AF352268

TRANSPORT PHENOMENA
IN POROUS MEDIA II

Elsevier Science Internet Homepage

http://www.elsevier.com

Consult the Elsevier Science Homepage for full catalogue information on all books, journals and electronic products and services.

Elsevier Titles of Related Interest

Journals
Free sample copies are available on-line, or can be requested by writing to:
Elsevier Science, The Boulevard, Langford Lane, Kidlington, Oxford, OX5 1GB, UK.

Applied Thermal Engineering
European Journal of Mechanics - B/Fluids
Experimental Thermal and Fluid Science
Flow Measurement and Instrumentation
Fluid Abstracts: Process Engineering
International Communications in Heat and Mass Transfer
International Journal of Heat and Fluid Flow
International Journal of Heat and Mass Transfer
International Journal of Multiphase Flow
International Journal of Refrigeration
International Journal of Thermal Sciences
Journal of Non-Newtonian Fluid Mechanics

Books

HESSELGREAVES: Compact Heat Exchangers
HOFFMAN: Unsteady-State Fluid Flow
INGHAM & POP: Transport Phenomena in Porous Media
KOTAKE & HIJIKATA: Numerical Simulations of Heat Transfer and Fluid Flow on a Personal
      Computer
POP & INGHAM: Convective Heat Transfer: Mathematical and Computational Modelling of Viscous
      Fluids and Porous Media

To Contact the Publisher

Elsevier Science welcomes enquiries concerning publishing proposals: books, journal special issues, conference proceedings, etc. All formats and media can be considered. Should you have a publishing proposal you wish to discuss, without obligation, please contact the publisher responsible for Elsevier's mechanical engineering programme:

Keith Lambert
Senior Publishing Editor
Elsevier Science
The Boulevard
Langford Lane
Kidlington, Oxford   Phone: +44 1865 843411
OX5 1GB   Fax: +44 1865 843920
UK   Email: k.lambert@elsevier.com

General enquiries, including placing orders, should be directed to Elsevier's Regional Sales Office – please access the homepage for full contact details.

TRANSPORT PHENOMENA
IN POROUS MEDIA II

Edited by

DEREK B. INGHAM
Department of Applied Mathematics
University of Leeds
Leeds LS2 9JT
United Kingdom

IOAN POP
Faculty of Mathematics
University of Cluj
R-3400 Cluj, CP 253
Romania

2002

PERGAMON
An imprint of Elsevier Science

Amsterdam - Boston - London - New York - Oxford - Paris
San Diego - San Francisco - Singapore - Sydney - Tokyo

ELSEVIER SCIENCE Ltd
The Boulevard, Langford Lane
Kidlington, Oxford OX5 1GB, UK

© 2002 Elsevier Science Ltd. All rights reserved.

This work is protected under copyright by Elsevier Science, and the following terms and conditions apply to its use:

Photocopying
Single photocopies of single chapters may be made for personal use as allowed by national copyright laws. Permission of the Publisher and payment of a fee is required for all other photocopying, including multiple or systematic copying, copying for advertising or promotional purposes, resale, and all forms of document delivery. Special rates are available for educational institutions that wish to make photocopies for non-profit educational classroom use.

Permissions may be sought directly from Elsevier Science Global Rights Department, PO Box 800, Oxford OX5 1DX, UK; phone: (+44) 1865 843830, fax: (+44) 1865 853333, e-mail: permissions@elsevier.co.uk. You may also contact Global Rights directly through Elsevier's home page (http://www.elsevier.com), by selecting 'Obtaining Permissions'.

In the USA, users may clear permissions and make payments through the Copyright Clearance Center, Inc., 222 Rosewood Drive, Danvers, MA 01923, USA; phone: (+1) (978) 7508400, fax: (+1) (978) 7504744, and in the UK through the Copyright Licensing Agency Rapid Clearance Service (CLARCS), 90 Tottenham Court Road, London W1P 0LP, UK; phone: (+44) 207 631 5555; fax: (+44) 207 631 5500. Other countries may have a local reprographic rights agency for payments.

Derivative Works
Tables of contents may be reproduced for internal circulation, but permission of Elsevier Science is required for external resale or distribution of such material.
Permission of the Publisher is required for all other derivative works, including compilations and translations.

Electronic Storage or Usage
Permission of the Publisher is required to store or use electronically any material contained in this work, including any chapter or part of a chapter.

Except as outlined above, no part of this work may be reproduced, stored in a retrieval system or transmitted in any form or by any means, electronic, mechanical, photocopying, recording or otherwise, without prior written permission of the Publisher.
Address permissions requests to: Elsevier Science Global Rights Department, at the mail, fax and e-mail addresses noted above.

Notice
No responsibility is assumed by the Publisher for any injury and/or damage to persons or property as a matter of products liability, negligence or otherwise, or from any use or operation of any methods, products, instructions or ideas contained in the material herein. Because of rapid advances in the medical sciences, in particular, independent verification of diagnoses and drug dosages should be made.

First edition 2002

Library of Congress Cataloging in Publication Data
A catalog record from the Library of Congress has been applied for.

British Library Cataloguing in Publication Data
A catalogue record from the British Library has been applied for.

ISBN: 0-08-043965-9

∞ The paper used in this publication meets the requirements of ANSI/NISO Z39.48-1992 (Permanence of Paper).
Printed in The Netherlands.

Preface

Transport phenomena in porous media continues to be an area of intensive research activity and this is primarily due to the fact that it plays an important role in a large variety of engineering and technological applications which span from the transport processes in biomechanical systems, such as blood flow in the pulmonary alveolar sheet, to the large scale circulation of brine in a geothermal reservoir. The acceleration in the progress in science and in the improvement in the design, efficiency and reliability of heat transfer equipment in power engineering, chemical, oil and gas industries are directly associated with the effective use of the modern tools of heat transfer analysis and measurement, predictive correlation equations, and with the sharing of the practical experience on the operation of all types of thermal equipment. This has caused a rapid expansion of research in diversified areas of heat transfer, including also porous media, and this has produced a huge amount of theoretical and experimental work.

The first volume of this book series, published in 1998, met with a very favourable reception within the porous media community, showing that there was a large demand for a series of books that emphasizes both the fundamentals and the applications of research in porous media. This has encouraged us to prepare the present second volume. In doing so, we have maintained the original concept of including a wide and diverse range of topics. In choosing the material we have also been influenced by the needs of the practical applications of porous media. Also, the book provides an up-to-date review of the current state-of-the-art in the various topics of heat transfer in porous media, presenting both the fundamental and experimental results of very active and internationally well-recognized authors. Thus, the book is primarily aimed at advanced researchers in porous media, and these may be applied mathematicians, physicists, geologists, chemists and practicing engineers.

All of the chapters in this book are very much interrelated. Further, some of the views expressed by some of the authors are contradictory and controversial, and therefore it was not easy to decide how best to order the chapters.

In Chapter 1, Nield has produced a very provoking and wide-ranging review of non-Darcy models, including comments on inertial effects, boundary friction, non-Newtonian fluids, viscous dissipation, rotation, magnetic field, radiation, and porous-medium/clear-fluid interface conditions. The solution of these model equations may be performed using many of the standard techniques, e.g., finite differences, finite elements, etc., but in

Chapter 2 Škerget and Jecl describe an alternative method which is becoming ever more popular, namely the boundary element method.

The work described in Chapters 1 and 2 assumes that the flows are stable and therefore in the next three chapters the stability of various fluid flows are investigated. In Chapter 3, Rees presents an overview of recent work on the onset of instabilities in thermal boundary-layer flows. In Chapter 4, Tyvand investigates the linear stability in three dimensions for the onset of Rayleigh–Bénard convection in finite porous bodies. In Chapter 5, Mamou looks at the stability of double-diffusive convection and finite amplitude flows in a tilted porous enclosure.

In Chapter 6, Howle investigates buoyancy-driven convection and notes the discrepancy that can exist between theoretical and laboratory results. This can be as a result of an instability or that the medium is not spatially homogeneous. Wang, in Chapter 7, further investigates the effects of the micromechanics of ordered, unidirectional heterogeneous materials.

In Chapters 8 and 9, the effects of turbulence are discussed. Lage, de Lemos and Nield, in Chapter 8, review four available methodologies for developing microscopic turbulence models for single-phase flow in rigid, fully-saturated porous media. Masuoka and Takatsu, in Chapter 9, experimentally and theoretically investigate the turbulence characteristics and discuss the mechanism for the production and thermal dissipation.

In Chapters 10 to 13, the effects of phase change, bubble growth, solidification and surface reactions are considered. In Chapter 10, Chang and Weng review coupled heat and moisture transfer in porous material with applications in the drying and storage of grain. In Chapter 11, Bories and Prat investigate the nucleation and bubble growth which are important in processes where there is pressure depletion and boiling. In Chapter 12, Riahi discusses the effects of rotation on convection adjacent to the solid–liquid interface during the solidification of a binary alloy. In Chapter 13, Pop, Merkin and Ingham investigate the effects of exothermic catalytic reactions which take place on a surface next to a porous medium.

In Chapters 14 to 16, problems involving porosity, either within or on the surface of the Earth, are considered. In Chapter 14, Bejan, Rocha and Cherry review the work that has been performed on the generation and flow of methane gas through a porous medium impregnated with solid clathrate hydrates. In Chapter 15, Woods develops a hierarchy of models to describe gravity driven flows in porous rocks, including the effects of layering, reaction, boiling and double advection. Finally, in Chapter 16, Lane and Hardy highlight the basic limitations of traditional treatments of complex bed geometries and vegetation in traditional models of river flows and the need for a more sophisticated porosity approach.

In the preparation of this volume, we have collaborated with a large number of researchers. Thus, our thanks go first to all the twenty-six authors, who have kindly accepted to contribute a chapter to the realization of the present volume, for their patience and dedicated effort. We would also like to express our sincere thanks to Professors P. J. Heggs and A. Nakayama who have throughout the preparation of the book given their support generously.

The formatting of this book and the preparation of the figures were performed by Dr Julie M. Harris and Dr Simon D. Harris and we are deeply indebted to them for all of their care and attention, and the patience that they have shown in both the preparation and the proof reading. Finally, but most sincerely, our grateful appreciation is extended to Mr Keith Lambert, Senior Publishing Editor, not only for his thoughtfulness, but also for his constant encouragement throughout the preparation of this book.

LEEDS/CLUJ   D. B. INGHAM & I. POP

DECEMBER, 2001

Contributors

A. BEJAN, Department of Mechanical Engineering and Materials Science, Duke University, Durham, NC 27708-0300, USA

S. BORIES, Institut de Mécanique des Fluides de Toulouse, UMR CNRS-INP/UPS N° 5502, Avenue du Professeur Camille Soula, F-31400 Toulouse, France

W. J. CHANG, Department of Mechanical Engineering, Kun Shan University of Technology, Tainan, Taiwan, ROC

R. S. CHERRY, Idaho National Engineering and Environmental Laboratory, P. O. Box 1625, Idaho Falls, ID 83415-2203, USA

M. J. S. DE LEMOS, Departamento de Energia, Instituto Tecnológico de Aeronáutica, 12228-900 São José dos Campos, SP, Brazil

R. J. HARDY, School of Geography, University of Leeds, Leeds, LS2 9JT, UK

L. E. HOWLE, Department of Mechanical Engineering and Materials Science, Center for Nonlinear and Complex Systems, Duke University, Durham, NC 27708-0300, USA

D. B. INGHAM, Department of Applied Mathematics, University of Leeds, Leeds, LS2 9JT, UK

R. JECL, University of Maribor, Faculty of Civil Engineering, Smetanova 17, 2000 Maribor, Slovenia

J. L. LAGE, Department of Mechanical Engineering, Southern Methodist University, Dallas, TX 75275-0337, USA

S. N. LANE, School of Geography, University of Leeds, Leeds, LS2 9JT, UK

M. MAMOU, Institute for Aerospace Research, National Research Council Canada, Ottawa, Ontario, K1A 0R6, Canada

T. MASUOKA, Department of Mechanical Engineering Science, Kyushu University, 6-10-1, Hakozaki, Higashi-ku, Fukuoka 812-8581, Japan

J. H. MERKIN, Department of Applied Mathematics, University of Leeds, Leeds, LS2 9JT, UK

D. A. NIELD, Department of Engineering Science, University of Auckland, Private Bag 92019, Auckland, New Zealand

I. POP, Faculty of Mathematics, University of Cluj, R-3400 Cluj, CP 253, Romania

M. PRAT, Institut de Mécanique des Fluides de Toulouse, UMR CNRS-INP/UPS N° 5502, Avenue du Professeur Camille Soula, F-31400 Toulouse, France

D. A. S. REES,   Department of Mechanical Engineering, University of Bath, Claverton Down, Bath, BA2 7AY, UK

D. N. RIAHI,   Department of Theoretical and Applied Mechanics, 216 Talbot Laboratory, University of Illinois at Urbana–Champaign, 104 South Wright Street, Urbana, IL 61801, USA

L. A. O. ROCHA,   Duke University, Department of Mechanical Engineering and Materials Science, Durham, NC 27708-0300, USA

L. ŠKERGET,   University of Maribor, Faculty of Mechanical Engineering, Smetanova 17, 2000 Maribor, Slovenia

Y. TAKATSU,   Department of Mechanical Engineering, Hiroshima Kokusai Gakuin University, 6-20-1, Nakano, Aki-ku, Hiroshima 739-0321, Japan

P. A. TYVAND,   Department of Agricultural Engineering, Agricultural University of Norway, 1432 Ås, Norway

C. Y. WANG,   Department of Mathematics, Michigan State University, East Lansing, MI 48824, USA

C. I. WENG,   Department of Mechanical Engineering, National Cheng Kung University, Tainan, Taiwan, ROC

A. W. WOODS,   BP Institute, University of Cambridge, Madingley Rise, Madingley Road, Cambridge, CB3 OEZ, UK

Contents

Contents

1 MODELLING FLUID FLOW IN SATURATED POROUS MEDIA AND AT INTERFACES

D. A. NIELD

Department of Engineering Science, University of Auckland, Private Bag 92019, Auckland, New Zealand

email: d.nield@auckland.ac.nz

Abstract

Since the days of Darcy, many refinements have been made to the equations used to model single-phase fluid flow and heat transfer in a saturated porous medium, to allow for such basic things as inertial effects, boundary friction and viscous dissipation, and additional effects such as those due to rotation or a magnetic field or due to radiative heat transfer. In this chapter all these developments are reviewed. Also reviewed are approaches to modelling flow at a porous-medium/clear-fluid interface.

Keywords: porous medium, interface, inertial effects, boundary friction, viscous dissipation, rotation, magnetic field, non-Newtonian fluid, radiative heat transfer

1.1 INTRODUCTION

Shenoy (1994) gives a two-page list of applications of the present subject under the headings Biomechanics, Ceramic engineering, Chemical engineering, Food technology, Geophysics, Groundwater hydrology, Industrial engineering, Mechanical engineering, Petroleum engineering, and Soil mechanics.

A porous medium is a fixed (or almost fixed) solid matrix with a connected void space through which a fluid can flow. The fraction of void space to total volume is called the porosity. Most naturally occurring porous media have porosities less than 0.6 (an exception is hair), but man-made materials, such as metallic foam, can have porosities up to 0.99.

The observations of Henry Darcy (1856) on the public water supply at Dijon and experiments on steady state unidirectional flow suggested Darcy's law, which in its refined modern form can be expressed as follows:

$$\frac{\partial p}{\partial x} = -\frac{\mu v}{k},$$ (1.1)

where $\partial p/\partial x$ is the pressure gradient, v is the filtration velocity, μ is the fluid viscosity and K is the permeability (the dimension are length squared). The filtration velocity v (velocity averaged over the medium) is related to the intrinsic velocity V (velocity averaged over the pore space) by $v = \phi V$, where ϕ is the porosity. In order to tie in with Darcy's results, the pressure here has to be an intrinsic quantity, i.e., the pressure is averaged only over the pore space. The permeability K depends on the pore size (or particle diameter) D_p, the porosity, and also on the detailed geometry. A useful estimate is given by the Carman–Kozeny relationship, derived for a packed bed of uniform spherical particles, namely

$$K = \frac{D_p^2 \phi^3}{180 \left(1 - \phi\right)^2}.$$ (1.2)

Darcy's law means that the drag is linearly proportional to the velocity. This holds for small velocities (namely when the Reynolds number, based on the pore scale, is less than unity). However, this 'law' breaks down for larger velocities. Dupuit (1863) and Forchheimer (1901) found empirically, for larger velocities, that the drag is a quadratic function of the velocity, and a detailed historical account has been given by Lage (1998), and what is commonly called the Forchheimer equation should really be called the Dupuit–Darcy equation.

1.2　THE BRINKMAN–FORCHHEIMER EQUATION

A modern refinement (see, for example, Hsu and Cheng, 1990 and Vafai and Kim, 1990) is the following equation:

$$\rho \left[\frac{1}{\phi}\frac{\partial v}{\partial t} + \frac{1}{\phi^2}\left(v \cdot \nabla v\right)\right] = -\nabla p + \mu_e \nabla^2 v - \frac{\mu}{K}v - \frac{c_F \rho}{K^{1/2}}vv.$$ (1.3)

This applies to an incompressible fluid of density ρ, where v denotes $|v|$, the magnitude of the Darcy velocity, while μ_e is an effective viscosity and c_F is a dimensionless Forchheimer coefficient. The inertial terms, on the left-hand side of equation (1.3), result from formal averaging. The first viscous term is the Brinkman term, and the last term is the Forchheimer term. We now consider the significance of the various terms in equation (1.3).

1.2.1　The local time-derivative inertial term

This is derived on the assumption that a spatial averaging process commutes with a derivative with respect to time and this breaks down when the porous medium has a macroscopic structure, such as a system of tubes. The decay of a transient is more rapid in narrow tubes than in wide tubes. Nield (1991) suggested in this case that the term $(1/\phi)\,\partial v/\partial t$ be replaced by $c_a \cdot \partial v/\partial t$, where c_a is a constant tensor, which is determined mainly by the nature of the pores of largest cross-sections. In any case, the ratio of the time-derivative term to the Darcy resistance is $c_a \rho K/\mu T$, where T is a characteristic time of the process being investigated, and this ratio is normally very small. However, it is essential to retain the time-derivative term when modelling certain convective instability problems, see Vadasz (1999a, 1999b).

1.2.2 Advective inertial term

Joseph *et al.* (1982) argued that, when modelling dense media, the advective (or convective) term involving $(v \cdot \nabla)\,v$ should be omitted because the inertial effects are already accounted for in the quadratic drag term involving vv. This arises as a result of the form drag on the solid particles. The drag is independent of the viscosity and acts in a direction opposite to v. Nield (1991) argued that the inclusion of the $(v \cdot \nabla)\,v$ term leads to the prediction that longitudinal momentum can, unimpeded by the fixed solid matrix, be transmitted transversely, but this conflicts with expectation based on basic physics.

This is related to the difficulty of spin-up which is achieved by just rotating a solid boundary, and the absence of true macroscopic turbulence (involving a cascade of energy from large eddies to smaller eddies), in a dense porous medium. The averaging process leads to misleading results because it leads to a loss of vital information about the way in which the geometry of the solid matrix affects the flow by reducing the coherence of the fluid momentum pattern.

From the vector identity $(v \cdot \nabla)\,v = \nabla\left(v^2/2\right) + v \times (\nabla \times v)$, it was noted by Nield (1994) that at least the irrotational part, $\nabla\left(v^2/2\right)$, of the term $(v \cdot \nabla)\,v$ needs to be retained in order to account for the phenomenon of choking in high speed flow of a compressible fluid, but he suggested that the rotational part, involving the intrinsic vorticity, should be deleted. His argument was based on the expectation that a medium of low porosity will allow scalar entities, such as fluid speed, to be freely advected, but it will inhibit the advection of vector quantities such as the vorticity. Nield and Bejan (1999) went a step further and suggested that even when the vorticity is being continuously produced, e.g., by the buoyancy force, one would expect that it would be destroyed by a momentum dispersion process due to the solid obstructions.

An argument providing further support for this point of view was presented by Nield (2001a). There are some subtleties about the effect of the inertial terms on the fluid motion in a porous medium. The power of the total drag force per unit volume is equal to the rate of viscous dissipation per unit volume, see below. Although the Forchheimer drag term appears to be independent of the viscosity, it does contribute to the viscous dissipation. The effect of inertia is mediated via a change in the pressure distribution and the fluid velocity distribution. The flip side of the coin is that when one closes the system of equations by introducing a Forchheimer drag term then one should not assume that the advective inertia term that remains in the momentum equation is identical with that obtained by formal volume-averaging. After integration, it should lead to the correct expression for the averaged kinetic energy, which involves the magnitude but not the direction of the fluid velocity, and this means that the irrotational part of the volume-averaged advective inertial term must be unchanged, but the rotational part is not determined by the averaging process, and there is no inconsistency in setting it to be zero as part of the closure process.

In the process of performing the closure, after volume-averaging, it has been traditional to adjust for the contribution to the overall drag force that includes a quadratic drag force, but to ignore the fact that one also needs to adjust for the fact that the overall moment of the force system has to be zero. Nield (2001c) suggested that an appropriate adjustment is simply to set to zero the rotational part of the volume-averaged advective inertial term.

It has sometimes been claimed that the retention of the advective inertial term is necessary in order to account for the formation of hydrodynamic boundary layers in channel flow, and in order to estimate the entrance length. However, this is not correct. The formation of such layers is primarily due to the action of viscous diffusion and the entrance length can be estimated by using the time-derivative inertial term.

In many practical situations it does not matter computationally whether the advective inertial term is included or not because, relative to the quadratic drag term, it is of the order of magnitude $K^{1/2}/c_F\phi^2 L$, where L is a characteristic length scale, and this expression is normally small. Lage (1992) verified numerically that the advective inertial term has a negligible effect on thermal convection in most cases of interest.

This topic is related to the question of how best to model turbulence in a porous medium. This is currently a controversial topic; see, for example, Nield (2001c) and the chapter in this book by Lage, de Lemos and Nield.

1.2.3 Brinkman viscous term

Brinkman (1947) introduced the Laplacian viscous term in a restricted context but its global use is due to other authors. The global treatment may fail to deal adequately with the distinctive features of flow in a porous medium. The ratio of the Brinkman term to the Darcy term is of the order $Da = K/L^2$, where L is the appropriate macroscopic length scale, and therefore $Da \to \infty$ corresponds to a fluid clear of solid material. In most practical cases the magnitude of Da will be very small and the Brinkman term will have a significant effect only in thin layers which are within a dimensional distance of the order $K^{1/2}$ from a solid wall. In many cases the reduction in the fluid velocity in this thin layer will be masked by an increase in velocity, namely the channeling effect, due to the increase in porosity near the wall where solid particles cannot pack as tightly as they can in the interior of the porous medium.

It is important to note that the Brinkman equation cannot be rigorously justified, except when the porosity is close to unity. The self consistent formulation of Brinkman breaks down when ϕ becomes less than 0.6 and there is an uncertainty about the effective viscosity μ_e. Brinkman simply took $\mu_e = \mu$, but the process of formal averaging, Bear and Bachmat (1990), leads to $\mu_e = \mu/\phi T$, where T is the tortuosity. Whitaker (1999, p. 173) ignores the tortuosity and he emphasizes that the Brinkman correction essentially involves the intrinsic velocity, so that when the correction is written in terms of the Darcy velocity then this immediately leads to $\mu_e = \mu/\phi$. Until comparatively recently it had not been possible to check the alternative estimates of μ_e against experiment because all the available experimental data pertained to media whose porosity was outside the range for which the theoretical results are valid. Givler and Altobelli (1994), using a Nuclear Magnetic Resonance technique, found $\mu_e \approx 8\mu$ for water flowing through a rigid foam material ($\phi = 0.972$), and it is clear that averaging is inadequate in this case.

It is worth noting that in the case of porous media of very high porosity then there is no theoretical limit on the value of the Darcy number K/L^2, but for such media the value of K that appears in the Brinkman equation cannot be determined solely from a simple Darcy-type experiment which involves a measurement of the ratio of the volume flux to the applied pressure gradient.

1.2.4 Dupuit–Forchheimer (form drag) term

The term $\left(c_F\rho/K^{1/2}\right)vv$ in equation (1.3) is in the form recommended by Joseph *et al.* (1982). The scalar form is due to Ward (1964), who thought that c_F might be a universal constant which takes the value 0.55 but subsequent experimenters found that c_F is approximately constant for a particular family of materials, e.g., $c_F = 0.1$ for foamed metal fibres. A semi-empirical derivation of an estimate for c_F was reported by Joseph *et al.* (1982). They emphasized that the drag is quadratic and in a direction opposed to v. This means that one cannot write down uncoupled equations for the x- and y-components of v. In fact, a cubic drag term arises in two circumstances. First, Lage *et al.* (1997) pointed out, when complications resulting from transition to turbulence are taken into account, that the coefficient of the quadratic term varies slowly with the velocity, and so the overall drag is effectively cubic in the velocity within a restricted range of Reynolds number. Second, Mei and Auriault (1991), and others, have pointed out that in the weak inertial regime, one for which the pore Reynolds number is less than unity, the variation from the linear term is in fact cubic, rather than quadratic. However, Lage and Antohe (2000) have shown that the experimental data are best fitted by a quadratic expression. This result indicates that as the Reynolds number increases then any cubic term is quickly dominated by a quadratic term.

Lage and Antohe (2000) have argued that the factor $K^{1/2}$ in the Dupuit–Forchheimer term is better replaced by another quantity which has the dimensions of length, namely a typical particle diameter. Accordingly, they replace the factor $c_F/K^{1/2}$ with a quantity C which has the dimensions of the reciprocal of the length. These authors note that the permeability relates essentially to the effective surface area of the solid porous matrix, whereas their form parameter C depends on the 'form' of the porous medium (defined as the variation of the cross-sectional area of the solid matrix). The present author prefers to retain the formulation in terms of the coefficient c_F with the understanding that c_F depends on the geometry of the porous medium. This choice is based simply on the fact that c_F is dimensionless. In other words, the factor $K^{1/2}$ is included simply because it is immediately available as a length scale for the quadratic drag term. The mathematical form of the expression for the Dupuit–Forchheimer term that appears in the equation (1.3) is not very important. What is important is making sure that one uses the most appropriate form of the Reynolds number in the correlation of the experimental results. The most important practical consideration is the magnitude of the Dupuit–Forchheimer form drag term relative to the Darcy term in the equation of motion. If U is a characteristic Darcy velocity then the relevant ratio is represented by the Reynolds number $c_F \rho K^{1/2} U / \nu$. It is important to note that many authors have used a Reynolds number defined with the factor c_F left out. In other words, they have used a Reynolds number that is based solely on the permeability of the medium.

Lage and Antohe (2000) have pointed out that there is a need to investigate how the geometry of the porous matrix affects the form drag effect, because very little is known in this regard. In numerical investigations, many authors have been content to use a relationship between c_F and K that is based on semi-empirical formulae, giving those two quantities in terms of the porosity, as obtained by Ergun (1952). This relationship leads to the formula

$$c_F = \frac{1.75}{150^{1/2}\phi^{3/2}}. \tag{1.4}$$

The reader should appreciate that the above expression is *ad hoc*. The Ergun expressions, and hence this one, are appropriate for a bed of spherical solid particles, but are less appropriate for other types of porous media. The reader is also warned that some authors have used equation (1.4) in association with a momentum equation in which an extra factor ϕ was incorporated in the Forchheimer term of the momentum equation, a procedure that is clearly inconsistent. This pitfall was pointed out by Nield (2001b).

There are several subtleties associated with the form drag term. One is the contribution to viscous dissipation, see below. Another is the contribution of the drag force on the kinetic energy of the fluid in a porous medium. There is a temptation to think of the drag as being just a surface effect, but in the case of a porous medium that is an oversimplification. The effect of viscosity acts throughout the fluid phase of the porous medium, and this means that the viscous dissipation, and hence the total drag, involves a volume integral over a representative elementary volume, not just a surface integral. This means that, in effect, the drag force acts on a particle having as its velocity the Darcy velocity, despite the no-slip condition holding on the pore boundaries.

Another point worthy of note is that the dichotomy into the Darcy drag and the form (Forchheimer) drag is not arbitrary, as might be thought from the way in which some early workers have added terms to an empirical relationship between pressure drop and velocity. Rather, the two terms arise naturally once a decision is made to include two terms, a term linear in the velocity and one quadratic in the velocity. It is then inevitable, on dimensional grounds, that the Darcy term will appear with the viscosity as a coefficient, whereas the Forchheimer term will not explicitly involve the viscosity. The absence of the viscosity as a factor in the Forchheimer term is inevitably related to the overall process of the production of increased drag by wake formation. Another point to note is that the Forchheimer drag term and the Darcy term are intrinsically linked. While it may make mathematical sense to treat the asymptotic case where the flow is so rapid that the Forchheimer term dominates the Darcy term, as several authors have done, the results may not have much physical significance, at least in the absence of careful interpretation.

In the light of the argument resolving the apparent paradox about viscous dissipation given below, a further observation can be made. For the flow past a single solid obstacle in an unbounded region, the size of the region in which velocity gradients are large can grow without limit, but in the case of a porous medium a limit is set by the pore size. For example, in the case of a porous medium with a spatially periodic structure, the limiting region is a period cell. This suggests, at very large velocities, that the rate of increase of the drag with an increase of the fluid velocity may fall below that predicted by the Forchheimer expression with an unchanged constant. The situation is complicated by the transition to unsteady and chaotic flow regimes, but it is interesting that Kaviany (1995, p. 49) refers to experimental results reported in Macdonald *et al.* (1979) and Dybbs and Edwards (1984) indicating an asymptotic behaviour in which the normalized pressure drop does not change with the Reynolds number.

1.3 MODELLING A POROUS-MEDIUM/CLEAR-FLUID INTERFACE

Since the differential equations for the two regions are of second-order in spatial derivatives, four matching conditions are needed if the Brinkman equation is employed. These involve the continuity of tangential velocity, normal velocity, tangential stress and normal stress. The velocity matching causes no problems, but with the stress matching it is different. Consider the matching of tangential stress. Over the pore portion of the interface the velocity shear, and hence the tangential stress, is continuous. Over the solid portion the tangential shear expression (viscosity times velocity gradient) is not continuous. It is clearly zero in the solid, but has some indeterminate nonzero value in the adjacent clear fluid (that is matched by the shear stress in the solid). Thus authors who have matched velocity shears have overdetermined the system of equations.

When one uses the Darcy equation (instead of the Brinkman equation) in the porous medium then the difficulty can be side-stepped. Now one needs only three matching conditions; two of these are the continuity of the tangential fluid velocity and the normal velocity, and the third is the Beavers–Joseph (Beavers and Joseph, 1967) boundary condition

$$\frac{\partial u_f}{\partial y} = \frac{\alpha_{BJ}}{K^{1/2}} \left(u_f - u_m \right).$$

(1.5)

Here the clear fluid occupies the region $y > 0$, and u_f is the fluid velocity, u_f and $\partial u_f / \partial y$ are evaluated at $y = 0^+$ and the Darcy velocity u_m is evaluated at some small distance from $y = 0$. The Beavers–Joseph constant α_{BJ} is dimensionless and independent of the fluid viscosity, but it depends on the structure of the porous material within the boundary region. Sahraoui and Kaviany (1992) have shown that the value of α_{BJ} depends on the flow direction at the interface, the Reynolds number, the precise choice of the interfacial location at which the boundary condition is applied, and nonuniformities in the arrangement of the solid material at the surface. In the present author's opinion α_{BJ} should be regarded as simply an empirical constant, to be determined experimentally. Its presence in the boundary condition provides the needed flexibility in modelling the tangential stress requirement.

The situation with respect to the normal stress is similar but there is an additional factor involved. The normal stress is the sum of the pressure and a viscous stress term. Some authors have argued that the pressure, being an intrinsic quantity, has to be continuous across the interface. They have failed to realize that the interface is an idealization of a thin layer in which the pressure can change substantially because of the presence of the solid material. In practice, the viscous term in the normal stress may be small compared with the pressure and in this case the continuity of normal stress does reduce to the approximate continuity of pressure. Also, for an incompressible fluid, the continuity of normal stress does reduce to the continuity of pressure if one takes the effective Brinkman viscosity equal to the fluid viscosity, as shown by Chen and Chen (1992). It should be noted that several authors who have formulated a problem in terms of stream function and vorticity have failed to deal properly with the normal stress boundary condition, see Nield (1997). For a more soundly based procedure for numerical simulation, and for a further discussion of this matter, the reader is referred to Gartling *et al.* (1996).

Ochoa-Tapia and Whitaker (1995a, 1995b) have expressly matched the Darcy and Stokes equations using the volume-averaging procedure. This approach produces a jump in the stress, but not in the fluid velocity, and involves a parameter which has to be fitted experimentally. They also explored the use of a variable porosity model as a substitute for the jump condition and they concluded that the latter approach does not lead to a successful representation of all the experimental data but it provides insight into the complexity of the interface region. Kuznetsov (1996) applied the jump condition to flows in parallel-plate and cylindrical channels which are partially filled with a porous medium. Kuznetsov (1997) reported an analytical solution for flow near an interface. Kuznetsov and Xiong (1999) have investigated the limitations of the 'single domain' approach, in which the same equations (but with different coefficients) are used in the fluid region as are used in the porous medium region. They concluded that the single-domain approach results in the correct matching of the shear stress only if the adjustable coefficient in the representation for the excess stress equals zero and if also the effective viscosity is equal to the fluid viscosity.

Salinger *et al.* (1994) found that a Darcy-slip, finite element formulation produced solutions which were more accurate and more economical to compute than those obtained using a Brinkman formulation.

For the case of forced convection in a composite channel between parallel plates, Alazmi and Vafai (2001) have reported the results, based on a Brinkman–Forchheimer formulation, of numerical calculations on the effects of varying the boundary conditions imposed at the interface. They found, as one would expect, that the changes had a more prominent effect on the velocity field, a less prominent effect on the temperature field, and thus an even less prominent effect on the Nusselt number. The effect of the changes increased with an increase in the Reynolds number and with an increase in the Darcy number.

Published studies of the interface problem have concentrated on situations where the basic flow is tangential to the interface. In fact, very little has been reported on situations in which the basic flow is normal to the interface. An exception is the dissertation by Fadda (1996), who investigated flow in a channel with a clear fluid upstream and a porous medium downstream. His numerical work confirmed the prediction of an analytic study by the present author (unpublished) for a porous medium of moderate or low Darcy number, namely that the presence of the porous medium would be felt several channel widths upstream of the interface, and he verified this with a flow visualization experiment. In this situation, one would expect that the precise form of the boundary condition at the interface would not be important, and Fadda confirmed that this was so.

1.4 NON-NEWTONIAN FLUID

Shenoy (1994) reviewed studies of flow of non-Newtonian fluids in porous media, which reveal that the various authors have concentrated on power-law fluids. Shenoy suggested, on the basis of volumetric averaging, that the Darcy term be replaced by $(\mu^*/K^*)\,v^{n-1}v$, the Brinkman term by

$$\frac{\mu^*}{\phi^n}\nabla\left\{\left|\sqrt{\left[\tfrac{1}{2}\boldsymbol{\Delta}:\boldsymbol{\Delta}\right]}\right|^{n-1}\boldsymbol{\Delta}\right\} \tag{1.6}$$

for an Ostwald–de Waele fluid, and the Forchheimer term be left unchanged, because it is independent of the viscosity. Here n is the power-law index, μ^* reflects the consistency of the fluid, K^* is a modified permeability, and $\boldsymbol{\Delta}$ is the deformation tensor. The present author agrees with Shenoy's suggestion, but in the Brinkman term he would replace μ^*/ϕ^n by an equivalent coefficient. Applications to natural, forced and mixed convection heat transfer have been discussed by Shenoy (1992, 1993) and others.

1.5 EFFECT OF ROTATION

The effect of rotation is to add two extra body-force terms to the momentum equation, reflecting the centrifugal and Coriolis effects. In the context of natural convection, this topic has been discussed in papers reviewed by Vadasz (1998). The left-hand side of equation (1.3) is now replaced by

$$\rho\left[\frac{1}{\phi}\frac{\partial v}{\partial t}+\frac{1}{\phi^2}\left(v\cdot\nabla v\right)+\frac{2}{\phi}\omega\times v+\omega\times\left(\omega\times x\right)\right], \tag{1.7}$$

where ω is the angular velocity of the rotating frame of reference and x is the position vector relative to that frame. The ratio of the Coriolis term to the Darcy term is of order

E^{-1}, where the Ekman–Darcy number E is given by

$$E = \frac{\phi Ek}{Da}, \quad \text{where} \quad Ek = \frac{\mu}{2\omega\rho L^2} \quad \text{and} \quad Da = \frac{K}{L^2}. \tag{1.8}$$

Here L is a characteristic length. In general, in most practical situations E is large since the Darcy number Da is small, and therefore the Coriolis term is usually not important. However, Vadasz points out, in the case of a heterogeneous medium, that the Coriolis acceleration distorts the direction of any existing flow and generates vortices in a plane perpendicular to the fluid flow. For the case of isothermal fluids, the centrifugal term, then being irrotational, merely affects a reduced pressure, but for free convection this term may be important.

Many authors have wrongly omitted the factor ϕ from the Coriolis term. As Nield (1999) pointed out, they have failed to realize that the pressure in Darcy's equation is an intrinsic quantity and hence the velocity appearing in an inertial term must also be an intrinsic quantity.

1.6 EFFECT OF A MAGNETIC FIELD

The technique of volume-averaging leads to the prediction that the effect of a magnetic field is to add a body force term $\sigma\left(v \times B\right) \times B/\phi$ to the right-hand side of equation (1.2). Here σ is the electrical conductivity of the fluid and B is the applied magnetic induction, see, for example, Raptis and Perdikis (1987). In the case of two-dimensional flow and with the magnetic induction in the plane of that flow, the extra body force reduces to $-\sigma B^2 v/\phi$. Thus the effect of the magnetic field is then simply to add an additional drag force. The ratio of the magnetic drag to the Darcy drag is $\sigma B^2 K/\mu\phi$, a parameter called the Chandrasekhar–Darcy number, and in most practical cases this number is very small, and therefore the effect of the magnetic field is negligible. Again, it should be noted that many authors have in error omitted the factor ϕ from their Chandrasekhar–Darcy number.

1.7 A REFORMULATION OF THE MOMENTUM EQUATION

Because so many people have been misled on the matter of modelling inertial terms, it is worthwhile considering in detail why the pressure in the traditional Darcy differential equation is necessarily an intrinsic quantity rather than a seepage quantity. It is also worthwhile investigating whether a reformulated equation would be less confusing. The reader should note that the original Darcy equation (relating to Darcy's experiments), in the form 'pressure drop divided by length of column equals constant times seepage velocity', leads directly to the modern differential equation for sufficiently slow flow. For the purpose of deriving this differential equation, the REV (representative elementary volume) is properly regarded as a miniature column to which Darcy's result can be applied.

As well as the assumptions of steady flow and incompressible fluid, three other important assumptions are made, namely,

(i) the porous medium is assumed to be homogeneous,

(ii) the macroscopic (REV scale) flow is assumed to be unidirectional,

(iii) a continuum assumption is made.

A consequence of the first assumption is that no distinction need be made between surface porosity and volume porosity. A consequence of the second assumption is that there is no flow out of the sides of the miniature column, so the situation in that column is the same as in the large column. A consequence of the third assumption is that it is permissible to consider the mathematical limit as the length of the miniature column tends to zero. In Darcy's experiments, the pressure drop was a *fluid* pressure drop measured in the usual way by a pair of manometers placed outside the porous medium. Each manometer measured the fluid pressure at the point at which it was placed and no cross-sectional average was involved. Rather, because of assumptions (i) and (ii) above, the pressure could be regarded as uniform, and so the pressure measured at one point was representative of the whole cross- section occupied by fluid, so the pressure measured by Darcy is an intrinsic pressure. It follows that, after the appropriate limit of length of the miniature column tending to zero is taken, the pressure in the modern differential equation is also an intrinsic quantity. This means that the pressure gradient at the REV level is effectively the average of the microscopic (pore scale) fluid pressure gradient averaged over just the fluid portion of the REV. Professor J. L. Lage has pointed out to the author in a personal communication that Darcy measured the pressure just outside the porous medium, rather than just inside it, and this means that an entrance and exit effect is involved, and this affects the permeability value measured in his experiments. In the argument presented here it is assumed that this effect is negligible.

Conversely, one can start with the differential equation and deduce an expression for the Darcy pressure drop. Because of assumptions (i) and (ii), macroscopic transverse pressure gradients are zero. Further, by mass conservation, the seepage velocity is independent of the longitudinal coordinate and so the same must be true of the longitudinal pressure gradient. For an integration over a volume, one can treat the medium as a continuum, in which no distinction need be made between the fluid and solid phases. This means that mathematically one can average over the entire cross-section of the Darcy column, and integrate between the ends of the column, and thereby recover the original Darcy expression for the pressure drop. In fact, since the pressure gradient is uniform over the whole column, the mathematics involved is very simple. In this process one starts with a fluid pressure gradient and ends up with a fluid pressure drop. Thus the argument is consistent.

There is overwhelming experimental evidence that the conventional differential equation, with P denoting an intrinsic quantity and K the standard permeability, is correct. If P were an REV averaged quantity but with K unchanged, then a factor ϕ would have to appear in the buoyancy term in the momentum equation, and consequently in the definition of Rayleigh number, see, for example, in the last term of equation (6.4) and equation (6.19) of Nield and Bejan (1999). There would then be a discrepancy with the well established experimental value for the critical Rayleigh number for the Horton–Rogers–Lapwood problem, see Nield and Bejan (1999, Section 6.9.1), to give just one example. Elder (1967) obtained the experimental value 40 with an estimated experimental error of 10%, in comparison with the theoretical value of 39.48. In his experiment he used a bed of packed spheres, and so the porosity was about 0.4. A critical Rayleigh number of 40/0.4 is incompatible with the experimental results.

When the conventional definition of permeability was introduced about 1920 (for references, see Lage, 1998), a definition which was popularized by the writings of Muskat (1937), workers had the option of invoking the Dupuit–Forchheimer relationship and expressing Darcy's law in terms of intrinsic velocity, effectively defining a different permeability which incorporates the factor ϕ, but they did not do so. The mixture of intrinsic pressure and seepage velocity in the conventional equation is unfortunate. Lage (1998) recognized this, and wrote an equation in terms of a seepage pressure and seepage velocity. Nield (1999) suggested that it is better to write the entire equation in terms of intrinsic quantities. In order to avoid possible confusion, he introduced a new quantity with a different name, symbol and dimensions from the permeability K. He wrote the Darcy differential equation in the form

$$\nabla P + \mu R V = 0, \tag{1.9}$$

where R denotes the 'retardability', which has dimensions $(\text{length})^{-2}$ and is thus measured in terms of the unit m^{-2}, and is defined in terms of the standard permeability K and porosity ϕ by $R = \phi/K$, while P is the intrinsic pressure and V is the intrinsic velocity. A major advantage of the new form is that it generalizes in a natural way to the Brinkman equation,

$$\nabla P + \mu R V - \mu_e \nabla^2 V = 0, \tag{1.10}$$

where μ_e is an effective viscosity. Volume averaging over an REV gives the estimate $\mu_e = \mu$ (rather than μ/ϕ). Clearly, the porosity ϕ does not appear explicitly in the equation, but is incorporated into the geometrical factor R. Other terms like the Coriolis term can be added and expressed in terms of V, and again the porosity does not appear in these terms. For the case of buoyancy, the term to be added to the left hand side of equation (1.9) is $-\rho g$, where g is the gravitational acceleration. A minor bonus is that the division solidus does not appear in the equation.

For the Brinkman–Forchheimer equation, Nield (1999) proposed the form

$$\nabla P + \mu R V - \mu_e \nabla^2 V + C_N \rho R^{1/2} V V = 0. \tag{1.11}$$

The new non-dimensional Forchheimer coefficient C_N is related to the Forchheimer coefficient c_F used in equation (1.3) by $C_N = \phi^{3/2} c_F$. There is evidence which suggests that C_N may be closer to being a universal constant than is c_F. For example, Beavers and Sparrow (1969) noted, for their fibrous foam metal materials, that c_F was about 0.1, compared with the value 0.55 obtained with beds of spheres (Beavers *et al.*, 1973).

Unfortunately the porosity values are not reported in Beavers and Sparrow (1969), but plausible ballpark values are 0.9 for the fibrous materials and 0.4 for the beds of spheres, and these values yield the C_N values 0.085 for the fibrous material and 0.14 for the beds of spheres.

1.8 VISCOUS DISSIPATION

For convection problems one must supplement the momentum equation by a thermal energy equation, which in steady state form is given by

$$\rho c_P \boldsymbol{v} \cdot \nabla T = \nabla \cdot (k_m \nabla T) + \Phi, \qquad (1.12)$$

where k_m is the effective thermal conductivity of the porous medium and Φ is the viscous dissipation term. This last term is generally negligible but, in general, it is given by the power of the drag force per unit volume, i.e., $\boldsymbol{v} \cdot \boldsymbol{F}$, where $\boldsymbol{F}$ is the drag force. In the case where the Forchheimer term is added to the Darcy term we have the following:

$$\Phi = \frac{\mu}{K}\boldsymbol{v} \cdot \boldsymbol{v} + \frac{c_F \rho}{K^{1/2}} |\boldsymbol{v}| \, \boldsymbol{v} \cdot \boldsymbol{v}. \qquad (1.13)$$

The remarkable thing is that the last term in equation (1.13) does not involve the viscosity as a factor, despite the fact that it contributes to the *viscous* dissipation term, and this paradox was resolved by Nield (2001a). The explanation of the apparent paradox lies in the recognition that, as pointed out by Joseph *et al.* (1982), the Forchheimer drag term models essentially a form drag effect, and involves the separation of boundary layers and wake formation behind solid obstacles. In fact, the basic idea goes back at least as far as Bakhmeteff and Feodoroff (1937). The inertial effects are mediated by the pressure distribution and this affects the velocity field and hence the drag in a complex fashion. The pore scale advective inertial effects contributing to the form drag lead to a substantial modification of the velocity field and, in particular, to an enlargement of the macroscopic region in which pore scale velocity gradients are large. This leads to an increase in the total viscous dissipation, which is summed over the whole region occupied by fluid and hence, because of the fundamental equality of viscous dissipation within a given volume and the power of the drag force on that volume, to the increase in the drag. Lage (1998) has emphasized the distinction between porous media of a bluff-body type and those of a conduit type. Therefore it is worthwhile pointing out that the above argument applies to both types of porous media. In the case of bluff bodies the separation of the boundary layers and wake formation occur behind the bodies, while in the case of conduits of converging–diverging type these phenomena occur downstream of the shoulders.

In their study of natural convection, Fand *et al.* (1994) argued, in the case of high flow rates, that the boundary layers are sufficiently thin as to render viscous dissipation negligible compared to conduction at the heated surface. We can now see that their argument is invalid, and that in their problem viscous dissipation should be important for both large values and small values of the Rayleigh number.

If the Brinkman model is employed, and so the drag is modelled by the term $(\mu/K)\,v - \mu_e \nabla^2 v$, then the viscous dissipation is given by

$$\Phi = \frac{\mu}{K} v \cdot v - \mu_e v \cdot \nabla^2 v. \tag{1.14}$$

(The reader should note that this expression is based on averages taken over a representative elementary volume and hence one should not expect to recover from equation (1.14) an expression for the viscous dissipation in a clear fluid in the limit of porosity tending to unity.)

Finally, it should be noted that the viscous dissipation term in the thermal energy equation is of the order of magnitude $\mu U^2/K c_P$. On the other hand, the thermal conduction term, that appears in the same equation, is of the order of magnitude $k\Delta T/L^2$, where U, L and ΔT are the characteristic velocity, length and temperature difference scales, respectively, and k is the thermal conductivity. Hence the viscous dissipation is negligible if

$$N \ll 1, \tag{1.15}$$

where

$$N = \frac{\mu U^2 L^2}{K k \Delta T} = \frac{Ec\,Pr}{Da} = \frac{Br}{Da}, \tag{1.16}$$

where Ec, Pr, Da and Br are the Eckert number, Prandtl number, Darcy number and Brinkman number defined respectively by

$$Ec = \frac{U^2}{c_P \Delta T}, \quad Pr = \frac{\mu/\rho}{k/\rho c_P}, \quad Da = \frac{K}{L^2}, \quad Br = \frac{\mu U^2}{k \Delta T}. \tag{1.17}$$

In most situations the Darcy number is small and therefore the viscous dissipation is important at even modest values of the Eckert number. The circumstances in which viscous dissipation is important are those involving flows of relatively large velocity and the author believes that the expression given in equation (1.13), or equation (1.14), is likely to be applicable in the context of particle bed nuclear reactors.

In the case of forced convection, a suitable choice for the characteristic velocity U is clear. In the case of natural convection, scale analysis arguments, such as those employed in Nield and Bejan (1999, Section 7.1.1), lead to the estimate that $U \sim (\alpha_m/L)\,Ra^{1/2}$, where α_m is the effective thermal diffusivity of the porous medium and Ra is the Rayleigh–Darcy number which is defined by

$$Ra = \frac{g\beta K L\Delta T}{\alpha_m \nu}, \tag{1.18}$$

where β is the coefficient of thermal expansion, and in this case we have

$$Ec \sim Pr^{-1} Da\,Ge, \tag{1.19}$$

where Ge is the Gebhart number, which is defined by

$$Ge = \frac{g\beta L}{c_P}, \tag{1.20}$$

and the condition (1.15) becomes

$$Ge \ll 1, \tag{1.21}$$

in agreement with the statement made by Nield and Bejan (1999, p. 25). The analysis of Nakayama and Pop (1989) and Murthy and Singh (1997) confirms that a Gebhart number, which was denoted in those papers by the symbol ε, is the pertinent group in natural convection problems, in the Darcy flow case. Murthy and Singh (1997) found that the Nusselt number decreases as a result of the viscous dissipation, by a fraction approximately equal to Ge, for small values (0.01 and 0.1) of Ge.

The above comments on forced convection are made on the assumption that the Péclet number $Pe = UL/\alpha_m$ is not large. If it is large, then the proper comparison is one between the magnitudes of the viscous dissipation term and the convective transport term in the thermal energy equation. This ratio is of the order of $Ec\,Pr/Da\,Pe = Ec/Da\,Re$, where the Reynolds number $Re = UL/\nu$.

Another significant paper in which viscous dissipation is correctly modelled is that by Ingham *et al.* (1990). The reader is warned that several papers have been published recently in which viscous dissipation has *not* been correctly modelled.

1.9 RADIATION

The topic of radiative heat transfer in porous media, and in particular in packed beds, is well covered in Kaviany (1995, Chapter 5) and recent work has been reviewed by Howell (2000). The methodology involved with radiation is somewhat different from that involved with convection, and only a limited number of investigations have been concerned with combined radiation and convection in porous media. Most porous media are opaque to radiation, and for them the effect of radiation is felt only in thin surface layers. Boundary layer analyses have been made by Chandrasekhara and Nagaraju (1988, 1993), Hossain and Pop (1997), Mansour (1997), Yih (1999) and Mohammadien and El-Amin (2000a, 2000b). In the recent papers the effect of radiation has been modelled by the addition of a particular volumetric distribution of heat sources. The present author would like to see more attention paid to volumetric heat sources having layered distributions of more general form.

1.10 CONCLUSION

A review has been made of the many refinements that over the years have been made to the equations used to model single-phase fluid flow and heat transfer in a saturated porous medium, to allow for such basic things as inertial effects, boundary friction and viscous dissipation, and additional effects such as those due to rotation or a magnetic field or due to radiative heat transfer. Approaches to modelling flow at a porous-medium/clear-fluid interface have also been reviewed. It has been demonstrated that the modelling of fluid flow in a porous medium involves some subtle matters, but substantial progress has been made in dealing with these. There are at least a couple of areas where further investigation is desirable. The first such area is the modelling of high speed fluid flow in a porous medium. In particular, the proposal about the advective inertial term made by Nield and Bejan (1999) has not been refuted, but neither has it been supported by other workers. Also, the author would like to see some experimental work done on the 'spin-up' problem in a porous medium, to provide results for comparison with alternative models. The second area where further work is desirable is the problem of modelling the interface between a clear fluid and a mushy zone, where the porosity varies with distance from the interface. Some preliminary analysis by David E. Loper of Florida State University (private communication) and some by the present author has been done. The analysis involves Airy functions or modified Bessel functions of order $1/3$.

Acknowledgements

This article is a revised and substantially expanded version of the paper by Nield (2000). The author has benefited from discussions with a number of people, and in particular from several with J. L. Lage, P. Vadasz and A. V. Kuznetsov.

REFERENCES

Alazmi, B. and Vafai, K. (2001). Analysis of fluid flow and heat transfer interfacial conditions between a porous medium and a fluid layer. *Int. J. Heat Mass Transfer* **44**, 1735–1749.

Bakhmeteff, B. A. and Feodoroff, N. V. (1937). Flow through granular beds. *J. Appl. Mech.* **4**, A97–A104.

Bear, J. and Bachmat, Y. (1990). *Introduction to Modeling of Transport Phenomena in Porous Media.* Kluwer, Dordrecht.

Beavers, G. S. and Joseph, D. D. (1967). Boundary conditions at a naturally permeable wall. *J. Fluid Mech.* **30**, 197–207.

Beavers, G. S. and Sparrow, E. M. (1969). Non-Darcy flow through fibrous porous media. *ASME J. Appl. Mech.* **36**, 711–714.

Beavers, G. S., Sparrow, E. M., and Rodenz, D. E. (1973). Influence of bed size on the flow characteristics and porosity of randomly packed beds of spheres. *ASME J. Appl. Mech.* **40**, 655–660.

Brinkman, H. C. (1947). A calculation of the viscous force exerted by a flowing fluid on a dense swarm of particles. *Appl. Sci. Res. A* **1**, 27–34.

Chandrasekhara, B. C. and Nagaraju, P. (1988). Composite heat transfer in the case of a steady laminar flow of a gray fluid with small optical density past a horizontal plate embedded in a saturated porous medium. *Wärme- und Stoffübertr.* **23**, 343–352.

Chandrasekhara, B. C. and Nagaraju, P. (1993). Composite heat transfer in a variable porosity medium bounded by an infinite flat plate. *Wärme- und Stoffübertr.* **28**, 449–456.

Chen, F. and Chen, C. F. (1992). Convection in superposed fluid and porous layers. *J. Fluid Mech.* **234**, 97–119.

Darcy, H. P. C. (1856). *Les Fontaines Publiques de la Ville de Dijon.* Victor Dalmont, Paris.

Dupuit, A. J. E. J. (1863). *Etudes Theoriques et Pratiques sur le Mouvement des Eaux dans les Canaux Découverts et a Travers les Terrains Perméables.* Victor Dalmont, Paris.

Dybbs, A. and Edwards, R. V. (1984). A new look at porous media fluid mechanics—Darcy to turbulent. In *Fundamentals of Transport Phenomena in Porous Media* (eds J. Bear and M. Y. Corapcioglu), pp. 199–256. Martinus Nijhoff, Dordrecht.

Elder, J. W. (1967). Steady free convection in a porous medium heated from below. *J. Fluid Mech.* **27**, 29–48.

Ergun, S. (1952). Fluid flow through packed columns. *Chem. Eng. Progress* **48**, 89–94.

Fadda, D. (1996). *Free Surface Flow with Heat Transfer in Clear and Porous Media.* Ph.D. thesis. Southern Methodist University, Dallas TX, USA.

Fand, R. M., Varahasamy, M., and Yamamoto, L. M. (1994). Heat transfer by natural convection from horizontal cylinders embedded in porous media whose matrices are composed of spheres: viscous dissipation. In *Heat Transfer 1994, Proceedings of 10th International Heat Transfer Conference,* Brighton, UK (ed. G. F. Hewitt), Vol. 5, pp. 237–242. Institute of Chemical Engineers, Rugby, UK.

Forchheimer, P. (1901). Wasserbewegung durch Boden. *VDI Z.* **45**, 1782–1788.

Gartling, D. K., Hickox, C. E., and Givler, R. C. (1996). Simulation of coupled viscous and porous flow problems. *Comp. Fluid Dyn.* **7**, 23–48.

Givler, R. C. and Altobelli, S. A. (1994). A determination of the effective viscosity for the Brinkman–Forchheimer flow model. *J. Fluid Mech.* **258**, 355–370.

Hossain, M. and Pop, I. (1997). Radiation effect on Darcy free convective flow along an inclined surface placed in porous media. *Heat Mass Transfer* **30**, 149–153.

Howell, J. R. (2000). Radiative transfer in porous media. In *Handbook of Porous Media* (ed. K. Vafai), pp. 663–698. Marcel Dekker, New York.

Hsu, C. T. and Cheng, P. (1990). Thermal dispersion in a porous medium. *Int. J. Heat Mass Transfer* **33**, 1587–1597.

Ingham, D. B., Pop, I., and Cheng, P. (1990). Combined free and forced convection in a porous medium between two vertical walls with viscous dissipation. *Transport in Porous Media* **5**, 381–398.

Joseph, D. D., Nield, D. A., and Papanicolaou, G. (1982). Nonlinear equation governing flow in a saturated porous medium. *Water Resources Res.* **18**, 1049–1052 and **19**, 591.

Kaviany, M. (1995). *Principles of Heat Transfer in Porous Media* (2nd edn). Springer–Verlag, New York.

Kuznetsov, A. V. (1996). Analytical investigation of the fluid flow in the interface region between a porous medium and a clear fluid in channels partially filled with a porous medium. *Appl. Sci. Res.* **56**, 53–57.

2 BOUNDARY ELEMENT METHOD FOR TRANSPORT PHENOMENA IN POROUS MEDIUM

L. ŠKERGET[*] and R. JECL[†]

[*]University of Maribor, Faculty of Mechanical Engineering, Smetanova 17, 2000 Maribor, Slovenia

email: leo@uni-mb.si

[†]University of Maribor, Faculty of Civil Engineering, Smetanova 17, 2000 Maribor, Slovenia

email: renata.jecl@uni-mb.si

Abstract

New computational methods and techniques have allowed us to model and simulate various phenomena in porous medium, and thus a deeper understanding of them is being gained on a perpetual basis. The aim of the present work is to obtain the numerical solution of the governing equations describing the flow of viscous incompressible fluid flow in a porous medium using an appropriate extension of the boundary element method (BEM). A basic description of the BEM is included based on a simple example of potential flow in porous medium. The results obtained on the basis of the Brinkman equations are discussed and the comparison and the suitability of the developed boundary domain integral method (BDIM) with other most commonly used numerical methods employed for this type of problem is evaluated.

Keywords: boundary element method, boundary domain integral method, porous medium, Brinkman equation, natural convection, velocity–vorticity formulation, diffusion–convective, subdomain technique

2.1 INTRODUCTION

Fluid transport phenomena in porous medium refers to the processes related to and accompanied with the transport of fluid momentum, mass and heat, through the given porous medium. These processes which are encountered in many different branches of science and technology, e.g., hydrology, geomechanics, and civil, petroleum, chemical and mechanical engineering, etc. are commonly subject to theoretical treatments which are based upon the methods traditionally developed in classical fluid dynamics. Over recent decades, fluid flows in porous medium have been studied both experimentally and theoretically. Different numerical methods were used for obtaining the solutions of some transport phenomena in porous media, e.g., the finite-difference method (FDM), finite element method (FEM), finite volume method (FVM), as well as the boundary element method (BEM). The main comparative advantage of the BEM, the application of which requires the given partial differential equation to be mathematically transformed into the equivalent integral equation representation, which is latter to be discretized, over the discrete approximative methods is demonstrated in cases where this procedure results in boundary integral equations only. This turns out to be possible only for potential problems, e.g., inviscid fluid flow, heat conduction, etc. In general, the procedure results in boundary domain integral equations and therefore several techniques were developed to extend the classical BEM. The dual reciprocity boundary element method (DRBEM) represents one of the possibilities for transforming the domain integrals into a finite series of boundary integrals, see for example Blobner *et al.* (2000) and Pérez-Gavilán and Aliabadi (2000). The key point of the DRBEM is the approximation of the field in the domain by a set of global approximation functions and the subsequent representation of the domain integrals of these global functions by boundary integrals. The discretization of the domain is represented only by grid points (poles of global approximation functions) in contrast to FDM meshes. However, the discretization of the geometry and fields on the boundary is piecewise polygonal, which gives the method greater flexibility over the FDM methods in coping with the boundary quantities. In the DRBEM all calculations reduce to the evaluation of boundary integrals only. Another more recent extension of the BEM is the so-called boundary domain integral method (BDIM), see Škerget *et al.* (1989, 1999) and Jecl *et al.* (2001). Here, the integral equations are given in terms of the variables on the integration boundaries as well as within the domain of the integration. This situation arises when we are dealing with strongly nonlinear problems with prevailing domain-based effects, for example diffusion–convection problems. The Navier–Stokes equations are commonly used as a framework for the solution of such a problems since they provide a mathematical model of physical conservation laws of mass, momentum and energy. The velocity–vorticity formulation of these equations results in the computational decoupling of the kinematics and kinetics of the fluid motion from the pressure computation, see Wu (1982). Since the pressure does not appear explicitly in the field functions conservation equations, the difficulty connected with the computation of the boundary pressure values is avoided. The main advantage of the BDIM, as compared to the classical domain type numerical techniques, is that it offers an effective way of dealing with boundary conditions on the solid walls when solving the vorticity equation. Namely, the boundary vorticity in the BDIM is computed directly from the kinematic part of the computation and not through the use of some approximate formulae. One of the few drawbacks of the BDIM are considerable CPU time and memory requirements, but they can be drastically reduced by the use of a subdomain technique, see Hriberšek and Škerget (1996). Convection-dominated fluid flows suffer from numerical instabilities. In domain-type numerical techniques upwinding schemes of different orders are used to suppress such instabilities while in BDIM the problem can be avoided by the use of Green's functions of the appropriate linear differential operators which results in a very stable and accurate numerical description of coupled diffusion–convective problems. There are no oscillations in the numerical solutions, which would have to be eliminated by using some artificial techniques, e.g., upwinding, artificial viscosity, as is the case with other approximation methods.

2.2 GOVERNING EQUATIONS

Due to the general complexity of the fluid transport process in porous medium, our work is based on a simplified mathematical model in which it is assumed that:

- the solid phase is homogeneous, non-deformable, and does not interact chemically with respect to the fluid,

- the fluid is single phase and Newtonian; its density does not depend on pressure variations, but only on variations of the temperature,

- the two average temperatures, T_s for the solid phase and T_f for the fluid phase are assumed to be identical and the porous medium is in thermodynamic equilibrium, that means it is described by a single equation for the average temperature $T = T_s = T_f$,

- no heat sources or sinks exist in the fluid; thermal radiation and Rayleigh dissipation are negligible,

- the natural convection effect is considered by using the Boussinesq approximation, where the temperature influence on the density is considered only in the term describing the body force, while in all the other terms the density is assumed to be constant.

Under these assumptions, having in mind that we are dealing with time dependent incompressible viscous fluid flow through porous medium, we may write the macroscopic set of equations, commonly called the modified Navier–Stokes equations or the Brinkman equations, as follows:

- continuity equation

$$\frac{\partial v_j}{\partial x_j} = 0, \qquad (2.1)$$

- momentum equation

$$\frac{1}{\phi}\frac{\partial v_i}{\partial t} + \frac{1}{\phi^2}\frac{\partial v_j v_i}{\partial x_j} = \underbrace{-\frac{1}{\rho}\frac{\partial P}{\partial x_j} + F g_i - \frac{\gamma}{K} v_i}_{\text{Darcy law}} + \underbrace{\frac{\gamma}{\phi}\frac{\partial^2 v_i}{\partial x_j \partial x_j}}_{\substack{\text{Brinkman} \\ \text{extension}}}, \qquad (2.2)$$

- energy equation

$$\sigma \frac{\partial T}{\partial t} + \frac{\partial v_j T}{\partial x_j} = \frac{\partial}{\partial x_j} \left(a \frac{\partial T}{\partial x_j} \right), \qquad (2.3)$$

where the vector field functions v_i, g_i and x_i are the filtration velocity, gravity and position, respectively. The scalar quantity $P = p - \rho g_i r_i$ is the modified pressure, while ρ and ϕ stand for the density and porosity. The quantities γ, K, a and σ are the kinematic viscosity coefficient, permeability, thermal diffusivity and the heat capacity ratio, which is defined by the expression

$$\sigma = \phi + \frac{\rho_s c_{p,s}}{\rho c_p} \left(1 - \phi \right), \qquad (2.4)$$

where the subscript s denotes the solid part of porous medium. The following general expression for the density variation can be used: $\rho = \rho_0 \left(1 + F \right)$, where the function F is frequently defined as follows:

$$F = -\beta_T \left(T - T_0 \right), \qquad (2.5)$$

β_T denotes the thermal volume expansion coefficient, and ρ_0 is the reference density at the temperature T_0.

The Brinkman extension expresses the viscous resistance or viscous drag force exerted by the solid phase on the flowing fluid at their contact surfaces. With the Brinkman equation one is able to satisfy the no-slip boundary conditions on an impermeable surfaces that bounds the porous medium, see for example Tong and Subramanian (1985), Lauriat and Prasad (1987), Kladias and Prasad (1989), Mamou *et al.* (1992), etc. It is important to stress that the Brinkman equation is essentially an interpolation scheme between the Navier–Stokes and the Darcy equations. It is well known, see Vasseur *et al.* (1990), that in the limit when the porosity approaches unity ($\phi \to 1$), and consequently the permeability tends to infinity ($K \to \infty$), then the Brinkman equation transforms into the classical Navier–Stokes equation for a pure fluid, while when the permeability tends to zero ($K \to 0$), the Brinkman term becomes negligible and the Darcy law is then recovered.

As we mentioned previously, our goal is to solve the Brinkman equations (2.1) – (2.3) using the boundary domain integral method (BDIM). To show the differences of the BDIM as compared with classical boundary element method (BEM) let us start with the basic description of the BEM.

2.3 BOUNDARY ELEMENT METHOD FOR POTENTIAL FLOW IN POROUS MEDIUM

The numerical methods used in continuum mechanics can be classified into three main approaches, e.g., finite-difference, finite element and boundary element approaches, see Becker (1992). The first two approaches may be treated as domain methods where the user is required to represent the geometry of the object under consideration by elements which themselves represent a volume. However, in the boundary element approach the user models only the surface (or boundary) of the body. So for two-dimensional applications, the elements are lines and for three-dimensional applications the elements are surfaces. The boundary of the body is modelled by joining these elements together edge-to-edge, as though they were being pasted onto the surface of the 3D body, until the entire body has been covered with non-overlapping elements. Since the user is required to model only the boundary, the amount of data which one has to supply is reduced substantially. By using lines to model 2D geometries and patches to model 3D geometries, the dimensionality of the problem is reduced by one in all cases. This greatly simplifies the data input and minimises the errors introduced at this early stage of the analysis. The boundary element approach implies that the governing differential equations may be transformed into integral equations which are applicable over the surface or boundary. These integrals are numerically solved over the boundary and provided that the boundary conditions are satisfied, a system of algebraic equations emerges for which a solution may be obtained.

2.3.1 Potential flow

In order to illustrate a typical boundary element derivation, all the necessary steps involved to obtain a solution are shown for a simple governing equation. Therefore we first consider the problem in which we may neglect the inertial force, the viscous resistance to the flow inside the fluid that saturate the porous medium domain, assume a constant fluid density and introduce the piezometric head ψ, namely

$$\psi = z + \frac{P}{\rho g}, \tag{2.6}$$

that expresses the mechanical energy due to gravity and the pressure of the fluid. In such a case, the momentum equation (2.2) transforms into the so-called Darcy's law, see Bear and Bachmat (1991), in the form

$$v_i = -K \frac{\partial \psi}{\partial x_i}, \tag{2.7}$$

where the coefficient K is a second-rank symmetrical tensor, called the hydraulic conductivity, defined as follows:

$$K = K \frac{g}{\gamma}, \tag{2.8}$$

and the term $\partial \psi / \partial x_i$ is called the hydraulic gradient. The hydraulic conductivity K depends on the properties of both the fluid phase ($1/\gamma$, often referred to as the fluidity, is equal to the reciprocal of the kinematic viscosity of the fluid), and the solid matrix (through the permeability K). Darcy's law in the form of equation (2.7) states that the relative specific discharge is proportional to the hydraulic gradient. For flow in an isotropic, homogeneous, fixed and non-deformable porous medium, $K = $ constant, and for $\rho = $ constant and $\gamma = $ constant the flow as defined by equation (2.7) is referred to as potential flow. The boundary element method is therefore an ideal method for solving this particular partial differential equation in order to find the solution of the unknown piezometric head in the porous medium domain. By taking the divergence of the Darcy equation (2.7), the Laplace equation is obtained in the form

$$\frac{\partial^2 \psi}{\partial x_i \partial x_i} = 0 \quad \text{in} \quad \Omega, \tag{2.9}$$

taking into account the mass conservation equation (2.1) and Ω being a two- or three-dimensional domain, see Figure 2.1. The solution of equation (2.9) is obtained by assuming that the boundary Γ consists of two parts, namely Γ_1 and Γ_2 on which the Dirichlet (essential) boundary conditions for the piezometric head and the Neumann (natural) boundary conditions or the normal derivative of the piezometric head, namely

$$\psi = \bar{\psi} \quad \text{on} \quad \Gamma_1, \qquad q = \frac{\partial \psi}{\partial n} = \bar{q} \quad \text{on} \quad \Gamma_2, \tag{2.10}$$

are known, with n being the outward normal vector to the boundary Γ ($\Gamma = \Gamma_1 + \Gamma_2$) and the bar indicates that values of ψ and q are known on Γ_1 and Γ_2, respectively.

2.3.2   Integral equation

The boundary element method is a weighted residual method for solving partial differential equations, characterised by choosing an appropriate fundamental solution as a weighting function and by using the Green's formula for the complete transform of one, or more, of the partial differential operators to the weighted function. The Laplace equation (2.9) is solved by using the weighted residual principle, see for example Banerjee and Butterfield (1981) and Brebbia and Dominquez (1992), to obtain

$$\int_\Omega \frac{\partial^2 \psi}{\partial x_i \partial x_i} u^* \, d\Omega = \int_{\Gamma_2} (q - \bar{q}) \, u^* \, d\Gamma - \int_{\Gamma_1} \left(\psi - \bar{\psi}\right) q^* \, d\Gamma, \tag{2.11}$$

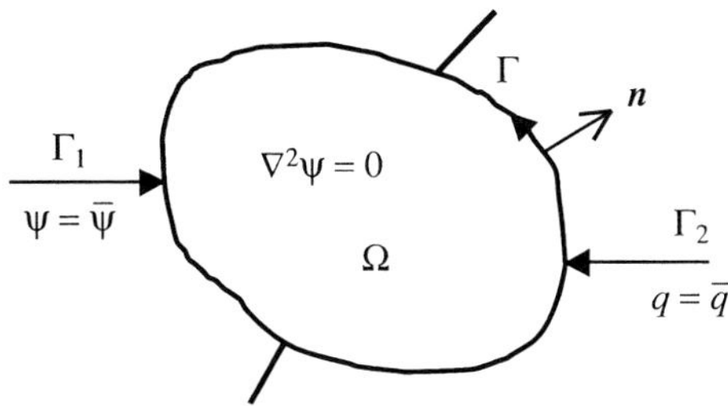

Figure 2.1   *Schematic diagram of the geometry and boundary conditions of the problem under consideration*

where u^* is a weighting function and $q = \partial u^*/\partial n$ its derivative on the boundary. Integrating twice by parts the left-hand side of the equation (2.11) we obtain

$$\int_\Omega \frac{\partial^2 u^*}{\partial x_i \partial x_i} \psi \, d\Omega = - \int_{\Gamma_2} \bar{q} u^* \, d\Gamma - \int_{\Gamma_1} q u^* \, d\Gamma + \int_{\Gamma_2} \psi q^* \, d\Gamma + \int_{\Gamma_1} \bar{\psi} q^* \, d\Gamma. \quad (2.12)$$

This equation is the starting point for the application of the boundary element method and our aim is now to render the equation (2.12) into a boundary integral equation. This is done by using a special type of weighting function u^*, called the fundamental solution, which satisfies the Laplace equation, namely u^* is the solution of the following equation:

$$\frac{\partial^2 u^*}{\partial x_i \partial x_i} + \delta(\xi, s) = 0. \quad (2.13)$$

The fundamental solution u^* represents the field generated by a concentrated unit charge acting at a source point ξ, while s denotes the reference point. The effect of this charge is transferred from source point to infinity without any consideration of the boundary conditions. Further, the function $\delta(\xi, s)$ represents the Dirac delta function which tends to infinity at the source point and is equal to zero everywhere else. The integral of a Dirac delta function multiplied by any other function is equal to the value of the latter at the point ξ. Therefore equation (2.12) can be written as follows:

$$c(\xi)\psi(\xi) + \int_{\Gamma_2} \psi q^* \, d\Gamma + \int_{\Gamma_1} \bar{\psi} q^* \, d\Gamma = \int_{\Gamma_2} \bar{q} u^* \, d\Gamma + \int_{\Gamma_1} q u^* \, d\Gamma, \quad (2.14)$$

or

$$c(\xi)\psi(\xi) + \int_\Gamma \psi q^* \, d\Gamma = \int_\Gamma U^* q \, d\Gamma, \quad (2.15)$$

which is valid for the whole boundary $\Gamma = \Gamma_1 + \Gamma_2$ before any boundary conditions have been applied. For an isotropic three-dimensional domain, the fundamental solution of equation (2.13) is given by

$$u^* = \frac{1}{4\pi r(\xi, s)}, \quad (2.16)$$

and for a two-dimensional isotropic domain it is given by

$$u^* = \frac{1}{2\pi} \ln \frac{1}{r(\xi, s)}, \quad (2.17)$$

where $r(\xi, s)$ is the distance from the source point ξ (from the point of application of the delta function) to the reference point s (any point under consideration), see Brebbia and Dominquez (1992). The equation (2.15) is valid for any point within the domain Ω because of the geometrically dependent free term $c(\xi)$, which accounts for the Cauchy type singularity of the integral on the left-hand side of equation (2.15). The geometrical coefficient $c(\xi)$ denotes the fundamental solution related coefficient depending on the position of the source point and it takes the following values:

$c(\xi) = 0$ when ξ is outside the domain Ω,

$c(\xi) = 1$ when ξ is inside the domain Ω,

$c(\xi) = 1/2$ when ξ lies on the smooth boundary Γ,

$c(\xi) = \beta/2\pi$ when ξ lies on the non-smooth boundary Γ where β is the internal angle of the boundary at the point ξ.

2.3.3 Discretization

Searching for an approximate numerical solution, the integral equation (2.15) is written in a discretized manner in which the integrals over the boundary are approximated by a sum of integrals over individual boundary elements. Therefore the boundary Γ is divided into N segments or elements. The points which define the geometry of these elements are called 'mesh points' which are always placed at the corners and midpoints of the elements. The points where the unknown values are considered are called 'nodes'. Unlike the finite element method, the mesh points do not necessarily coincide with the nodes. The mesh points define only the geometry and the nodes define the values of the unknown function, or its derivative, on the element. Two different types of elements are basically used, namely continuous and discontinuous elements. Continuous elements have nodes on the edges and corners of the element, which are shared with neighbouring elements. Discontinuous elements do not share nodes with neighbouring elements, and the nodes are therefore not at the edges and corners of the element, but are displaced towards the element centroid. The number of nodes on an element varies according to the user requirements. For constant elements there is only one node (taken to be in the middle of the element) for each element, linear elements are those for which the nodes are at the ends of the elements, and quadratic elements (curved elements) are those for which a further mid-element node is required. In what follows, the description of different continuous elements is given. For constant elements, the values of the unknown function, in our case the piezometric head ψ, and the values of the derivative of this function are assumed to be constant over each element and equal to the value at the mid-element node, see Figure 2.2.

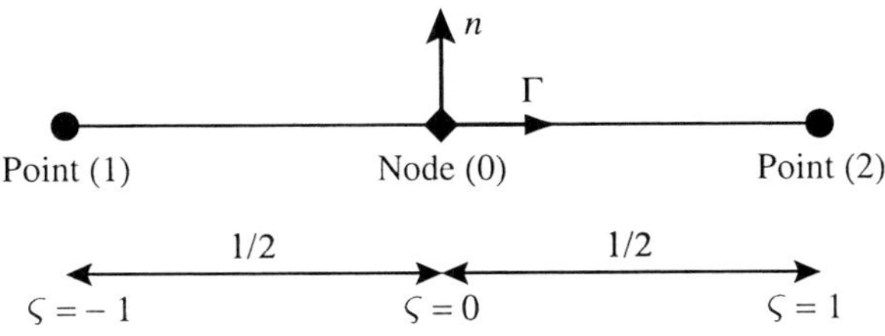

Figure 2.2 *Constant element definitions*

to the left-hand side of the equation then we can write

$$AX = F, \qquad (2.25)$$

where X is a vector of the unknowns of the functions ψ and q. It should be mentioned that the vector F may be determined by multiplying the corresponding columns of the equation (2.25) by the known values of ψ or q. It is interesting to note that the unknowns are now a mixture of the potential and its derivative, rather than the potential only as in the finite element method. This is a consequence of the boundary element being of a 'mixed' formulation and this gives an important advantage to the boundary element method over the finite element method. Equation (2.25) can now be solved and all the boundary values are then known. Once this is done then it is possible to calculate any internal value of ψ, or its derivative q, using equation (2.15) for $c(\xi) = 1$.

Integrals, such as $\hat{h}_j$ and g_j in the expression (2.22), can be calculated analytically or by using a numerical integration formulae, such as the Gauss quadrature scheme, for the case $\xi \neq j$. For the element $\xi - j$, the presence of the singularity due to the fundamental solution requires a more accurate integration. For these integrals it is recommended to use a higher-order integration scheme or a special formula, such as the logarithmic transformation. However, when other types of boundary elements are used in the discretization, then the integrals are more difficult to evaluate as the known function and its derivative vary linearly or quadratically over each part of the boundary Γ_j. Hence it is not possible to take them out of the integrals as done in equation (2.21), see for example Banerjee and Butterfield (1981) or Brebbia and Dominquez (1992).

Here the procedure of the classical BEM has been described for the most simple governing equation, namely the Laplace equation. This equation is often used for water seepage through porous medium, for which the examples can be found in the literature, see Cheng (1984) and da Veiga *et al.* (1994).

2.4   BOUNDARY DOMAIN INTEGRAL METHOD

So far we have explained the basic principles of the boundary element method (BEM), which is an ideal tool for solving potential problems involving the Laplace or Poisson equation. However, when the problems are more complicated, such as the diffusion–convective equations then the method must be extended in a way that we need to carry out integrals in the domain as well as on the boundary. In what follows we give all the necessary steps which are required to determine the solution of such an equation, e.g., a system of diffusion–convective equations, using the boundary domain integral method.

2.4.1   Velocity–vorticity formulation

Following the concept of BEM, the governing equations (2.1) – (2.3) are further transformed by using the velocity–vorticity variable formulation (VVF), according to Guj and

Stella (1993). Defining the vorticity vector ω_i, which represents the curl of the velocity field, as follows:

$$\omega_i = e_{ijk}\frac{\partial v_k}{\partial x_j}, \tag{2.26}$$

where e_{ijk} is the unit tensor, and applying the velocity–vorticity approach, the fluid motion computation scheme is partitioned into its kinematic and kinetic parts so that the continuity and the momentum equations are replaced by the equations of kinematics and kinetics, see Škerget *et al.* (1989). With a slight modification of the momentum equation, i.e., introducing the new variable τ_v ($\tau_v = t/\phi$), and taking the curl of equation (2.2), then the kinetics is governed by the following parabolic kinetic equation, namely the vorticity transport equation:

$$\frac{\partial \omega_i}{\partial \tau_v} + v_j\frac{\partial \omega_i}{\partial x_j} = \omega_j\frac{\partial v_i}{\partial x_j} + \phi\gamma\frac{\partial^2 \omega_i}{\partial x_j\partial x_j} + \phi^2 e_{ijk}g_k\frac{\partial F}{\partial x_j} - \frac{\phi^2\gamma}{K}\omega_i, \tag{2.27}$$

which describes the redistribution of the vorticity vector in the fluid flow field. The quantity τ_v is the so-called modified vorticity time step which is introduced only as a necessary mathematical step allowing one to use the VVF principle in the momentum equation (2.2).

Applying the curl operator directly to the vorticity defined in equation (2.26), and using the continuity equation (2.1), then the kinematics can be formulated in the form of an elliptic Poisson equation for the velocity vector, namely

$$\frac{\partial^2 v_i}{\partial x_j\partial x_j} + e_{ijk}\frac{\partial \omega_k}{\partial x_j} = 0, \tag{2.28}$$

which represents the compatibility, or the restriction condition between the velocity and the vorticity fields, at any given point in space and time. In order to improve the convergence of the coupled velocity–vorticity iterative scheme, then instead of using equation (2.28), we have used the following parabolized kinematic equation, see for example Guj and Stella (1993):

$$\frac{\partial^2 v_i}{\partial x_j\partial x_j} - \frac{1}{\alpha}\frac{\partial v_i}{\partial t} + e_{ijk}\frac{\partial \omega_k}{\partial x_j} = 0, \tag{2.29}$$

which is extended with the fictitious velocity accumulation term which smoothes out the numerical oscillations, and α is a relaxation parameter. The velocity equation, as given by equation (2.28), is exactly satisfied in the steady state ($t \to \infty$) when the time derivatives vanish. Defining the modified temperature time step τ_T ($\tau_T = t/\sigma$), the energy equation (2.3) may be rewritten in the form of a diffusion–convective equation which represents the energy transport equation

$$\frac{\partial T}{\partial \tau_T} + v_j\frac{\partial T}{\partial x_j} = a\frac{\partial^2 T}{\partial x_j\partial x_j}. \tag{2.30}$$

Applying the integral formulation given by equation (2.49), we obtain the boundary domain integral formulation for the vorticity kinetics as follows:

$$
c\left(\xi\right)\omega_i\left(\xi\right) + \int_\Gamma \omega_i \frac{\partial U^*}{\partial n}\, \mathrm{d}\Gamma =
$$

$$
\frac{1}{\phi\gamma}\int_\Gamma \left(\phi\gamma\frac{\partial\omega_i}{\partial n} - \omega_i v_n - v_i\omega_n + \phi^2 e_{ijk}g_k F n_j\right) U^*\, \mathrm{d}\Gamma
$$

$$
+ \frac{1}{\phi\gamma}\int_\Omega \left(\omega_i\tilde{v}_j - v_i\omega_j - \phi^2 e_{ijk}g_k F\right)\frac{\partial U^*}{\partial x_j}\, \mathrm{d}\Omega
$$

$$
- \frac{\phi}{K}\int_\Omega \omega_i U^*\, \mathrm{d}\Omega + \beta \int_\Omega \omega_{i,F-1} U^*\, \mathrm{d}\Omega,
$$

$$
(2.56)
$$

where $U^* = \phi\gamma u^*$, β is now given by $\beta = 1/\phi\gamma\Delta\tau_v$ and u^* is the elliptic diffusion–convective fundamental solution.

Considering that the temperature T, as described by equation (2.30), satisfies the non-homogeneous equation (2.46), subject to the boundary conditions as given by equation (2.33), is the solution of the following equation:

$$
\bar{a}_p \frac{\partial^2 T}{\partial x_j \partial x_j} - \frac{\partial \bar{v}_j T}{\partial x_j} - \frac{T}{\Delta\tau_T} + b = 0, \tag{2.57}
$$

where the pseudo body force b, which includes convection due to the perturbed part of the velocity, then the nonlinear diffusion part and the initial conditions is given by

$$
b = -\frac{\partial \tilde{v}_j T}{\partial x_j} + \frac{\partial}{\partial x_j}\left(\tilde{a}_p \frac{\partial T}{\partial x_j}\right) + \frac{T_{F-1}}{\Delta\tau_T}. \tag{2.58}
$$

Applying the integral formulation, given by equation (2.49), we obtain the boundary domain integral formulation for the temperature kinetics as follows:

$$
c\left(\xi\right)T\left(\xi\right) + \int_\Gamma T\frac{\partial U^*}{\partial n}\, \mathrm{d}\Gamma = \frac{1}{\bar{a}_p}\int_\Gamma \left(a_p\frac{\partial T}{\partial n} - T v_n\right) U^*\, \mathrm{d}\Gamma
$$

$$
+ \frac{1}{\bar{a}_p}\int_\Omega \left(\tilde{v}_j T - \tilde{a}_p\frac{\partial T}{\partial x_j}\right)\frac{\partial U^*}{\partial x_j}\, \mathrm{d}\Omega \tag{2.59}
$$

$$
+ \beta \int_\Omega T_{F-1} U^*\, \mathrm{d}\Omega,
$$

where $U^* = \bar{a}_p u^*$, β is now the parameter defined as $\beta = 1/\bar{a}_p\Delta\tau_T$, and u^* is the elliptic diffusion–convective fundamental solution. The coefficient $\bar{a}_p$ is the constant part of the thermal diffusivity and $\tilde{a}_p$ is the perturbed part of thermal diffusivity, such that $a_p = \bar{a}_p + \tilde{a}_p$, where the thermal diffusivity is calculated as $a_p = \lambda_p/\rho c$, and λ_p is the heat conductivity of the porous medium, defined as $\lambda_p = (1 - \phi)\lambda_s + \phi\lambda$ with λ_s denoting the heat conductivity of the solid.

2.4.5 Numerical solution of the diffusion–convective integral equation

Comparing the equations (2.44) and (2.49), we can observe that they are almost identical and that is the reason why we present the discretization procedure only once. The governing partial differential equations for the velocity, vorticity and temperature was previously written in an integral formulation but the solution of those equation is still confined to simple cases, e.g., elementary geometries, basic forms of the boundary conditions, etc. The lack of an appropriate analytical solutions of those equations is the main reason for the use of the boundary element method. For the numerical approximate solution of the field functions, namely the velocity, vorticity, and temperature, then the corresponding boundary domain integral representation is written in a discretized manner in which the integrals over the boundary and domain are approximated by a sum of the integrals over E boundary elements and over C internal cells. The regions of integration, called cells, can now be employed to compute the domain integrals. The cells are surface elements which cover the boundary of the solution domain. They are usually of two types: triangular or quadrilateral and both can be flat or curved. The potential and its derivative, and functions which describe the geometry of the cell can be constant over the element or vary linearly, being second-order functions and others which produce a curved element. For example, the zero-order constant quadrilateral cell is defined with one point in the centre of the cell which is bounded by four boundary elements, the second-order bilinear quadrilateral cell is defined with four points placed in the corners of the cell, etc., see Brebbia and Dominquez (1992) for more details.

The integral equation (2.44) can be written as the sum of the boundary and domain integrals as follows:

$$c\left(\xi\right) u\left(\xi\right) + \sum_{e=1}^{E} \int_{\Gamma_e} u \frac{\partial u^*}{\partial n}\, \mathrm{d}\Gamma = \frac{1}{\bar{a}} \sum_{e=1}^{E} \int_{\Gamma_e} \left(a \frac{\partial u}{\partial n} - uv_n\right) u^*\, \mathrm{d}\Gamma$$

$$+ \frac{1}{\bar{a}} \sum_{c=1}^{C} \int_{\Omega_c} \left(uv_j - \tilde{a} \frac{\partial u}{\partial x_j}\right) \frac{\partial u^*}{\partial x_j}\, \mathrm{d}\Omega$$

$$+ \frac{1}{\bar{a}} \sum_{c=1}^{C} \int_{\Omega_c} I_u u^*\, \mathrm{d}\Omega + \beta \sum_{c=1}^{C} \int_{\Omega_c} u_{F-1} u^*\, \mathrm{d}\Omega. \tag{2.60}$$

The variation of the field functions, or their products within each boundary element or internal cell, is defined in terms of their nodal values and by the use of the interpolation polynomials 'φ' or 'Φ' with respect to the boundary or domain nodal function values, which are similar to those obtained for the potential problem, e.g., we have

$$u = \{\varphi\}^{\top} \{u\}^n, \quad a \frac{\partial u}{\partial n} = \{\varphi\}^{\top} \left\{a \frac{\partial u}{\partial n}\right\}^n, \quad uv_n = \{\varphi\}^{\top} \{uv_n\}^n, \tag{2.61}$$

(2.72) – (2.74) can be rewritten in the form to determine the boundary normal flux, or boundary field function nodal values, and the field function domain nodal values in the form of equation (2.25), see Škerget *et al.* (1999) for more details.

Since the implicit set of equations is written simultaneously for all the boundary and internal nodes, this results in a very large, fully populated system matrix which is influenced by the diffusion and the convection. The consequence is a very stable and accurate numerical scheme with substantial computer time and memory demands. In order to improve the economics of the computation, and thus widen the applicability of the proposed numerical algorithm, the subdomain technique has to be used, see Hribersek and Škerget (1996). The idea is to partition the entire solution domain into subdomains to which the same discretized numerical procedure can be applied. The final system of equations for the entire domain is then obtained by adding the sets of equations for each subdomain, considering the compatibility and equilibrium conditions between their interfaces. This results in a much more sparse matrix system, which is suitable to be solved by iterative techniques. On the interface Γ_1, between the subdomains Ω_1 and Ω_2, the following compatibility and equilibrium conditions can be applied for each conservation equation:

- kinematic equation

$$v_i|_I^1 = v_i|_I^2, \quad \frac{\partial v_i}{\partial n}\bigg|_I^1 = -\frac{\partial v_i}{\partial n}\bigg|_I^2, \tag{2.75}$$

- vorticity equation

$$\omega|_I^1 = \omega|_I^2, \quad \left(\gamma\frac{\partial \omega}{\partial n}\right)\bigg|_I^1 = -\left(\gamma\frac{\partial \omega}{\partial n}\right)\bigg|_I^2, \tag{2.76}$$

- energy equation

$$T|_I^1 = T|_I^2, \quad \left(\lambda_p\frac{\partial T}{\partial n}\right)\bigg|_I^1 = -\left(\lambda_p\frac{\partial T}{\partial n}\right)\bigg|_I^2. \tag{2.77}$$

The discrete model is based on the substructure technique following the concept of the finite volume, e.g., that each quadrilateral internal cell represents one subdomain bounded by four boundary elements, see Škerget *et al.* (1999). The geometrical singularities are overcome by using 3-node discontinuous quadratic boundary elements combined with 9-node corner continuous internal cells, see Figure 2.5.

The quadratic interpolation functions have been used since a good approximation of the boundary values of the velocity gradients ensure an accurate evaluation of the boundary vorticity values, which strongly influence the stability of the proposed method. The interpolation polynomials are given by the expressions φ for discontinuous three-noded quadratic boundary elements and by Φ for continuous nine-node quadratic internal cells,

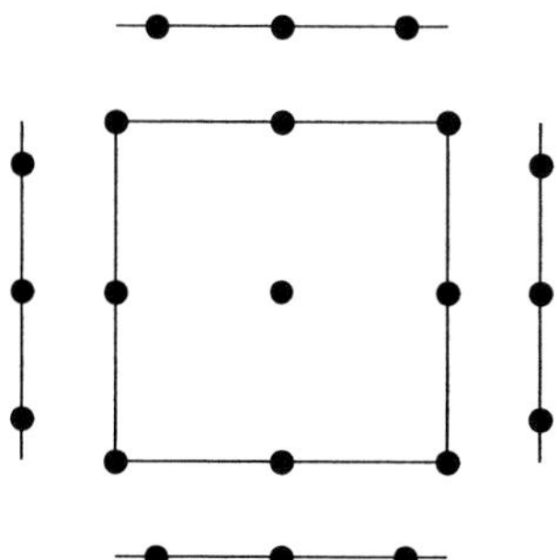

Figure 2.5 *Discrete model: layout of a subdomain*

where ς and η are the local coordinate, as follows:

$$\Phi^1 = \tfrac{1}{4}\left(\eta - \eta^2\right)\left(\varsigma - \varsigma^2\right), \quad \Phi^2 = \tfrac{1}{2}\left(1 - \eta^2\right)\left(\varsigma^2 - \varsigma\right), \quad \Phi^3 = \tfrac{1}{4}\left(\eta + \eta^2\right)\left(\varsigma^2 - \varsigma\right),$$

$$\Phi^4 = \tfrac{1}{2}\left(\eta + \eta^2\right)\left(1 - \varsigma^2\right), \quad \Phi^5 = \tfrac{1}{4}\left(\eta + \eta^2\right)\left(\varsigma + \varsigma^2\right), \quad \Phi^6 = \tfrac{1}{2}\left(1 - \eta^2\right)\left(\varsigma^2 + \varsigma\right),$$

$$\Phi^7 = \tfrac{1}{4}\left(\eta^2 - \eta\right)\left(\varsigma + \varsigma^2\right), \quad \Phi^8 = \tfrac{1}{2}\left(\eta - \eta^2\right)\left(\varsigma^2 - 1\right), \quad \Phi^9 = \left(1 - \eta^2\right)\left(1 - \varsigma^2\right),$$

$$\varphi^1 = \tfrac{1}{9}\left(-6\varsigma + 8\varsigma^2\right), \qquad \varphi^2 = \tfrac{1}{9}\left(9 - 16\varsigma^2\right), \qquad \varphi^3 = \tfrac{1}{9}\left(6\varsigma + 8\varsigma^2\right).$$

$$(2.78)$$

The kinematics, given by equation (2.72) and the velocity boundary conditions prescribed by equations (2.31) and (2.32), cannot assure the solenoidality of the velocity field for an arbitrary vorticity distribution, and therefore this property may be fulfilled only by coupling the kinetic and kinematic equations. Thus, the solenoidality conditions of the velocity and vorticity field requires a coupled iterative solution of the nonlinear dynamical system of equations (2.72) – (2.74) with the corresponding boundary conditions described by equations (2.33) – (2.35). To obtain a solution of the fluid motion problem, the following iterative steps have to be performed:

(i) Start with some initial values for the vorticity distribution.

(ii) Kinematic computational part (velocity field):

- solve the implicit sets for the boundary velocity or the velocity normal flux values, equation (2.72);
- transform the new function values from the element nodes to the cell nodes;
- compute the gradient of the velocity components;
- determine the new boundary vorticity values, equation (2.34);
- determine the new boundary domain integral kinetic matrices, if the constant velocity vector is perturbed more than the prescribed tolerance.

(iii) Energy kinetic computational part (temperature field):

- solve the implicit set for the boundary and domain values, equation (2.74);

- transform the new function values from the element nodes to the cell nodes.

(iv) Vorticity kinetic computational part (vorticity field):

- solve the implicit set for the unknown boundary vorticity flux and internal domain vorticity values, equation (2.73);

- transform the new function values from the element nodes to the cell nodes.

(v) Relax all the new values and the test for convergence. If the convergence criterion is satisfied then stop the iterative procedure, otherwise go to step (ii).

2.5   TEST EXAMPLE

2.5.1   Natural convection in a porous cavity saturated with a Newtonian fluid

When the temperature of the saturating fluid phase in a porous medium is not uniform then flows which are induced by buoyancy effects occur. These flows, which depend on the density differences due to the temperature gradients and the pertinent boundary conditions, are commonly called free or natural convection. Due to their numerous applications in energy related engineering problems, natural convection has been receiving an increasing amount of interest and it has become one of the most commonly studied transport phenomena in porous medium. Studies have been reported, for different geometries and a variety of heating conditions. For example, a vertical cavity in which a horizontal temperature gradient is induced by side walls being maintained at different temperatures has been analysed by several researchers, see Lauriat and Prasad (1987), Nakayama and Pop (1989) and Vasseur *et al.* (1990). Others have examined the natural convection in a porous layer heated from below, for example Kladias and Prasad (1989) and Mamou *et al.* (1992). In all these studies, use has been made of the Brinkman-extended Darcy formulation as the governing momentum equation. This is because it was established in the earlier work on flows through porous medium that the simple Darcy law does not give satisfactory results when we wish to take into account the no-slip boundary condition, see for example Tong and Subramanian (1985) and Lage (1998).

To check the validity of the proposed numerical procedure, we discus the problem of natural convection in a vertical porous cavity. The description of the physical problem is shown in Figure 2.6, which represents a two-dimensional, vertical cavity filled with an isotropic, homogeneous, Newtonian fluid-saturated porous medium, with one vertical wall being isothermally heated and the other is isothermally cooled, while the horizontal walls are adiabatic.

The thermophysical properties of the solid and the fluid phases are assumed to be constant, except for the density variation in the body force term. Assuming that both the solid and the fluid in the porous medium are in thermal equilibrium then the governing equations are written in the form of equations (2.1) – (2.3). Computations have been carried out for the complete Brinkman-extended Darcy model with the transport term in the momentum

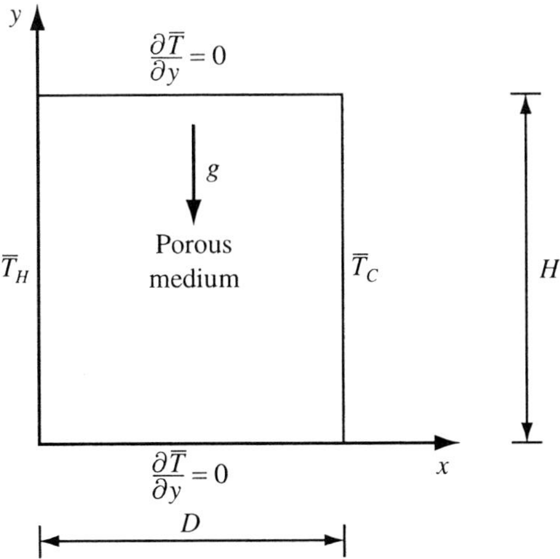

Figure 2.6 *Geometry and boundary conditions for the porous cavity*

equation included. Whenever we consider the Brinkman term, we have to deal with the so-called Darcy number Da, see Lauriat and Prasad (1987), which is defined as the ratio between the permeability and the characteristic length multiplied by the viscosity ratio Λ, which in our case is equal to the reciprocity of the porosity ($\Lambda = 1/\phi$), see Jecl *et al.* (2001). The governing parameters for the present problem are: the porosity ϕ, modified Rayleigh number $Ra^* = g\beta_T K D \Delta T / \gamma a_p$, Darcy number as $Da = (1/\phi)\left(K/D^2\right)$, aspect ratio $A = H/D$, and the ratio of the volumetric heat capacity of the solid and fluid phase σ. Here D, H and ΔT are the width of the cavity, the height of the cavity and the temperature difference between hot and cold walls, respectively.

We have tested our numerical model for several different parameters and therefore we can confirm that the effect of an increase in the Darcy number Da appears to be very similar at all the values of the Rayleigh number Ra we have considered, namely $100 \leqslant Ra^* \leqslant 1000$. However, it is well known that the effect of the viscous (Brinkman) term becomes more important at high modified Rayleigh numbers, see Lauriat and Prasad (1987). The proposed BDIM scheme has been verified for a square porous cavity with aspect ratio $A = 1$. Because of the above mentioned similarity, we graphically present only one example, namely $Ra^* = 500$, in order to outline the relevant characteristics that are common at all modified Rayleigh numbers. The boundary conditions for the computed test examples are as follows:

$$\begin{aligned}
\bar{v}_x = \bar{v}_y = 0 \qquad &\text{for} \quad x = 0, D \quad \text{and} \quad y = 0, H, \\
\frac{\partial \overline{T}}{\partial y} = 0 \qquad &\text{for} \quad y = 0, H, \\
\overline{T} = \overline{T}_H = 0.5 \qquad &\text{for} \quad x = 0, \\
\overline{T} = \overline{T}_C = -0.5 \qquad &\text{for} \quad x = D.
\end{aligned} \tag{2.79}$$

The streamlines and isotherms are presented in Figures 2.7 and 2.8 for $Da = 10^{-1}, 10^{-2}$, 10^{-3}, 10^{-4} and 10^{-5}. A computational mesh of 10×10 subdomains is used and time steps ranging from $\Delta t = 10^{16}$ (steady state) for $Da = 10^{-1}$, $\Delta t = 10^{-1}$ for $Da = 10^{-2}$, $\Delta t = 10^{-2}$ for $Da = 10^{-3}$ to $\Delta t = 10^{-3}$ for $Da = 10^{-4}$, 10^{-5} have been employed. The convergence criterion was selected as $\varepsilon = 10^{-5}$ and if the difference between all the computed values at iteration n and iteration $n - 1$ is greater than this value the next iteration is required, otherwise the computation is finished. Further, in order to illustrate the typical results obtained we have taken the porosity to be $\phi = 0.5$ and the heat capacity ratio to be $\sigma = 1$.

The streamlines in Figure 2.7(a) are observed to be closely spaced near the solid boundaries. This configuration indicates, as expected, that the fluid velocity has a maximum near the boundary since in the limit when $Da = 0$ (Darcy law) the velocity has a maximum on the boundaries. In this case Da is small enough that the viscous term which is responsible for the boundary effects becomes negligible and the Darcy law correctly describes the fluid flow behaviour. Figures 2.7(b) to (d) illustrate typical results obtained on the basis of the Brinkman model for various values of Da. It is evident that when the Darcy number increases then the influence of the boundary effects on the flow field becomes significant and the streamlines are observed to become relatively more and more sparsely spaced near the solid boundaries. This is due to the fact that the viscous–Brinkman effect becomes

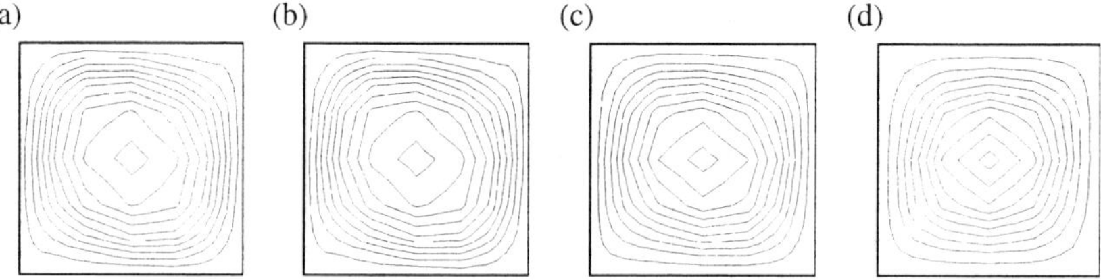

Figure 2.7   *Streamlines for $A = 1$, $\phi = 0.5$, $\Delta T = 1$ and $Ra^* = 500$ for (a) $Da = 10^{-4}$, (b) $Da = 10^{-3}$, (c) $Da = 10^{-2}$, and (d) $Da = 10^{-1}$*

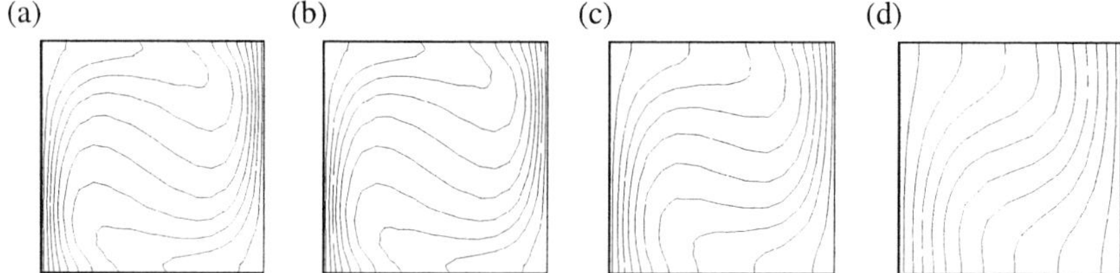

Figure 2.8   *Isotherms for $A = 1$, $\phi = 0.5$, $\Delta T = 1$ and $Ra^* = 500$ for (a) $Da = 10^{-4}$, (b) $Da = 10^{-3}$, (c) $Da = 10^{-2}$, and (d) $Da = 10^{-1}$*

gradually more important and slows down the fluid in the neighbourhood of the solid walls. It is also observed that the region where the flow has a maximum velocity, as indicated by the closely spaced streamlines, moves away from the walls towards the core region of the cavity as Da increases.

Similarly, we can observe the effects of the Darcy number on the isotherms and on the temperature field, see Figure 2.8. When the Darcy number is small, see Figure 2.8(a), the convective motion inside the cavity is strong and the isotherms are considerably distorted. The flat isotherms in the core indicate a negligible lateral conduction. As Da increases then the viscous effects become more important and slow down the buoyancy induced flow inside the cavity. The isotherm profiles become more linear and the heat transfer across the cavity results from the combined action of conduction and convection.

From the above results we can clearly observe that the streamlines and isotherms redistribution are almost identical for small values of the Darcy number, $Da = 10^{-4}$ and $Da = 10^{-3}$, but with a further increase in the Darcy number, e.g., by increasing the permeability K, the velocity and temperature fields start to become significantly modified.

The rate of heat transfer is expressed by the average Nusselt number which is defined as follows:

$$\overline{Nu} = \int_0^1 \left(\frac{\partial T}{\partial n} \right) dy. \tag{2.80}$$

The variation of the Nusselt number for different Rayleigh numbers is shown in Table 2.1, where in the brackets the results of Lauriat and Prasad (1987) are shown. A direct comparison is not fully possible because in the above mentioned study the authors have calculated the Nusselt number considering the Brinkman momentum equation in which the transport term is assumed to be negligible, while in our work the computations have been made on the basis of the complete Brinkman equation in the form as given by equation

Table 2.1 *Values of the average Nusselt number $\overline{Nu}$ (values in brackets are those obtained by Lauriat and Prasad, 1987)*

Da/Ra^*	100	200	500	1000
10^{-1}	1.026	1.061	1.370	1.815
	–	–	–	–
10^{-2}	1.479	2.016	2.823	3.691
	(1.46)	(1.70)	(2.58)	(3.30)
10^{-3}	1.816	2.666	4.030	7.410
	(1.88)	(2.41)	(3.80)	(5.42)
10^{-4}	1.895	2.718	4.370	7.921
	(2.14)	(2.84)	(4.87)	(7.37)
10^{-5}	2.010	2.765	4.474	8.200
	(2.15)	(3.02)	(5.37)	(8.41)

(2.2). Also we have assumed that the viscosity ratio Λ is equal to the reciprocity value of the porosity ϕ ($\Lambda = 1/\phi$), according to Nield and Bejan (1992), while Lauriat and Prasad (1987) have taken $\Lambda = 1$. Therefore, their results can be used by replacing the modified Rayleigh number Ra^* in their formulation with Ra^* multiplied by a factor 2.

The variation of the average Nusselt number with Da is presented in Figures 2.9 and 2.10 for $A = 1$, $Ra^* = 100, 200, 500$ and 1000. As expected, the average Nusselt number approaches the conduction value ($\overline{Nu} = 1$) when Ra^* approaches zero ($Ra^* \to 0$). Also, the average Nusselt number always increases with increasing Ra^*, but the effect of the Darcy number is just the reverse, namely from Figure 2.9 we can clearly observe that

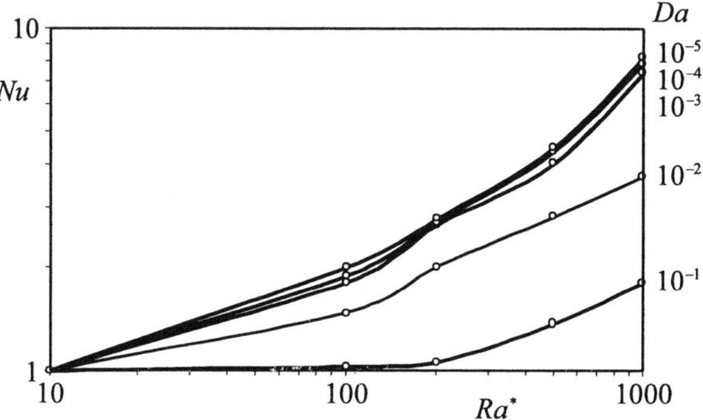

Figure 2.9   *Variation of the average Nusselt number with Ra^* for $10^{-5} \leqslant Da \leqslant 10^{-1}$*

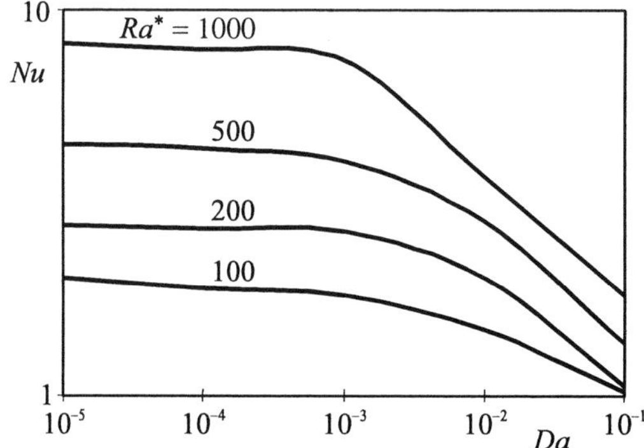

Figure 2.10   *Variation of the average Nusselt number with Da for $Ra^* = 100, 200, 500$ and 1000*

the effect of the viscous–Brinkman term becomes negligible when $Da < 10^{-3}$, which is in complete agreement with the observation of Lauriat and Prasad (1987) who have solved the same problem using the finite-difference method. Our results are obtained on a relatively small computational mesh of 10×10 subdomains, while Lauriat and Prasad (1987) have performed their computations on a 41×41 grid.

Using the boundary domain integral method, we can therefore confirm the basic fact that, when using the Brinkman momentum equation which should be used when considering the problem of natural convection in configurations that are bounded by a solid walls, the effect of the viscous term becomes negligible when $Da < 10^{-3}$. Further, if the Brinkman equation is employed then the no-slip boundary condition on the impermeable walls that bound the porous medium domain may be enforced and this gives physically more realistic results, especially when the Darcy number is small, than if the classical Darcy law is used.

2.6 CONCLUSION

A numerical approach, which is based on the boundary domain integral method (BDIM), which is an extension of the boundary element method (BEM), has been applied to the solution of the transport equations in porous medium. The modified Navier–Stokes equations (Brinkman-extended Darcy formulation with the inertial term included) have been employed to describe the fluid motion in porous medium. The solution is based on the velocity–vorticity formulation of the governing equations which allows the separation of the computational scheme into its kinematic and kinetic parts. An elliptic modified Helmholtz fundamental solution is used for the kinematic part of the computation, while an elliptic diffusion–convective fundamental solution is employed for the kinetic part. The subdomain technique, in its limited version, e.g., each subdomain is being constructed of four discontinuous 3-node quadratic boundary elements and one continuous 9-node corner continuous quadratic cell, is applied. The proposed numerical procedure is applied to the case of natural convection in a porous cavity saturated with a Newtonian fluid, which is heated from the side, for different values of the Rayleigh and Darcy numbers. It can be stated that the boundary domain integral method, extended in a way which enables the investigation of the fluid transport phenomena also in a porous medium, appears to posses the potential to become a very powerful alternative over existing numerical methods, e.g., finite differences or finite elements, as a means for obtaining solutions to the most complex systems of nonlinear partial differential equations, when attacking some unsolved problems in engineering practice.

REFERENCES

Banerjee, P. K. and Butterfield, R. (1981). *Boundary Element Methods in Engineering Science.* McGraw–Hill, Bristol.

3 RECENT ADVANCES IN THE INSTABILITY OF FREE CONVECTIVE BOUNDARY LAYERS IN POROUS MEDIA

D. A. S. REES

Department of Mechanical Engineering, University of Bath, Claverton Down, Bath, BA2 7AY, UK

email: D.A.S.Rees@bath.ac.uk

Abstract

This chapter presents an overview of recent work by the author on the onset and evolution of instabilities in thermal boundary-layer flows in porous media. Attention is confined to those cases where the heated surface is almost vertical, for in that mathematical limit the onset of convection takes place at a very large distance from the leading edge, and therefore it is possible to analyse both the linear and nonlinear instability characteristics of the flow within the framework of the boundary-layer approximation. The chapter covers the following four topics: linear instability, the nonlinear evolution of vortices, secondary instabilities of vortices, and the effect of inertia on linear instability. After the vortex disturbance has been seeded into the boundary-layer we find that vortices enjoy only a finite region of growth before they ultimately decay. The region of growth depends not only on the wavenumber of the vortex, but also on where it is introduced and on its initial amplitude. The eventual decay of the nonlinear vortex signals the possibility of secondary instabilities since the local Rayleigh number, which is proportional to the local boundary-layer thickness, continues to grow with distance from the leading edge, and therefore the basic flow becomes increasingly susceptible to instability. We present two types of secondary instability, a wavelength doubling instability, and a modal cascade reminiscent of the Eckhaus or sideband instability. Finally we touch on the effect of form drag inertia, as modelled by the Forchheimer terms, on the linear stability characteristics of the flow. We find that the presence of inertia serves to stabilise the boundary-layer slightly, although the increasing thickness of the basic flow as the inertia parameter increases causes a large change in the critical wavenumber of the most unstable vortex.

Keywords: porous media, free convection, vortices, linear stability, nonlinear stability, secondary instability

3.1 INTRODUCTION

The aim of this chapter is to present the latest analyses of the onset, nonlinear development and eventual secondary instability of vortices within an inclined thermal boundary-layer flow in porous media. In particular we consider the situation which arises when the heated surface is nearly vertical, for in that limit instability occurs asymptotically far from the leading edge and therefore the boundary-layer approximation may be used as the sole approximation.

There are presently 29 papers in print which consider the instability of thermal boundary-layer flows in porous media—see the references indicated by an asterisk, especially the review by Rees (1998). Many of these papers use the approximate method known as the parallel flow approximation to determine stability criteria. Unfortunately there is a mathematical inconsistency inherent in the method in that the boundary-layer approximation is an asymptotic analysis with $x \gg 1$ but the final result is a finite value of x beyond which disturbances grow. In the porous medium context such values of x are often very small indeed (for example $x \sim 33$ for the horizontal boundary-layer) rendering the boundary-layer approximation invalid at the point of instability. The paper by Storesletten and Rees (1998) investigates this in great detail and concludes that progress may only be made by either using a numerical simulation of the full equations, or by looking at the near vertical limit. The aim of this chapter is to report on recent efforts in the latter area.

3.2 THE GOVERNING EQUATIONS AND BASIC FLOW

The flow configuration we consider is as shown in Figure 3.1. A generally inclined and upward-facing surface is set at a uniform temperature which is above the ambient temperature of the medium. The coordinate directions are as defined in Figure 3.1, and z, the spanwise coordinate, forms a right-handed orthogonal system with x and y.

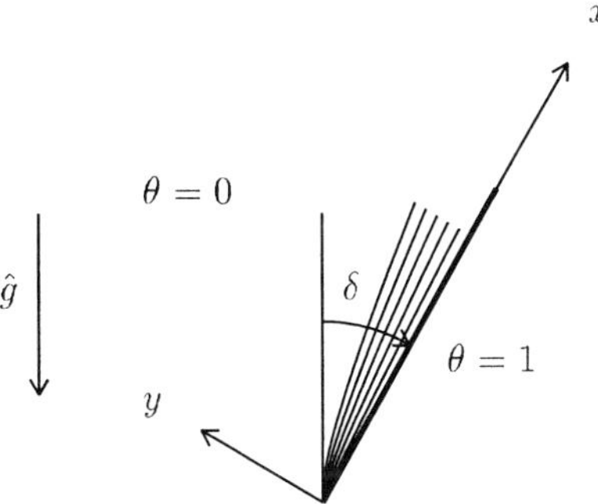

Figure 3.1 *Sketch of the flow domain and boundary conditions*

Assuming the simplest possible situation, namely, Darcy's law is valid, the Boussinesq approximation applies, the porous matrix and the fluid are in local thermal equilibrium, then the governing non-dimensional equations are given by

$$\frac{\partial u}{\partial x} + \frac{\partial v}{\partial y} + \frac{\partial w}{\partial z} = 0, \tag{3.1a}$$

$$u = -\frac{\partial p}{\partial x} + \theta \cos \delta, \tag{3.1b}$$

$$v = -\frac{\partial p}{\partial y} + \theta \sin \delta, \tag{3.1c}$$

$$w = -\frac{\partial p}{\partial z}, \tag{3.1d}$$

$$\frac{\partial \theta}{\partial t} + u\frac{\partial \theta}{\partial x} + v\frac{\partial \theta}{\partial y} + w\frac{\partial \theta}{\partial z} = \frac{\partial^2 \theta}{\partial x^2} + \frac{\partial^2 \theta}{\partial y^2} + \frac{\partial^2 \theta}{\partial z^2}. \tag{3.1e}$$

Here u, v and w are the fluid flux velocities in the x, y and z directions, respectively, p is the pressure and θ the temperature. The inclination from the vertical of the upward-facing heated surface is δ. The resulting steady two-dimensional flow may be studied by introducing a streamfunction in the form

$$u = \frac{\partial \psi}{\partial y}, \quad v = -\frac{\partial \psi}{\partial x}, \quad w = 0. \tag{3.2}$$

For the basic flow there are no z variations and equations (3.1) become

$$\frac{\partial^2 \psi}{\partial x^2} + \frac{\partial^2 \psi}{\partial y^2} = \frac{\partial \theta}{\partial y}\cos \delta - \frac{\partial \theta}{\partial x}\sin \delta, \tag{3.3a}$$

$$\frac{\partial^2 \theta}{\partial x^2} + \frac{\partial^2 \theta}{\partial y^2} = \frac{\partial \psi}{\partial y}\frac{\partial \theta}{\partial x} - \frac{\partial \psi}{\partial x}\frac{\partial \theta}{\partial y}. \tag{3.3b}$$

The heated surface is assumed to be semi-infinite in extent in this idealised problem. Therefore there is no external physical length scale which may be used to non-dimensionalise the equations, and hence the porous medium Rayleigh number has been set equal to unity. The boundary-layer approximation now consists in treating x as an asymptotically large variable with $x \gg y$.

At general inclinations ($0 \leqslant \delta < \pi/2$), the boundary-layer approximation reduces equations (3.3) to the form

$$\frac{\partial^2 \psi}{\partial y^2} = \frac{\partial \theta}{\partial y}\cos \delta, \tag{3.4a}$$

$$\frac{\partial^2 \theta}{\partial y^2} = \frac{\partial \psi}{\partial y}\frac{\partial \theta}{\partial x} - \frac{\partial \psi}{\partial x}\frac{\partial \theta}{\partial y}, \tag{3.4b}$$

for which there exists a similarity solution with the form

$$\psi = x^{1/2} f(\eta), \quad \theta = g(\eta), \quad \text{where} \quad \eta = y/x^{1/2}, \tag{3.5}$$

where f and g satisfy

$$f'' = g' \cos \delta, \tag{3.6a}$$

$$g'' + \frac{1}{2} f g' = 0, \tag{3.6b}$$

subject to

$$\begin{aligned} f = 0, \quad g = 1 \quad &\text{on} \quad \eta = 0, \\ f', g \to 0 \quad &\text{as} \quad \eta \to \infty. \end{aligned} \tag{3.7}$$

This flow was first considered by Cheng and Minkowycz (1977) for the vertical case, $\delta = 0$.

3.3   PERTURBATION EQUATIONS

We will be dealing with the onset and development of vortex disturbances and therefore it is necessary to use the fully three-dimensional equations of motion. We do not consider unsteady effects, but rather we determine the downstream effect of steady disturbances placed within the boundary-layer. A scaling analysis of equations (3.1), subject to the requirement that the vortices have an $O(1)$ cross-section, yields the following transformations:

$$x = \left(\frac{\cos \delta}{\sin^2 \delta} \right) x^*, \quad y = \left(\frac{1}{\sin \delta} \right) y^*, \quad z = \left(\frac{1}{\sin \delta} \right) z^*, \tag{3.8}$$

$$u = (\cos \delta) u^*, \quad v = (\sin \delta) v^*, \quad w = (\sin \delta) w^*, \quad p = P^*, \quad \theta = \Theta.$$

On substitution into equations (3.1) we obtain

$$u_x + v_y + w_z = 0, \tag{3.9a}$$

$$u = \Theta - \left(\frac{\sin^2 \delta}{\cos^2 \delta} \right) P_x, \tag{3.9b}$$

$$v = \Theta - P_y, \tag{3.9c}$$

$$w = -P_z, \tag{3.9d}$$

$$u\Theta_x + v\Theta_y + w\Theta_z = \left(\frac{\sin^2 \delta}{\cos^2 \delta} \right) \Theta_{xx} + \Theta_{yy} + \Theta_{zz}, \tag{3.9e}$$

where the asterisk superscripts have been omitted. The limiting case, $\delta \to 0$, may now be seen as being equivalent to invoking the boundary-layer approximation since, in this limit, the length scale represented by $x^* = 1$ is asymptotically greater than that represented by $y^* = 1$, see equations (3.8).

We now eliminate the velocity components in (3.9) to obtain the pressure/temperature formulation of the equations, formally let $\delta \to 0$ and employ the new coordinate system, (ξ, η, z), where

$$\xi = x^{1/2} \quad \text{and} \quad \eta = y/x^{1/2}. \tag{3.10}$$

We obtain the equations

$$P'' + \xi^2 P_{zz} = \frac{1}{2}\left(\xi\Theta_\xi - \eta\Theta'\right) + \xi\Theta', \tag{3.11a}$$

$$\Theta'' + \xi^2 \Theta_{zz} = \frac{1}{2}\left(\xi\Theta_\xi - \eta\Theta'\right)\Theta - P'\Theta' + \xi\Theta\Theta' - \xi^2 P_z\Theta_z, \tag{3.11b}$$

subject to the boundary conditions

$$\begin{aligned}
P' = 1, \quad \Theta = 1 \quad &\text{on} \quad \eta = 0, \\
P, \Theta \to 0 \quad &\text{as} \quad \eta \to \infty.
\end{aligned} \tag{3.11c}$$

Finally, we perturb about the basic flow by setting

$$P = q_1(\eta) + \xi q_2(\eta) + \hat{p}(\xi, \eta, z), \tag{3.12a}$$

$$\Theta = g(\eta) + \hat{\theta}(\xi, \eta, z, t), \tag{3.12b}$$

where

$$q_1' = \frac{1}{2}\left(f - \eta f'\right), \quad q_2' = g, \tag{3.13}$$

to obtain the following full disturbance equations:

$$\hat{p}'' + \xi^2 \hat{p}_{zz} = \frac{1}{2}\left(\xi\hat{\theta}_\xi - \eta\hat{\theta}'\right) + \xi\hat{\theta}', \tag{3.14a}$$

$$\begin{aligned}
\hat{\theta}'' + \xi^2 \hat{\theta}_{zz} = {}&\left(\frac{1}{2}f'\right)\xi\hat{\theta}_\xi - \left(\frac{1}{2}f\right)\hat{\theta}' + (f'')\left(\xi - \frac{1}{2}\eta\right)\hat{\theta} \\
&- (f'')\hat{p}' - \xi^2\hat{p}_z\hat{\theta}_z + \frac{1}{2}\left(\xi\hat{\theta}_\xi - \eta\hat{\theta}'\right)\hat{\theta} - \hat{p}'\hat{\theta}' + \xi\hat{\theta}\hat{\theta}',
\end{aligned} \tag{3.14b}$$

where all the nonlinear terms have been retained, and primes denote derivatives with respect to η. The boundary conditions are that

$$\begin{aligned}
\hat{p}' = 0, \quad \hat{\theta} = 0 \quad &\text{on} \quad \eta = 0, \\
\hat{p}, \hat{\theta} \to 0 \quad &\text{as} \quad \eta \to \infty.
\end{aligned} \tag{3.14c}$$

3.4   LINEAR EVOLUTION OF VORTICES

Equations (3.14) may be linearised by simply neglecting the terms involving products. The evolution with ξ of a vortex disturbance of wavenumber k may obtained by first introducing

$$\hat{p}\left(\xi,\eta,z,t\right) = p\left(\xi,\eta\right)\mathrm{e}^{\mathrm{i}kz}, \quad \hat{\theta}\left(\xi,\eta,z,t\right) = \theta\left(\xi,\eta\right)\mathrm{e}^{\mathrm{i}kz}, \tag{3.15}$$

and hence equations (3.14) reduce to

$$p'' - k^2\xi^2 p = \frac{1}{2}\left(\xi\theta_\xi - \eta\theta'\right) + \xi\theta', \tag{3.16a}$$

$$\theta'' - k^2\xi^2\theta = \left(\frac{1}{2}f'\right)\xi\theta_\xi - \left(\frac{1}{2}f\right)\theta' + \left(\xi - \frac{1}{2}\eta\right)g'\theta - g'p'. \tag{3.16b}$$

These equations have single ξ-derivatives which implies that the system is parabolic, and that the appropriate way of analysing instability must be to introduce a disturbance at some place within the boundary-layer and to monitor its evolution with ξ. However, to set such analyses into context we first perform a 'parallel flow approximation' stability analysis by simply neglecting the ξ-derivatives. Clearly this forms a constraint on the disturbance which is unlikely to occur in practice, and therefore stability criteria obtained in this way will not be accurate, but are nevertheless likely to give good qualitative results. Equations (3.16) with the ξ-derivative terms neglected form an ordinary differential eigensystem with ξ as the eigenvalue and k as a parameter. The result of solving this eigensystem is shown in Figure 3.2.

The minimum value of ξ on the neutral curve is 10.479 (equivalent to $x = 108.82$) at which point $k = 0.05744$. The maximum wavenumber for which neutral stability occurs is $k = 0.09236$. This figure may be interpreted as meaning that disturbances of any chosen

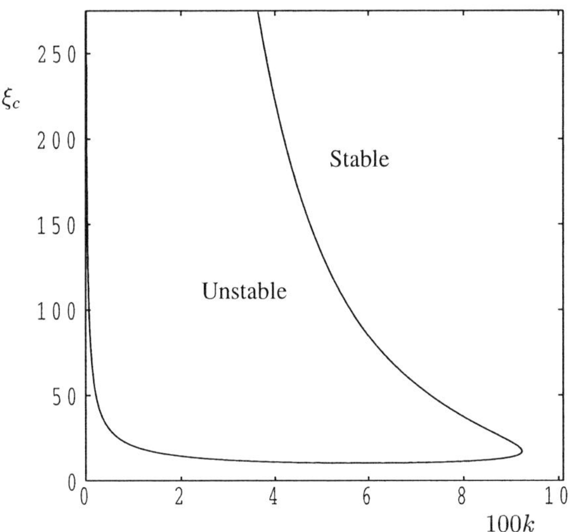

Figure 3.2 *Neutral stability curve: ξ_c against k*

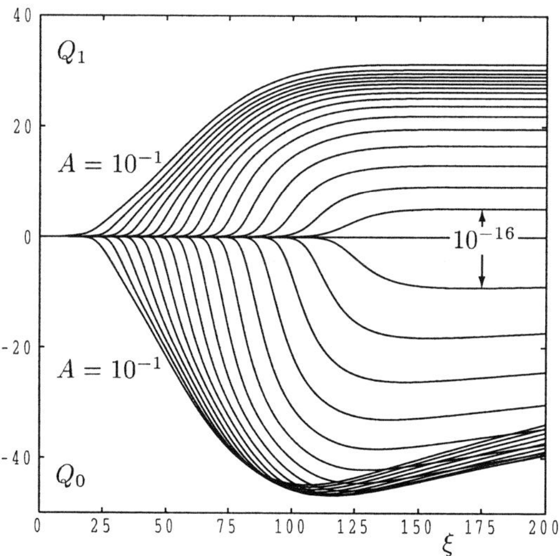

Figure 3.10 *Variation of Q_0 and Q_1 with ξ for the initial disturbance amplitudes $A = 10^{-1}, 10^{-2}, 10^{-3}, \ldots, 10^{-16}$*

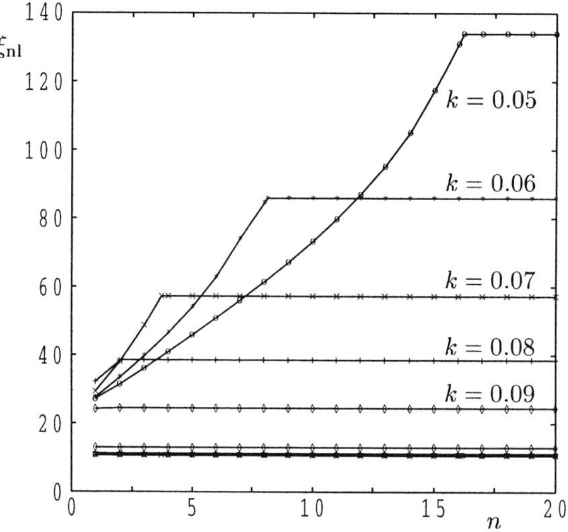

Figure 3.11 *Variation in the values of the nonlinear neutral points, ξ_{nl}, with $n = -\log_{10} A$ for different vortex wavenumbers. The lines placed near to $\xi_{nl} = 11$ correspond to onset of instability, while the others correspond to the beginning of decay*

$k = 0.06$ nonlinear effects become significant when $A < 10^{-8}$. Such a small amplitude is very likely to be exceeded in practice, and therefore we can conclude that nonlinear effects due to very small amplitude perturbations should be very significant.

We now consider the effect of varying ξ_0, the position where the disturbance is introduced. We set $A = 0.1$ in the computations displayed in Figure 3.12. In this case we find that the most dangerous location for the disturbance in terms of its local response is at approximately $\xi_0 = 50$. Once more the local response eventually becomes positive as ξ increases. The maximum local response in terms of $\partial\theta_0/\partial y|_{y=0}$ takes place nearer the leading edge at roughly $\xi_0 = 30$. The corresponding cumulative rates of heat transfer display the same characteristics as in Figure 3.10 except that the cumulative responses, Q_0 and Q_1, vary monotonically with increasing ξ_0.

Figure 3.13 displays values of ξ_{nl} as a function of ξ_0 for $k = 0.06$ for the following values of A: $10^{-1}, 10^{-3}, 10^{-5}, 10^{-7}$ and 10^{-9}. The horizontal lines correspond to the positions linear neutral stability, while the diagonal line indicates where the disturbance has been introduced into the boundary-layer. The onset of convection is seen to correspond roughly to the linear value until ξ_0 gets close to 10 whereupon convection begins to occur just after ξ_0. Conversely, for $A \leqslant 10^{-9}$, stability is re-established just after the upper value of ξ_c given by linear theory. However, for larger values of A, nonlinear effects become increasingly significant and the onset of decay is very premature when $A = 10^{-1}$.

Space limits the presentation of further results (see Rees, 2001c for more comprehensive details), but it is sufficient to comment briefly on the flow when the wavenumber takes other values. When k is larger, but nevertheless within the range where instabilities may

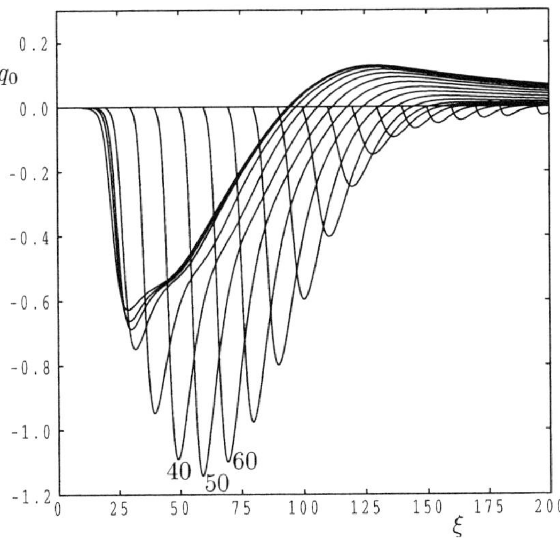

Figure 3.12 *Variation of q_0 with ξ for the initial disturbance locations, $\xi_0 = 8, 10, 15, 20, 30, \ldots$*

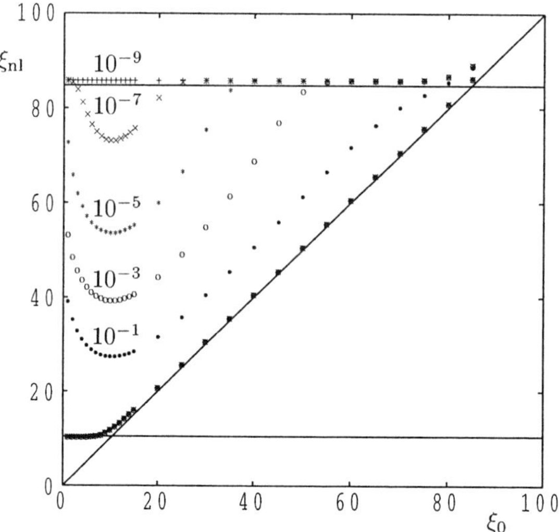

Figure 3.13 *Variation in the values of the nonlinear neutral points, ξ_{nl}, with the point of introduction, ξ_0, of the vortex for $k = 0.06$. The initial vortex amplitudes are $A = 10^{-1}, 10^{-3}, 10^{-5}, 10^{-7}$ and 10^{-9}. The lowest data points for each ξ_0 correspond to onset for all values of A, while the rest mark the beginning of decay*

occur, we find that the range of values of ξ over which the local response grows is much reduced and the cumulative rate of heat transfer is also reduced for any given value of ξ. Conversely, when k takes smaller values, the strength of the vortex system becomes greater and persists in a growing state for much longer. In some circumstances we find that the use of five modes ($N = 5$) does not give good resolution and it is necessary to increase the value of N and to decrease the η step length to proceed further with sufficiently accurate results.

3.6 SECONDARY INSTABILITIES

The vortex instabilities we have been considering are thermoconvective in origin, rather than hydrodynamic, since Darcy's law does not have advective terms. The boundary-layer whose stability is being analysed grows in thickness with distance from the leading edge and therefore the local Rayleigh number (which is a measure of the relative size of destabilising buoyancy forces and stabilising viscous forces) increases. Thus we would expect the boundary-layer to become increasingly susceptible to instability as ξ increases. The results presented in the last section indicate that this is true for a limited range of values of ξ. Irrespective of the value of A or of ξ_0, the local response to a disturbance always decays eventually. Therefore it is natural to expect that this decaying disturbance will itself be subject to instability. Although the wavelength of the above vortices remains constant, the wavelength compared with the local boundary-layer thickness becomes small and the aspect ratio of the vortices becomes large within the decaying region. It is frequently the case that the most dangerous disturbances have a roughly square aspect ratio, and therefore any destabilisation of the vortex systems already computed is likely to be initiated by disturbances of larger wavelength or of smaller wavenumber. The purpose of this section is to investigate this argument using numerical means. It is essential to point out that the results presented are not yet complete and that we present a small selection of destabilisations to indicate that destabilisation does indeed occur with respect to vortex disturbances of other wavelengths.

If the vortex disturbances considered are restricted to those which may be described using the expansions (3.20), then the same numerical code can be used to investigate secondary instabilities. The effect of out-of-phase disturbances would necessitate the use of sine terms in addition to the cosines used in (3.20) and this would reduce considerably the speed of the code, but it is intended to report on this aspect in the future. Here we choose a fundamental wavenumber, k, and introduce disturbances of similar form to expression (3.21) but applicable to more than one mode. Specifically we allow for all cosine modes to have an initial disturbance

$$\theta_n = A_n \eta e^{-\eta}. \tag{3.23}$$

In Figures 3.14, 3.15 and 3.16 we display the results of computing with the amplitudes

$$A_0 = 0, \quad A_1 = 0.01, \quad A_2 = 0.3, \quad A_n = 0 \quad \text{for} \quad n \geqslant 3, \tag{3.24}$$

for the respective wavenumbers $k = 0.05, 0.04$ and 0.03. Thus the second mode may be regarded as the primary mode of instability while the first mode forms the subharmonic mode of secondary instability. Given that the response of the boundary-layer to the primary mode of instability depends strongly on the amplitude and initial location of the disturbance, then clearly its destabilisation will depend not only on these parameters but also on the amplitude of the secondary mode. We therefore offer these figures as being illustrative of what is likely to happen in a realistic situation.

The primary mode shown in Figure 3.14 has a wavenumber which lies outside of the unstable range and it therefore decays after introduction. The secondary mode grows quite quickly and soon takes over as the strongest mode present. However, in line with the results of the last section, it becomes highly nonlinear and soon begins to decay itself. It is highly likely that this latter mode may itself be destabilised by a smaller wavenumber disturbance if that were present. The decay of the primary mode ($n = 2$) is seen easily in Figure 3.14(a) together with the growth of the secondary mode ($n = 1$). A temporary increase in amplitude of the primary mode near to $\xi = 30$ is caused by the nonlinear interaction of the subharmonic mode with itself. The corresponding surface heat transfer profiles are shown in Figure 3.14(b) and these emphasise how quickly the primary mode is destabilised.

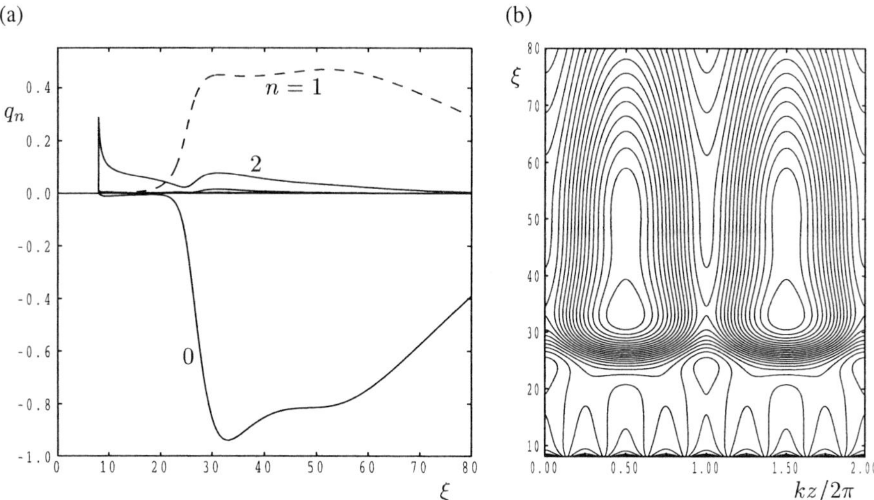

Figure 3.14   *For $k = 0.05$: (a) depicts the variation with ξ of the surface rate of heat transfer, q_n; (b) isolines of the surface rate of heat transfer beginning with eight vortices at $\xi = 8$*

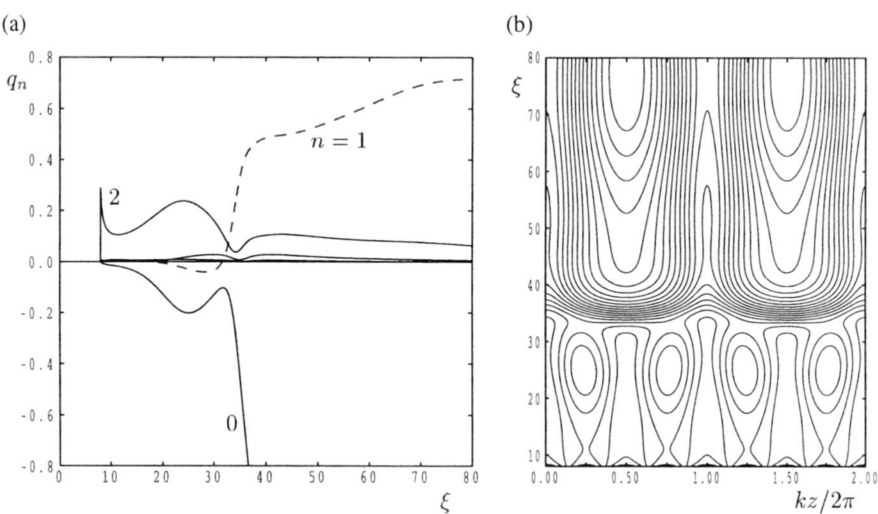

Figure 3.15   *For $k = 0.04$: (a) depicts the variation with ξ of the surface rate of heat transfer, q_n; (b) isolines of the surface rate of heat transfer beginning with eight vortices at $\xi = 8$*

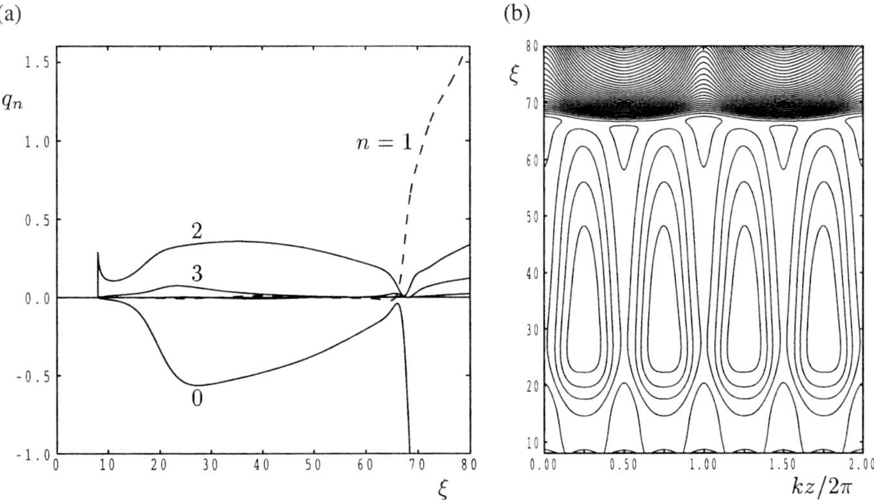

Figure 3.16 *For $k = 0.03$: (a) depicts the variation with ξ of the surface rate of heat transfer, q_n; (b) isolines of the surface rate of heat transfer beginning with eight vortices at $\xi = 8$*

Figures 3.15 and 3.16 correspond to primary modes with wavenumber within the unstable region of Figure 3.2 and therefore both these instabilities grow initially although the latter does so for a greater distance from the leading edge. Otherwise we have the same qualitative phenomena existing in these cases too, namely, rapid destabilisation and growth with an eventual slow decay of the secondary mode. We find that the smaller the wavenumber the larger is the distance from the leading edge at which these different events occur, and this is linked directly to the increasing strength of the primary vortex.

In Figure 3.17 we show cross-sections of the temperature perturbations at various values of ξ to indicate how the number of vortices changes with ξ. We see that the central vortex in the frame corresponding to $\xi = 8$ eventually decays and its two neighbouring vortices merge in order to reduce a four-vortex set to a two-vortex set.

In other contexts, such as the Darcy–Bénard or Bénard problems, flows consisting of vortices with wavenumbers which are slightly too small to be stable to disturbances aligned in the same direction but with a different wavenumber, have as their most unstable disturbance a mode with wavenumber which is very close to the original wavenumber. The nonlinear interactions between the primary flow and the destabilising vortex often causes a cascade of vortex interactions with many wavenumbers which differ by an integer multiple of the original difference between the primary and the destabilising vortex. Eventually a single steady vortex system is obtained which has a stable wavenumber. With this in mind, the subharmonic mode of instability shown in Figures 3.14 to 3.16 may not represent the most dangerous secondary instability mechanism, and therefore we attempted an

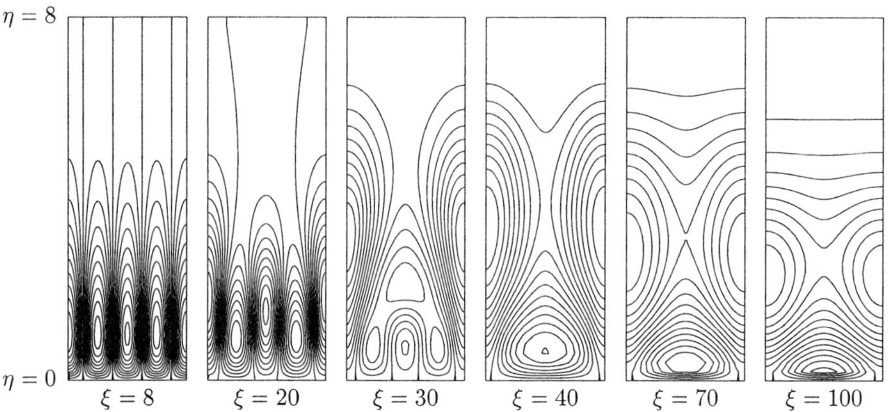

Figure 3.17 *For $k = 0.04$: cross-section of the vortex system corresponding to the perturbation temperature profiles. This figure corresponds to the solutions shown in Figure 3.15*

alternative scenario to investigate whether other wavenumbers could destabilise a primary vortex. Again, we emphasise that this is an isolated numerical simulation which indicates possibilities, and therefore definite statements about most unstable disturbances can only be made after many more numerical simulations are undertaken.

Figure 3.18 considers $k = 0.03$ as the fundamental wavenumber and uses the amplitudes

$$A_0 = A_1 = 0, \quad A_2 = 10^{-4}, \quad A_3 = 10^{-2}, \quad A_4 = A_5 = 0. \tag{3.25}$$

The primary mode ($n = 3$) has wavenumber $k = 0.09$ and its initial growth phase is difficult to see in Figure 3.18(a) due to the vertical scale of the graph. The secondary mode ($n = 2$) corresponds to $k = 0.06$ and the growth of this mode becomes evident near $\xi = 25$. However, the interaction between the $n = 2$ and $n = 3$ modes yields an $n = 1$ disturbance whose growth may be first seen near $\xi = 30$. Eventually the growth of the $n = 1$ mode dominates both the $n = 2$ and $n = 3$ modes, although nonlinear interactions causes their magnitudes to exceed their respective peak values when forming the locally dominant mode. The surface heat transfer profile is shown in Figure 3.18(b) and it is possible to discern the brief interval over which the $n = 2$ mode dominates between $\xi = 28$ and $\xi = 34$.

We redisplay in Figure 3.19 the numerical results shown in Figure 3.18 as cross-sections of the temperature perturbation profiles. The initial effect of this 'modal cascade' instability takes the form of a spanwise wavy modification of the local amplitude of the vortex system; this is seen most clearly in the profiles corresponding to $\xi = 25$. As in the subharmonic instability of Figure 3.17, the central vortex is weakened relative to the others displayed, and then disappears followed by the merging of its two neighbouring vortices; this takes place between $\xi = 25$ and $\xi = 30$. However, the resulting central vortex also shrinks and

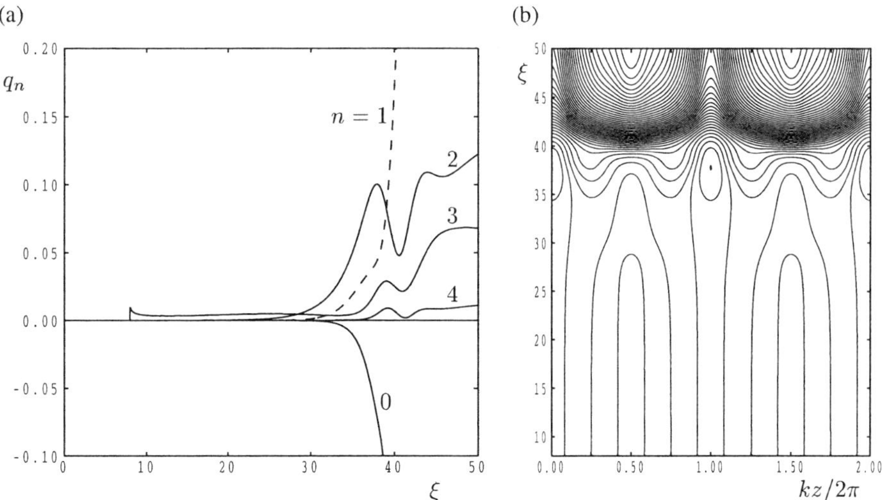

(a)

(b)

Figure 3.18   *For $k = 0.03$: (a) depicts the variation with ξ of the surface rate of heat transfer, q_n; (b) isolines of the surface rate of heat transfer beginning with twelve vortices at $\xi = 8$*

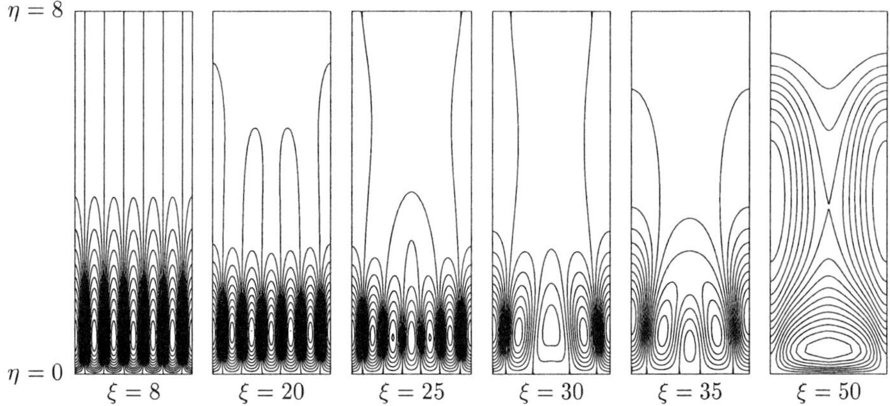

Figure 3.19   *For $k = 0.03$: cross-section of the vortex system corresponding to the perturbation temperature profiles. This figure corresponds to the solutions shown in Figure 3.18*

undergoes the same fate as the original central vortex by disappearing. Thus we have a mechanism which reduces eliminates rolls by reducing three into one.

It is only possible to study modal cascades formed by disturbances at nearer wavenumbers to that of the primary mode by increasing the number of Fourier modes in the numerical code. Once more, we intend to report on this in future work.

Finally, it would also be of some considerable interest to study the effect of a localised disturbance on the evolving vortex pattern. By 'localised' is meant a disturbance which has finite extent in the spanwise (z) direction. We would expect there to be a spreading out of the disturbance as it travels downstream. However, such a disturbance would by its nature contain components with many different spanwise wavenumbers and therefore the spreading out of the disturbance will also be accompanied by a complicated modal cascade sequence. Such a numerical simulation might require an entirely different solution strategy.

3.7 THE EFFECT OF INERTIA ON LINEAR STABILITY

Finally, we return to linearised theory and determine the effect which form drag has on the stability of boundary-layer flow. Given the large number of papers which have been published which deal with linear theory, it is surprising that none to date have tackled the effect of inertia on an inclined surface.

We assume that form drag manifests itself mathematically by the presence of quadratic terms in Darcy's law. In non-dimensional terms (see Riley and Rees, 1985) equations (3.1b) – (3.1d) are now replaced by

$$u\left(1 + Gq\right) = -\frac{\partial p}{\partial x} + \theta\cos\delta, \quad v\left(1 + Gq\right) = -\frac{\partial p}{\partial y} + \theta\sin\delta, \quad w\left(1 + Gq\right) = -\frac{\partial p}{\partial z},$$

$$\tag{3.26}$$

where

$$G = \left(\frac{\rho}{\mu}\right)^2 K\tilde{K}g\beta\,\Delta T \quad \text{and} \quad q = \sqrt{u^2 + v^2 + w^2} \tag{3.27}$$

are the inertia parameter and fluid flux speed; the meaning of the various terms in the definition of G are well known and are given in Banu and Rees (2000). As above we assume that the inclination, δ, is asymptotically small.

The basic boundary-layer flow is again two-dimensional and is given by $\psi = x^{1/2} f\left(\eta\right)$ and $\theta = g\left(\eta\right)$ where ψ is defined by (3.2) and f and g satisfy the equations

$$f''\left(1 + 2Gf'\right) = g', \quad g'' + \frac{1}{2}fg' = 0, \tag{3.28}$$

subject to the boundary conditions (3.7). We have followed precisely the earlier reduction of the disturbance equations to pressure/temperature form and have linearised them to obtain the following equations (which replace equations (3.16)):

$$p'' - \left[\frac{Gf''}{1+Gf'}\right] p' - k^2 \xi^2 p = \frac{Gf''}{(1+2Gf')^2} \left[\frac{1}{2}\eta + G\left(\eta f' + f\right)\right.$$

$$\left. - \frac{1}{2}\left(f - \eta f'\right)\frac{1+2Gf'}{1+Gf'}\right]\theta$$

$$+ \frac{1}{2}\left[\frac{1+Gf'}{1+2Gf'}\right]\left(\xi\theta_\xi - \eta\theta'\right) + \left[\xi + \frac{G\left(f - \eta f'\right)}{2\left(1+2Gf'\right)}\right]\theta',$$

$$\tag{3.29a}$$

$$\theta'' - k^2 \xi^2 \theta = \left(\frac{1}{2}f'\right)\xi\theta_\xi - \left(\frac{1}{2}f\right)\theta'$$

$$+ \frac{g'}{1+Gf'}\left[\xi - \frac{\eta\left(1+Gf'\right)}{2\left(1+2Gf'\right)} + \frac{G\left(f - \eta f'\right)}{2\left(1+2Gf'\right)}\right]\theta - g'p'.$$

$$\tag{3.29b}$$

Clearly these equations reduce to those given by (3.16) when $G = 0$.

Figure 3.20 shows the neutral stability curves which are obtained by setting to zero the ξ-derivative terms in (3.29) and by solving the resulting eigenvalue problem for ξ as a function of both k and G. It is interesting to note two main features of these curves as the inertia parameter, G, increases. First, the critical value of ξ (or, equivalently, x) increases only slightly as G increases from 0 (Darcy flow) to 10 (inertia-dominated flow), and second, the wavenumber decreases quite substantially; these are depicted in Figure 3.21.

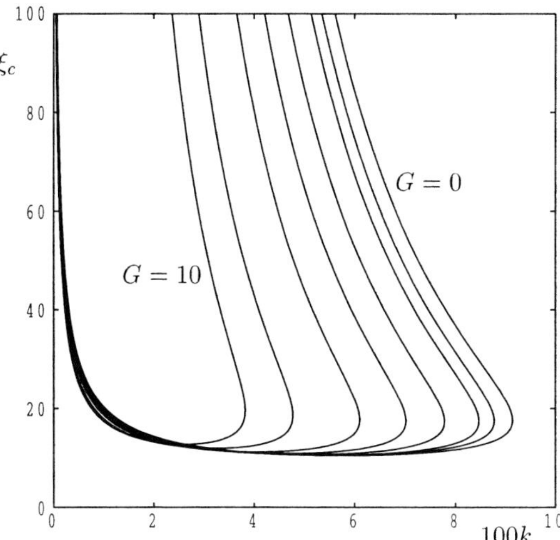

Figure 3.20 *Neutral stability curves depicting the variation of ξ_c with k for $G = 0, 0.1, 0.2, 0.5, 1, 2, 5$ and 10*

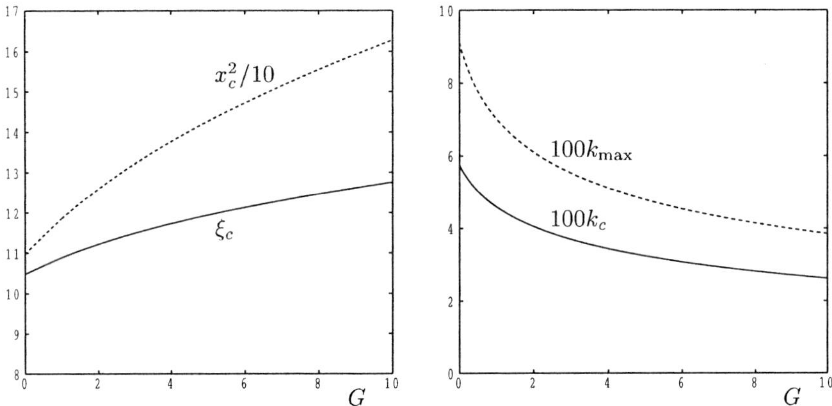

Figure 3.21   *Variation of ξ_c, x_c, k_c and $k_{\max}$ against G*

In trying to give good physical reasons why ξ_c and k_c behave in this way, it is important to note that there are two competing effects as G increases: the increasing resistance to flow due to inertia (which should raise the value of ξ_c) and the increasing boundary-layer thickness, or equivalently the local Rayleigh number (which destabilises the flow and would decrease the value of ξ_c). In the related field of Darcy–Bénard convection, He and Georgiadis (1990) showed that inertia does not affect the linear stability criterion for free convection, but that the critical Rayleigh number is raised in the presence of mixed convection (Rees, 1997). Thus we conclude, at least tentatively, that the increasing resistance to motion as G increases has a greater effect on stability than has the increasing boundary-layer thickness. However, if we were to assume that convective vortices tend to maintain a roughly unit aspect ratio, then the increasing boundary-layer thickness as G increases corresponds to decreasing critical wavenumbers, as computed.

3.8   CONCLUSION

We have found that the concept of a neutral position beyond which disturbances grow is not as straightforward as it is for uniform configurations such as the Bénard problem. We have shown that neutral locations depend on the form and position of the introduction of the disturbance and on its amplitude. The nonlinear evolution of vortices is counter-intuitive for after an initial interval of growth they decay despite the fact that the boundary-layer has increasing local Rayleigh number and hence it becomes more unstably stratified with x. We have also shown that secondary instabilities may take more than one form.

Clearly there remains a substantial amount of work to be undertaken on this topic in order to quantify where one might expect primary and secondary disturbances to appear. Furthermore, it is also possible that the vortex patterns presented here could be destabilised by transverse or oblique wave-like instabilities. If this is true, then it is likely that there will exist a streamwise wavenumber comparable with the local boundary-layer thickness, and this would require an alternative numerical strategy perhaps along the lines of the method of parabolized stability equations (see Herbert, 1997). There also exists the possibility of unsteadiness in the developing vortices; again if the analogy with Darcy–Bénard convection holds (for which unsteady flow first appears in an aspect ratio 1 box at $9Ra_c$) then we might expect unsteady vortices at or beyond $\xi \simeq 100$. Furthermore, when other effects, such as those considered by Jang, Chang and co-workers, are included in the analysis, it could very well be the case that qualitative results, such as the form of the most dangerous disturbance, may change. Indeed this is quite likely to be the case for Darcy–Brinkman flow, as the resulting equations are much closer to the familiar Navier–Stokes equations than are those for Darcy flow, and as the vertical thermal boundary-layer flow of a clear fluid is destabilised by waves, rather than by vortex disturbances.

REFERENCES

Banu, N. and Rees, D. A. S. (2000). The effect of inertia on vertical free convection boundary-layer flow from a heated surface in porous media with suction. *Int. Comm. Heat Mass Transfer* **27**, 775–783.

Bassom, A. P. and Rees, D. A. S.* (1995). The linear vortex instability of flow induced by a horizontal heated surface in a porous medium. *Quart. J. Mech. Appl. Math.* **48**, 1–19.

Chang, W. J. and Jang, J. Y.* (1989a). Non-Darcian effects on vortex instability of a horizontal natural convection flow in a porous medium. *Int. J. Heat Mass Transfer* **32**, 529–539.

Chang, W. J. and Jang, J. Y.* (1989b). Inertia effects on vortex instability of a horizontal natural convection flow in a saturated porous medium. *Int. J. Heat Mass Transfer* **32**, 541–550.

Cheng, P. and Minkowycz, W. J. (1977). Free convection about a vertical flat plate embedded in a porous medium with application to heat transfer from a dyke. *J. Geophys. Res.* **82**, 2040–2044.

He, X. S. and Georgiadis, J. G. (1990). Natural convection in porous media: effect of weak dispersion on bifurcation. *J. Fluid Mech.* **216**, 285–298.

Herbert, Th. (1997). Parabolized stability equations. *Ann. Rev. Fluid Mech.* **29**, 245–283.

Hsu, C. T. and Cheng, P.* (1979). Vortex instability in buoyancy induced flow over inclined heated surfaces in porous media. *J. Heat Transfer* **101**, 660–665.

Hsu, C. T. and Cheng, P.* (1980a). Vortex instability of mixed convection flow in a semi-infinite porous medium bounded by a horizontal surface. *Int. J. Heat Mass Transfer* **23**, 789–798.

Hsu, C. T. and Cheng, P.* (1980b). The onset of longitudinal vortices in mixed convective flow over an inclined surface in a porous medium. *Trans. ASME J. Heat Transfer* **102**, 544–549.

Hsu, C. T., Cheng, P., and Homsy, G. M.* (1978). Instability of free convection flow over a horizontal impermeable surface in a porous medium. *Int. J. Heat Mass Transfer* **21**, 1221–1228.

Jang, J. Y. and Chang, W. J.* (1987). Vortex instability of inclined buoyant layer in porous media saturated with cold water. *Int. Comm. Heat Mass Transfer* **14**, 405–416.

Jang, J. Y. and Chang, W. J.* (1988a). Vortex instability of buoyancy-induced inclined boundary-layer flow in a saturated porous medium. *Int. J. Heat Mass Transfer* **31**, 759–767.

Jang, J. Y. and Chang, W. J.* (1988b). The flow and vortex instability of horizontal natural convection in a porous medium resulting from combined heat and mass buoyancy effects. *Int. J. Heat Mass Transfer* **31**, 769–777.

Jang, J. Y. and Chang, W. J.* (1989). Maximum density effects on vortex instability of horizontal and inclined buoyancy-induced flows in porous media. *Trans. ASME J. Heat Transfer* **111**, 572–574.

Jang, J. Y. and Chen, J. L.* (1993a). Thermal dispersion and inertia effects on vortex instability of a horizontal mixed convection flow in a saturated porous medium. *Int. J. Heat Mass Transfer* **36**, 383–389.

Jang, J. Y. and Chen, J. L.* (1993b). Variable porosity effect on vortex instability of a horizontal mixed convection flow in a saturated porous medium. *Int. J. Heat Mass Transfer* **36**, 1573–1582.

Jang, J. Y. and Chen, J. L.* (1994). Variable porosity and thermal dispersion effects on vortex instability of a horizontal natural convection flow in a saturated porous medium. *Wärme- und Stoffübertr.* **29**, 153–160.

Jang, J. Y. and Leu, J. S.* (1993). Variable viscosity effects on the vortex instability of free-convection boundary-layer flow over a horizontal surface in a porous medium. *Int. J. Heat Mass Transfer* **36**, 1287–1294.

Jang, J. Y. and Lie, K. N.* (1992). Vortex instability of mixed convection flow over horizontal and inclined surfaces in a porous medium. *Int. J. Heat Mass Transfer* **35**, 2077–2085.

Jang, J. Y., Lie, K. N., and Chen, J. L.* (1995). The influence of surface mass flux on vortex instability of a horizontal mixed convection flow in a saturated porous medium. *Int. J. Heat Mass Transfer* **38**, 3305–3311.

Keller, H. B. and Cebeci, T. (1971). Accurate numerical methods for boundary-layer flows. 1. Two-dimensional flows. In *Proceedings of the International Conference on Numerical Methods in Fluid Dynamics. Lecture Notes in Physics.* Springer–Verlag, New York.

Lee, D. H., Yoon, D. Y., and Choi, C. K.* (2000). The onset of vortex instability in laminar convection flow over an inclined plate embedded in a porous medium. *Int. J. Heat Mass Transfer* **43**, 2895–2908.

Leu, J. S. and Jang, J. Y.* (1993). Variable viscosity effects on the vortex instability of the convective boundary-layer flow over a horizontal uniform heat flux surface in a saturated porous medium. In *Proceedings of the 6th International Symposium on Transport Phenomena and Thermal Engineering*, Seoul, Korea, Vol. 1, pp. 203–208.

Lewis, S., Bassom, A. P., and Rees, D. A. S.* (1995). The stability of vertical thermal boundary-layer flow in a porous medium. *Eur. J. Mech. B – Fluids* **14**, 395–408.

Lie, K. N. and Jang, J. Y.* (1993). Boundary and inertia effects on vortex instability of a horizontal mixed convection flow in a porous medium. *Numer. Heat Transfer, Part A* **14**, 361–378.

Rees, D. A. S.* (1993a). Nonlinear wave stability of vertical thermal boundary-layer flow in a porous medium. *J. Appl. Math. Phys. (ZAMP)* **44**, 306–313.

Rees, D. A. S.* (1993b). Numerical simulation of thermal boundary-layer instabilities in porous media. In *Proceedings of the IUTAM Symposium on Nonlinear Instabilities of Non-Parallel Flows*, Clarkson University, New York, pp. 242–247.

Rees, D. A. S. (1997). The effect of inertia on the onset of mixed convection in a porous layer heated from below. *Int. Comm. Heat Mass Transfer* **24**, 277–283.

Rees, D. A. S.* (1998). The instability of free convection boundary-layer flows in porous media: a critical review. In *Transport Phenomena in Porous Media* (eds D. B. Ingham and I. Pop), pp. 233–259. Pergamon, Oxford.

Rees, D. A. S.* (2001a). Vortex instability from a near-vertical heated surface in porous media. I. Linear instability. *Proc. Roy. Soc. Lond.* In press.

Rees, D. A. S.* (2001b). Vertical free convective boundary-layer flow in a porous medium using a thermal non-equilibrium model: elliptical effects. *J. Appl. Math. Phys. (ZAMP).* In press.

Rees, D. A. S. (2001c). Vortex instability from a near-vertical heated surface in porous media. II. Nonlinear evolution. *Proc. Roy. Soc. Lond.* Submitted.

Rees, D. A. S. and Bassom, A. P.* (1993). The nonlinear non-parallel wave instability of boundary-layer flow induced by a horizontal heated surface in porous media. *J. Fluid Mech.* **253**, 267–295.

Rees, D. A. S. and Bassom, A. P.* (1994). The linear wave instability of boundary-layer flow induced by a horizontal heated surface in porous media. *Int. Comm. Heat Mass Transfer* **21**, 143–150.

Riley, D. S. and Rees, D. A. S. (1985). Non-Darcy natural convection from arbitrarily inclined heated surfaces in porous media. *Quart. J. Mech. Appl. Math.* **38**, 277–295.

Storesletten, L. and Rees, D. A. S.* (1998). The influence of higher-order effects on the linear instability of thermal boundary-layer flow in porous media. *Int. J. Heat Mass Transfer* **41**, 1833–1843.

4 ONSET OF RAYLEIGH–BÉNARD CONVECTION IN POROUS BODIES

P. A. TYVAND

Department of Agricultural Engineering, Agricultural University of Norway, 1432 Ås, Norway

email: `peder.tyvand@itf.nlh.no`

Abstract

The onset of Rayleigh–Bénard convection in finite porous bodies is investigated theoretically. The linear stability problem is formulated in three dimensions. The temperature distribution along the boundaries is prescribed as a linearly decreasing function of height and this makes the basic state motionless with uniform conduction transport of heat upwards. The thermal condition along the body contour is chosen either as zero perturbation temperature (conducting boundaries) or zero normal derivative of the perturbation temperature (insulating boundaries). The kinematic condition along the body contour is chosen either as zero normal velocity (impermeable or closed boundaries) or zero tangential velocity (open boundaries). Reference is given to existing solutions for porous cylinders and rectangular boxes. Some simplified results are found for thin porous shells.

Keywords: Rayleigh–Bénard, porous medium, thermal instability, convection, eigenvalue problem, Rayleigh number

4.1 INTRODUCTION

The theory of thermal instability in a horizontal fluid layer heated from below was founded by Lord Rayleigh (1916). This phenomenon of buoyancy-induced instability is now called Rayleigh–Bénard convection, even though the original experiments by Bénard (1900) involved variable surface tension as the driving mechanism, see Pearson (1958).

Here we consider Rayleigh–Bénard convection in the strict sense that the flow is entirely due to buoyancy-induced instability. This is true for a steady and motionless basic state, which exists only if the basic temperature gradient is constant and strictly vertical, and then there is uniform conductive heat transport in the vertical direction. In a horizontal layer of fluid this motionless state of conduction is easily achieved by enforcing constant temperatures along the top and bottom boundaries. If the fluid domain is constricted by a more complex geometry, the temperature distribution along the boundaries must be given as linearly decreasing with the height. Then the basic temperature field is uniform in x and y at each level $z =$ constant, as any nonzero horizontal temperature gradient would induce a flow in the unperturbed state.

The Rayleigh–Bénard instability in a saturated porous medium was first investigated by Horton and Rogers (1945), and later by Lapwood (1948). They found the critical Rayleigh number to be $4\pi^2$ for the onset of convection in an infinitely wide horizontal porous layer. The porous medium was assumed homogeneous and isotropic and the boundaries were taken to be impermeable and perfect heat conductors. The preferred mode of instability consists of straight rolls with a cell width equal to the height of the porous layer and the cell width is half the spatial period, since every two neighbouring cells have opposite signs of the circulation. For a layer which is unlimited in the horizontal x- and y-directions the preferred mode of motion is degenerate both in the orientation of the roll pattern and in the location of each individual cell.

In the presence of a finite porous box with vertical lateral walls, the degeneracy of the most unstable mode is removed. If the geometry of the box is rectangular, the preferred mode of motion consists of a uniform roll pattern aligned along one of these lateral walls. However, this requires the assumption of thermally insulating lateral walls. This traditional boundary condition at the lateral boundaries is chosen for mathematical convenience because these lateral walls behave as cell walls within the roll structure itself. Beck (1972) was the first who investigated the selection of these roll patterns in rectangular porous boxes according to linear theory. A number of studies have been performed on nonlinear convection in porous boxes. In the present chapter we only consider the linear stability problem for the onset of convection, but various nonlinear instability phenomena in porous media have been reviewed recently by Rees (1998, 2000).

Most research papers on convection in finite porous boxes assume thermally insulating and impermeable lateral walls but in the present chapter we allow for more general boundary conditions. The porous body may be confined by impermeable walls, or it may have an open surface where the saturating fluid may flow freely in and out. The porous body may have zero perturbation temperature along its boundary, or the normal derivative of the perturbation velocity may be zero.

Nield (1968) was the first to consider the variety of possible boundary conditions for the top and bottom of an infinitely wide horizontal porous layer of uniform thickness. He determined the criteria for the onset of convection for all different combinations of open, closed, conducting and insulating boundaries, see also Nield and Bejan (1999, p. 181), but no such investigation exists of the possible combinations of lateral boundary conditions for finite porous boxes. During the last two decades, a number of papers have been written which take into account heat conduction in the lateral walls of a porous box and Nilsen and Storesletten (1990) has found an exact analytical solution for a two-dimensional box with conducting lateral walls. However, there are still a number of challenging onset problems for convection in finite three-dimensional porous bodies with different boundary

Our second thermal boundary condition is the assumption that the normal derivative of the temperature perturbation is zero along the boundary, i.e.,

$$n \cdot \nabla \Theta = 0 \quad \text{along the boundary.} \tag{4.13}$$

In general this second thermal condition means that the basic upward heat flux $k_m \, |dT_s/dz|$ (with dimension) is fixed at the boundary and it is not affected by the temperature perturbation $\Theta\,(x, y, z, t)$. Therefore condition (4.13) is the condition of constant heat flux. One special case deserves special treatment, namely $n \cdot k = 0$. This is generally the case where the body is cylindrical with vertical walls. Then the above interpretation of constant heat flux is still valid but this flux is zero. Therefore condition (4.13) is the condition of thermally insulating walls when the walls are vertical. The insulating boundary condition is the most common condition in the literature since it corresponds to internal cell walls within the flow domain, see Beck (1972) and Zebib (1978).

In general we consider two different kinematic conditions. The first condition is that of a closed (impermeable) boundary, namely

$$n \cdot v = 0 \quad \text{along the boundary.} \tag{4.14}$$

This is the standard condition of an impermeable wall separating the porous body from its surroundings, where n is the unit normal vector at the surface of the body. The second condition is the condition of an open boundary, namely

$$n \times v = 0 \quad \text{along the boundary.} \tag{4.15}$$

This is known as the constant pressure condition, even though a more appropriate description is that the surrounding fluid is hydrostatic. For simplicity we classify it as a kinematic condition even though it has a hydrodynamic (or at least hydrostatic) explanation. This condition of an open boundary arises from the assumption that the porous body is surrounded by its saturating fluid in a condition of hydrostatic pressure. This surrounding fluid either occupies a pure fluid domain or it saturates a neighbouring porous medium of much greater permeability than that of the porous body where the convection takes place.

Let us derive the open boundary condition (4.15). The tangential component of equation (4.7) at the boundary gives

$$n \times v + n \times \nabla P' - Ra\,\Theta n \times k = 0 \quad \text{just inside the boundary.} \tag{4.16}$$

Outside the porous body we take an equation of hydrostatic balance, namely

$$\nabla p' - Ra\,\theta k = 0, \tag{4.17}$$

and this means that the Darcian resistance is neglected. Here p' denotes the dimensionless pressure perturbation whereas θ denotes the temperature perturbation outside the porous body. The tangential component of this equation gives

$$n \times \nabla p' - Ra\,\theta n \times k = 0 \quad \text{just outside the boundary.} \tag{4.18}$$

The continuity of pressure through an open boundary implies the continuity of tangential derivatives, namely

$$n \times \nabla p' = n \times \nabla P' \quad \text{along the boundary.} \tag{4.19}$$

The temperature field is assumed continuous ($\Theta = \theta$) through the open boundary. Combining these pressure and temperature conditions with equations (4.16) – (4.18) finally yields the condition (4.15) for an open boundary. The continuity of temperature through the open boundary is a condition for the surrounding fluid and it does not put any restrictions on the flow inside the porous body.

4.3   A TWO-DIMENSIONAL CASE: THE RECTANGLE

For the two-dimensional problem in x and z, we may introduce the streamfunction $\Psi(x, z)$ and obtain the equations replacing equations (4.10) and (4.11) as follows:

$$\nabla^2 \Psi + Ra \frac{\partial \Theta}{\partial x} = 0, \tag{4.20}$$

$$\nabla^2 \Theta - \frac{\partial \Psi}{\partial x} = 0. \tag{4.21}$$

We may eliminate the temperature and obtain an equation for the streamfunction alone, namely

$$\left(\nabla^4 + Ra \frac{\partial^2}{\partial x^2} \right) \Psi = 0, \tag{4.22}$$

and a similar equation for the perturbation temperature.

The simplest two-dimensional geometry is the rectangle with vertical sidewalls and horizontal top and bottom walls. The rectangle is confined by

$$-\frac{a}{2} < x < \frac{a}{2}, \quad 0 < z < 1. \tag{4.23}$$

We wish to demonstrate the effects of different lateral boundary conditions, and therefore restrict our attention to the case of conducting and closed (impermeable) top and bottom boundaries, namely

$$\Psi = \Theta = 0 \quad \text{on} \quad z = 0 \quad \text{and} \quad z = 1. \tag{4.24}$$

We consider sidewalls that are kinematically either closed ($\Psi = 0$) or open ($\partial \Psi / \partial x = 0$), and thermally either conducting ($\Theta = 0$) or insulating ($\partial \Theta / \partial x = 0$). These options give four different combinations. In Figures 4.1 to 4.4 we show the preferred streamlines and isotherms for these four cases at the onset of convection for a square box where $a = 1$. The isotherms are plotted with an arbitrary amplitude for the perturbations, chosen for the purpose of illustration.

The streamfunction for the displayed case $a = 1$ is given by

$$\Psi = \sin \pi x \sin \pi z, \tag{4.29}$$

and this preferred mode of instability is triggered at the critical Rayleigh number $Ra_c = 4\pi^2$. The general formula for the Rayleigh number as a function of the rectangle width a is given by formula (4.27), as in the previous case.

Figure 4.3 shows the case of closed and conducting lateral sides, which was solved by Nilsen and Storesletten (1990), namely

$$\Psi = \Theta = 0 \quad \text{on} \quad x = \pm\frac{a}{2}. \tag{4.30}$$

There are two different eigenfunctions. The first streamfunction is shown in Figure 4.3(a) and is composed of two convection cells flowing symmetrically with respect to the z-axis. Its general formula is given by

$$\Psi_1 = \sin\left(\sqrt{a^2 + 1}\,\frac{\pi x}{a}\right) \cos\frac{\pi x}{a} \sin \pi z. \tag{4.31}$$

The second eigenfunction for the streamfunction is shown in Figure 4.3(b). It is composed of three convection cells flowing antisymmetrically with respect to the z-axis and is generally given by

$$\Psi_2 = \cos\left(\sqrt{a^2 + 1}\,\frac{\pi x}{a}\right) \cos\frac{\pi x}{a} \sin \pi z. \tag{4.32}$$

When Ψ_1 is the eigenfunction for the streamfunction, the eigenfunction for the temperature perturbation is proportional to Ψ_2, and vice versa. Both of these eigenfunctions are triggered at the same critical Rayleigh number, which is given by

$$Ra_c = 4\pi^2\left(1 + \frac{1}{a^2}\right). \tag{4.33}$$

This means that the general mode for the onset of instability is a linear combination of Ψ_1 and Ψ_2. There is no buoyancy along a conducting wall. Without buoyancy the motion along a closed wall cannot be supported, and the instability has to be driven in the interior of the fluid domain. Therefore conducting walls are stabilizing compared with insulating walls, assuming these walls are closed and vertical.

Figure 4.4 shows the case of open and insulating lateral sides, namely

$$\frac{\partial \Psi}{\partial x} = \frac{\partial \Theta}{\partial x} = 0 \quad \text{on} \quad x = \pm\frac{a}{2}. \tag{4.34}$$

Physically this case can be approximated by surrounding the porous rectangle laterally by a different medium of much greater permeability and much smaller conductivity. To achieve this, the porous body in which the convection takes place could be made of a highly conductive metal and be saturated by a poorly conducting fluid.

There are two different eigenfunctions for the displayed case $a = 1$ and they are derived by integrating up the eigenfunctions of the previous case, see Figure 4.3, in the x-direction. The first of these eigenfunctions for the streamfunction is shown in Figure 4.4(a) and is given by

$$\Psi_1 = \left\{ \frac{\cos\left[\left(\sqrt{a^2+1}+1\right)\pi x/a\right]}{\sqrt{a^2+1}+1} + \frac{\cos\left[\left(\sqrt{a^2+1}-1\right)\pi x/a\right]}{\sqrt{a^2+1}-1} \right\} \sin \pi z. \quad (4.35)$$

This streamline pattern forms one single convection cell and this is 43.7% of the volume flux recirculating within the cell. The second eigenfunction is shown in Figure 4.4(b) and is composed of two symmetrically flowing convection cells, namely

$$\Psi_2 = \left\{ \frac{\sin\left[\left(\sqrt{a^2+1}+1\right)\pi x/a\right]}{\sqrt{a^2+1}+1} + \frac{\sin\left[\left(\sqrt{a^2+1}-1\right)\pi x/a\right]}{\sqrt{a^2+1}-1} \right\} \sin \pi z. \quad (4.36)$$

In this solution only 3.5% of the volme flux in each of these cells recirculates within the cell. The flow pattern is very different from that of the first eigenfunction shown in Figure 4.4(a). Figure 4.4(b) is more similar to Figure 4.2, where an open boundary was assumed conducting instead of insulating, and where there was a full exchange of fluid with the surroundings (zero recirculation).

Comparing Figures 4.2 and 4.4(b), we see that in the latter case the streamlines are much more concentrated in the middle. The flow shown in Figure 4.2 is a pure upwelling flow, where there is positive buoyancy in all points inside the fluid domain. The whole cell at the onset of convection is effectively a rising plume, with no recirculation and no downward motion. (Since the flow direction is arbitrary, the flow could also have occurred as pure downwelling.)

In Figure 4.4(b) there is a small region of negative buoyancy so that 3.5% of the volume flux recirculates. This partly recirculating flow can be compared with the case of 6.17% recirculating flux for the onset of convection in a uniform porous layer heated from below with a closed/conducting bottom and an open/conducting top, see McKibbin *et al.* (1984). We note that the streamline patterns are similar in these two cases, in spite of the different orientations. In the present case the open boundaries are on the sides and not the top and these boundaries are insulating instead of conducting.

In the present case the amount of recirculation is highly different for the two different modes of instabilities (4.35) and (4.36). Both of these eigenfunctions are triggered at the critical Rayleigh number, which is given by

$$Ra_c = 4\pi^2 \left(1 + \frac{1}{a^2} \right). \quad (4.37)$$

Thus the critical Rayleigh number for open and insulating lateral boundaries is the same as for closed and conducting lateral boundaries. The general unstable mode is a linear combination of Ψ_1 and Ψ_2. These are eigenfunctions for the flow only, not the temperature, but the associated thermal eigenfunctions are simply proportional to Ψ_2 and Ψ_1, respectively.

Along an insulating lateral boundary, the buoyancy tends to be preserved. If such a vertical lateral boundary is closed, as was the case in Figure 4.1, the tangential vertical flow will be strong. This vertical flow is driven locally by positive or negative buoyancy along the wall, which is a dominating driving mechanism for the whole convection instability. However, the situation is very different in the present case, Figure 4.4, where an insulating vertical wall is open. The kinematic condition of an open lateral boundary forces the flow to be horizontal along the wall. The local buoyancy cannot support a horizontal flow other than indirectly through the pressure field in the porous medium. Consequently, the conditions of open and insulating vertical walls are stabilizing compared with the conditions of closed and insulating vertical walls.

4.3.1   The rectangle with asymmetric lateral conditions

So far, we have only considered rectangles with the same conditions at both end walls. This symmetry restriction allows four different combinations out of the two different kinematic and the two different thermal conditions. Two of these four combinations of wall conditions allow physical extensions beyond the wall:

(i) The combination closed/insulating at a wall corresponds to extending the mathematical solutions for the flow field as well as the temperature field symmetrically on the other side of the wall.

(ii) The open/conducting combination at a wall corresponds to extending the flow field as well as the temperature perturbation antisymmetrically on the other side of the wall.

The last two combinations do not allow any extension on the other side of the wall:

(iii) The closed/conducting wall.

(iv) The open/insulating wall.

Combining the last two conditions produces the single asymmetric case which cannot be deduced from the above analysis: a closed/conducting wall in combination with an open/insulating wall. A solution may be developed along the lines of Nilsen and Storesletten (1990) but this is not straightforward.

Here we show how the solution of the eigenvalue problem is deduced for all the other asymmetric cases. The porous rectangle under consideration is now $0 < x < a/2$, $0 < z < 1$. It is chosen to have half the width of the rectangles (squares) with symmetric conditions that have been solved above.

Figures 4.3 and 4.4 illustrated the degeneracy of the eigenfunctions when both walls are either closed/conducting or open/insulating. Asymmetric conditions remove the degeneracy. Figures 4.5(a) to (e) illustrate the unique eigenfunctions in five different cases of

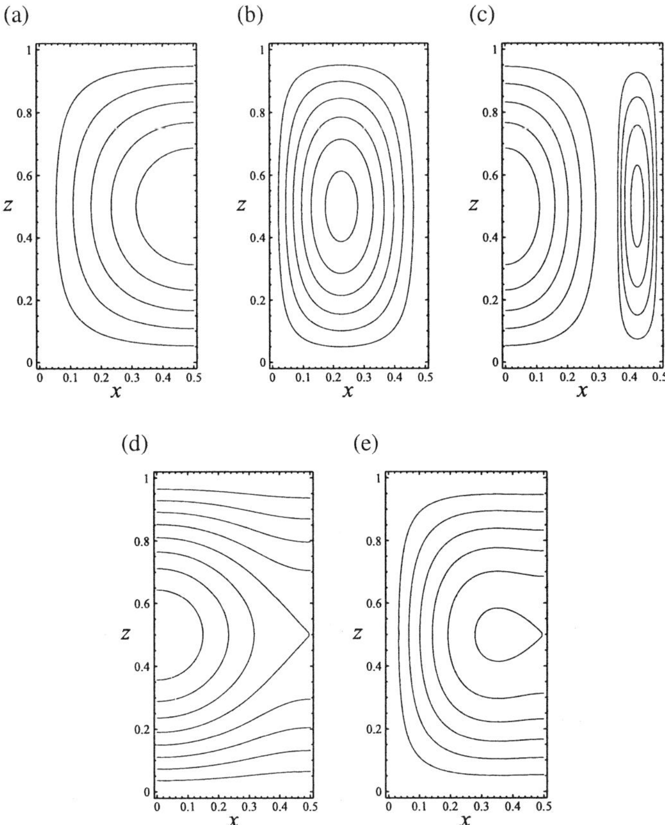

Figure 4.5 *Streamlines at the onset of convection in a rectangle with aspect ratio 1/2 having asymmetric lateral boundary conditions. Top and bottom are closed and conducting. (a) One closed/insulating wall and one open/conducting wall. (b) One closed/insulating wall and one closed/conducting wall. (c) One open/conducting wall and one closed/conducting wall. (d) One open/conducting wall and one open/insulating wall. (e) One closed/insulating wall and one open/insulating wall*

asymmetric boundary conditions. As these solutions correspond to dividing each of the previous figures in two equal parts, we keep the right-hand wall at $x = a/2$ as before and introduce a left-hand wall at $x = 0$. The formulas for the critical Rayleigh number are still given by equations (4.27) and (4.33).

In Table 4.1 we tabulate all possible combinations of lateral boundary conditions for a rectangle of height H and width L (now with dimension). The critical Rayleigh number is given as a function of height and width. The first four rows represent the

Table 4.1 *Critical Rayleigh number Ra_c for the onset of convection in a porous rectangle of height H and width L. First column: conditions at left-hand wall. Second column: conditions at right-hand wall. IMP = impermeable (closed), FRE = free (open), CON = conducting, INS = insulating. The top and bottom are assumed closed and conducting. Last column refers to Figures 4.1 to 4.5*

Kinematic/thermal	Kinematic/thermal	Ra_c	Figure
IMP/INS	IMP/INS	$\pi^2 \min\left(\frac{nH}{L} + \frac{L}{nH}\right)^2$	4.1
FRE/CON	FRE/CON	$\pi^2 \min\left(\frac{nH}{L} + \frac{L}{nH}\right)^2$	4.2
IMP/CON	IMP/CON	$4\pi^2\left(1 + \frac{H^2}{L^2}\right)$	4.3
FRE/INS	FRE/INS	$4\pi^2\left(1 + \frac{H^2}{L^2}\right)$	4.4
IMP/INS	FRE/CON	$\pi^2 \min\left(\frac{nH}{2L} + \frac{2L}{nH}\right)^2$	4.5(a)
IMP/INS	IMP/CON	$4\pi^2\left(1 + \frac{H^2}{4L^2}\right)$	4.5(b)
FRE/CON	IMP/CON	$4\pi^2\left(1 + \frac{H^2}{4L^2}\right)$	4.5(c)
FRE/CON	FRE/INS	$4\pi^2\left(1 + \frac{H^2}{4L^2}\right)$	4.5(d)
IMP/INS	FRE/INS	$4\pi^2\left(1 + \frac{H^2}{4L^2}\right)$	4.5(e)
FRE/INS	IMP/CON	unknown	void

different symmetric conditions, and the last six rows represent the different asymmetric conditions. The five first asymmetric problems follow from the symmetric ones, modified only by dividing the width by two. The sixth asymmetric problem, yet unsolved, is an open/insulating wall in combination with a closed/conducting wall.

Table 4.1 is our counterpart to Nield and Bejan (1999, Table 6.1, p. 181), and we adopt their notation. This systematic investigation of all top and bottom conditions was first performed by Nield (1968) and he considered infinite lateral extent. This is equivalent to having a rectangle of the preferred size constricted by closed and insulating lateral walls. In the present work we fix the top and bottom conditions and investigate all possible lateral conditions.

4.4 THE RECTANGULAR BOX

The simplest three-dimensional geometry is the rectangular box, and here we consider only the case where the box has vertical sidewalls and horizontal top and bottom walls. The box is confined by

$$0 < x < a, \quad 0 < y < b, \quad 0 < z < 1, \tag{4.38}$$

and the top and bottom are assumed conducting and closed (impermeable). The nth mode of disturbance in the vertical direction can then be expressed as follows:

$$\Theta = F(x, y) \sin n\pi z, \tag{4.39}$$

$$u = U(x, y) \cos n\pi z, \tag{4.40}$$

$$v = V(x, y) \cos n\pi z, \tag{4.41}$$

$$w = W(x, y) \sin n\pi z. \tag{4.42}$$

If we eliminate the vertical velocity and introduce the differential operator

$$\mathbb{L} = \frac{\partial^2}{\partial x^2} + \frac{\partial^2}{\partial y^2} - n^2 \pi^2, \tag{4.43}$$

then the eigenvalue problem is governed by the equations

$$n\pi \mathbb{L} F = \frac{\partial U}{\partial x} + \frac{\partial V}{\partial y} = \nabla \cdot \boldsymbol{V}, \tag{4.44}$$

$$\mathbb{L} \boldsymbol{V} + n\pi Ra \, \nabla F = 0. \tag{4.45}$$

Here the horizontal vector $\boldsymbol{V} = (U, V)$ has been introduced. Let us eliminate it to derive a fourth-order equation for the thermal eigenfunction, namely

$$\left(\mathbb{L}^2 + Ra \, \nabla^2 \right) F = 0. \tag{4.46}$$

This eigenvalue problem has been solved for the case of closed and insulating walls, see Beck (1972), and the result can be written as follows:

$$Ra_c = \pi^2 \min \left(f + \frac{1}{f} \right)^2, \quad f = \sqrt{\left(\frac{m}{a} \right)^2 + \left(\frac{n}{b} \right)^2}. \tag{4.47}$$

The minimum is to be taken with respect to all non-negative integers m and n, with the restriction that at least one of them must be nonzero. This formula gives an extension of the classical theory by Horton and Rogers (1945) and Lapwood (1948). It is straightforward because the lateral walls are mathematically equivalent to internal cell walls in the flow field.

Tewari and Torrance (1981) considered the same problem as Beck (1972) with the only exception being that they assumed the upper boundary to be open instead of closed. The results were qualitatively very similar to those obtained by Beck (1972) and Tewari and Torrance (1981) who applied the conventional conditions of closed and insulating lateral walls.

The eigenvalue problem is much more difficult when we consider the conditions of closed and conducting lateral walls, namely

$$F(0, y) = F(a, y) = F(x, 0) = F(x, b) = 0, \tag{4.48}$$

$$U(0, y) = U(a, y) = 0, \tag{4.49}$$

$$V(x, 0) = V(x, b) = 0, \tag{4.50}$$

and these lateral boundary conditions do not correspond mathematically to cell walls. This eigenvalue problem can only be solved in its original coupled version involving all three eigenfunctions $F(x, y)$, $U(x, y)$ and $V(x, y)$. The plausible method of solution is double Fourier expansions in x and y but this is troublesome because the problem does not allow full separation of the x and y dependence. No work appears to have been published on this problem.

The governing equation (4.46) is valid for a vertical cylinder with arbitrary cross-section. In the asymptotic limit of a very slender cylinder, the differential operator $\mathbb{L}$ reduces to ∇^2. This implies that the fourth-order governing equation degenerates to the second-order Helmholtz equation

$$\left(\nabla^2 + Ra\right) F = 0. \tag{4.51}$$

Then the onset of convection will be independent of the kinematic conditions along the wall. The Helmholtz equation is the standard equation for acoustic eigenmodes in cavities as well as standing oscillations in elastic membranes and in shallow water basins.

4.4.1   Two conducting and two insulating vertical walls

We now consider the situation where the two parallel vertical walls $x = 0$ and $x = a$ are conducting and the other two parallel vertical walls $y = 0$ and $y = b$ are insulating. All these four walls are assumed impermeable and the horizontal planes $z = 0$ and $z = 1$ are still taken as impermeable and conducting. This problem was first considered by Zebib and Kassoy (1977) and Lowell and Shyu (1978). This and similar problems were investigated further by Murphy (1979), Kassoy and Cotte (1985), Weidman and Kassoy (1986), Chelghoum $et\ al.$ (1987) and Wang $et\ al.$ (1987), and they applied numerical and asymptotic methods.

The two-dimensional version of the original problem with conducting lateral walls has been solved analytically by Nilsen and Storesletten (1990). The early numerical work by Lowell and Shyu (1978) included this two-dimensional problem as a special case. However, their two-dimensional results are not in full agreement with the simple general formula (4.33) found by Nilsen and Storesletten (1990). There is agreement for large values of a only. Nilsen and Storesletten's formula (4.33) is in agreement with the asymptotic theory by Kassoy and Cotte (1985) for small values of a.

Let us take a closer look at the three-dimensional onset problem for two conducting and two insulating walls. The boundary conditions (4.48) – (4.50) are then partly changed to give

$$F\left(0,y\right) = F\left(a,y\right) = \frac{\partial F}{\partial y}\left(x,0\right) = \frac{\partial F}{\partial y}\left(x,b\right) = 0, \tag{4.52}$$

$$U\left(0,y\right) = U\left(a,y\right) = 0, \tag{4.53}$$

$$V\left(x,0\right) = V\left(x,b\right) = 0. \tag{4.54}$$

We still assume the kinematic condition of closed boundaries everywhere and in the limit $b \ll a$ the preferred mode of convection will be two-dimensional in x and z, as assumed by Nilsen and Storesletten (1990). Along a conducting boundary, the motion is impeded because buoyancy is diffusing into the wall and it is easier for the fluid to flow along an insulating wall where its buoyancy is being maintained by the wall. Therefore it is clear that the two-dimensional mode found by Nilsen and Storesletten (1990) is not preferred for $b > a$, which restricts the validity of their formula (4.33).

Let us derive the critical Rayleigh number in the following asymptotic limit of a narrow gap between the two conducting boundaries:

$$a \ll \min\left(b,1\right). \tag{4.55}$$

In this asymptotic limit (4.55), the thickness a is half a wavelength of the preferred mode in the x-direction and we can write

$$\frac{\partial^2 F}{\partial x^2} = -\frac{\pi^2}{a^2} F. \tag{4.56}$$

The thermal eigenfunction F is given by

$$F\left(x,y\right) = \sin \pi x a \cos \frac{m\pi y}{b}, \tag{4.57}$$

where m denotes a positive integer. Expression (4.57) introduced into the governing equations produces the following formula for the Rayleigh number at the onset of convection:

$$Ra = \frac{\left(\pi^2/a^2 + m^2\pi^2/b^2 + n^2\pi^2\right)^2}{\pi^2/a^2 + m^2\pi^2/b^2}. \tag{4.58}$$

As we noted in connection with equation (4.51), the onset problem in this asymptotic thin-layer limit is governed solely by thermal conditions at the walls. We find that $m = n = 1$ gives the critical Rayleigh number. Our asymptotic formula for the critical Rayleigh number is given by

$$Ra_c = \frac{\pi^2}{a^2} + \left(\frac{\pi^2}{b^2} + 2\pi^2\right) + O\left(\pi^2 a^2\right). \tag{4.59}$$

This result appears as a singular perturbation in terms of a^2 starting with the minus first power, see Kassoy and Cotte (1985) who have performed a more rigorous asymptotic expansion. To the leading order in the expansion we have

$$Ra_c = \frac{\pi^2}{a^2},\tag{4.60}$$

and this is only $1/4$ of the corresponding limit for two-dimensional disturbances according to equation (4.33). The critical temperature gradient is only $1/4$ of the corresponding critical gradient prescribed by a two-dimensional mode of disturbance. We refer to the critical temperature gradient instead of the critical Rayleigh number because the asymptotic onset criterion (4.60) is essentially independent of the height of the porous layer. Gravity and temperature gradient are the only quantities in these asymptotic onset criteria that refer to the vertical direction. The vertical length scale (unity) disappears and is replaced by the horizontal gap thickness a. The narrow-gap limit (4.60) for the critical Rayleigh number can be rewritten as

$$\widetilde{Ra}_c = \frac{g\beta K \left|\nabla T_s\right| L^2}{\nu\alpha_m} = \pi^2.\tag{4.61}$$

Here a modified Rayleigh number $\widetilde{Ra} = Ra\, L^2 H^{-2}$ has been introduced, where $L = aH$ is the gap width (with dimension) and $|\nabla T_s|$ denotes the unperturbed temperature gradient.

4.5   THE HORIZONTAL CIRCULAR CYLINDER

Lyubimov (1975) was the first to consider Rayleigh–Bénard convection in a horizontal porous cylinder but he did not carry the onset problem far enough to identify preferred modes and critical Rayleigh number. This was done by Storesletten and Tveitereid (1991), who solved the onset problem for a vertical circle with conducting cylinder walls. This represents the two-dimensional modes of disturbance for a horizontal circular cylinder with insulating end walls. The Rayleigh number definition according to Storesletten and Tveitereid (1991) applies the radius as the unit length, whereas our definition of Ra, equation (4.5), takes as unit length the diameter of the cylinder. The critical Rayleigh number Ra_c is then redefined as twice the value found by Storesletten and Tveitereid (1991), namely

$$Ra_c = 92.53.\tag{4.62}$$

This critical value is 17% greater than the value $8\pi^2$ for a square body, given by equation (4.33), and this is a plausible result since 21.5% of the unit area is being removed when a unit square is replaced by its inscribed circle with unit diameter. It is a stabilizing effect to have less space available for the flow. The critical Rayleigh number given by equation (4.62) occurs for two different modes of motion. The first mode consists of two counter-rotating cells with strictly vertical motion (upward or downward) in the middle. The second mode consists of three cells: one dominating roll occupying most of the area, with two smaller rolls filling out the space to the left and to the right of the one big central roll.

Storesletten and Tveitereid (1991) also investigated a three-dimensional mode of disturbance, which is preferred for cylinders with a length of the order of their diameter, or larger. They assumed thermally insulating end walls so that the axial dependence in the solution is one single Fourier mode. When the aspect ratio of the horizontal cylinder (its length divided by its diameter) is shorter than about 0.43, the preferred mode of convection is still two-dimensional. For all larger aspect ratios, the preferred disturbance is three-dimensional, and its structure in a vertical cross-section consists of two cells. The critical Rayleigh number for this three-dimensional flow takes the following minimum value:

$$Ra_c = 62.48. \tag{4.63}$$

This value was found with a preferred wave number of 1.76 for the axial flow component. This lowest critical Rayleigh number therefore repeats itself for discrete aspect ratios which coincide with an integer number of critical wavelengths. The preferred wave number gives 3.57 as the wavelength with the radius as unity and there are two cells in each wavelength. This means that each cell will have the width 0.892 in the x-direction, dimensionless with the cylinder diameter as length unity.

4.5.1 General linearized equations in cylindrical coordinates

We introduce the cylindrical coordinate system (x, r, ϕ) for a porous cylinder with axis along the x-direction. The transformations between Cartesian and cylindrical coordinates are as follows:

$$(x, y, z) = (x, r \cos \phi, r \sin \phi) \tag{4.64}$$

and the velocity components are denoted by (u, v_r, v_ϕ).

After eliminating the pressure, the governing equations (4.7) – (4.9) can be expressed as follows:

$$\frac{\partial}{\partial r}(rv_\phi) - \frac{\partial v_r}{\partial \phi} = Ra\left(r\frac{\partial \Theta}{\partial r}\cos\phi - \frac{\partial \Theta}{\partial \phi}\sin\phi\right), \tag{4.65}$$

$$\frac{1}{r}\frac{\partial u}{\partial \phi} - \frac{\partial v_\phi}{\partial x} = -Ra\cos\phi\frac{\partial \Theta}{\partial x}, \tag{4.66}$$

$$\frac{\partial v_r}{\partial x} - \frac{\partial u}{\partial r} = Ra\sin\phi\frac{\partial \Theta}{\partial x}, \tag{4.67}$$

$$\frac{\partial u}{\partial x} + \frac{1}{r}\frac{\partial}{\partial r}(rv_r) + \frac{1}{r}\frac{\partial v_\phi}{\partial \phi} = 0, \tag{4.68}$$

$$v_r \sin\phi + v_\phi \cos\phi + \frac{1}{r}\frac{\partial}{\partial r}\left(r\frac{\partial \Theta}{\partial r}\right) + \frac{1}{r^2}\frac{\partial^2 \Theta}{\partial \phi^2} + \frac{\partial^2 \Theta}{\partial x^2} = 0. \tag{4.69}$$

In order to solve the challenging problem of zero perturbation temperature along all boundaries, including the end walls, one would have to start with these general equations.

4.5.2 A thin horizontal cylindrical shell with closed walls

We now consider the onset of convection in the small-gap porous shell defined by

$$a < r < a + \epsilon. \tag{4.70}$$

Here $\epsilon \ll \min(a, L)$ and L is the length of the cylindrical shell in the x-direction.

We choose the most natural thermal condition which is a prescribed boundary temperature, i.e., zero perturbation temperature, namely

$$\Theta = 0 \quad \text{on} \quad r = a \quad \text{and} \quad r = a + \epsilon. \tag{4.71}$$

Also we assume the walls of the cylinder shell to be closed (impermeable), i.e.,

$$v_r = 0 \quad \text{on} \quad r = a \quad \text{and} \quad r = a + \epsilon, \tag{4.72}$$

and asymptotically for small ϵ we may take $v_r = 0$. Then we can reduce the eigenvalue problem to being governed by two coupled equations for v_ϕ and Θ, namely

$$\frac{\partial}{\partial r}(r v_\phi) = Ra \frac{\partial}{\partial r}(r\Theta)\cos\phi - Ra\frac{\partial}{\partial \phi}(\Theta\sin\phi), \tag{4.73}$$

$$v_\phi \cos\phi + \frac{1}{r}\frac{\partial}{\partial r}\left(r\frac{\partial\Theta}{\partial r}\right) + \frac{1}{r^2}\frac{\partial^2\Theta}{\partial\phi^2} + \frac{\partial^2\Theta}{\partial x^2} = 0, \tag{4.74}$$

subject to boundary conditions (4.71). To the leading order of ϵ, we replace the factor r by a. Moreover, we cancel a couple of terms by taking into account that $\partial/\partial r \gg 1$, while a and $\partial/\partial\phi$ are of order unity.

Equation (4.73) can then be simplified and integrated to give

$$v_\phi = Ra\,\Theta\cos\phi. \tag{4.75}$$

In the heat equation (4.74), the effects of curvature are neglected in this thin-layer approximation and therefore we have

$$v_\phi \cos\phi + \frac{\partial^2\Theta}{\partial r^2} + \frac{1}{a^2}\frac{\partial^2\Theta}{\partial\phi^2} + \frac{\partial^2\Theta}{\partial x^2} = 0. \tag{4.76}$$

We combine equations (4.75) and (4.76) to find the governing equation for the temperature perturbation, namely

$$\frac{\partial^2\Theta}{\partial r^2} + \frac{1}{a^2}\frac{\partial^2\Theta}{\partial\phi^2} + \frac{\partial^2\Theta}{\partial x^2} + Ra\,\Theta\cos^2\phi = 0. \tag{4.77}$$

This thin-shell approximation allows us to write the eigenfunction as a Fourier mode in the r- and x-directions, namely

$$\Theta = F(\phi)\sin k(r - a)\,\mathrm{e}^{\mathrm{i}px}, \tag{4.78}$$

where the physical mode is represented by the real part. So far, nothing has been said about the boundary conditions at the two end walls $x = 0$ and $x = L$ of the cylinder. According to equation (4.78) we now assume closed (impermeable) and thermally insulating walls, which is the same as the conditions between neighbouring cell walls. Therefore, instead of considering the parameter L representing the cylinder length, we choose to consider the parameter p representing the wave number of the convection cell in the x-direction. We also consider the wave number k in the radial direction, and for the preferred mode of motion we have

$$pL = k\epsilon = \pi. \tag{4.79}$$

The governing equation is now given by

$$\frac{F''(\phi)}{a^2} + \left(Ra\cos^2\phi - k^2 - p^2\right) F(\phi) = 0, \tag{4.80}$$

and only two thermal conditions are thermal in this eigenvalue problem. We would in principle state these conditions at the top ($\phi = \pi/2$) and at the bottom ($\phi = \pi/2$) and at the bottom of the porous shell.

According to our general definition of the Rayleigh number, equation (4.5), we choose the outer shell diameter as length unit, i.e.,

$$a = \frac{1}{2} - \epsilon. \tag{4.81}$$

We solve the eigenvalue problem numerically by the shooting method, see Keller (1968), assuming that the preferred mode is symmetrical with respect to $\phi = 0$. Therefore we start the integration from $\phi = 0$ with the shooting conditions

$$F(0) = 1, \quad F'(0) = 0, \tag{4.82}$$

and these conditions are fixed, so the only variable to be iterated on is the Rayleigh number. In the computation of equation (4.80), we neglect p compared with k, since the only role of p in the thin-shell limit is to assure that there is just one cell width in the x-direction. Thus neglecting p means that we must consider a cylinder which is long compared with the thickness of the porous shell.

We start by choosing a small value of ϵ and evaluate the wave number $k = \pi/\epsilon$. The iteration for Ra starts with the value k^2, which has been found above for a thin vertical layer, see equation (4.60).

In Table 4.2 we show some numerical computations of the critical Rayleigh number Ra_c for several values of ϵ and the equation (4.80) is solved using MATHEMATICA. The first choice of layer thickness $\epsilon = 0.15$ is chosen for the purpose of illustration, as it is somewhat too large for the asymptotic thin-layer theory to be valid. We see that Ra_c for this curved shell converges quickly to its asymptotic limit k^2 for a thin vertical layer as the layer thickness is reduced.

Table 4.2 *Results for the onset of convection in the thin-shell limit of a cylindrical porous shell. The cylindrical shell has a horizontal axis and conducting impermeable boundaries. For various thicknesses ϵ, the critical Rayleigh number Ra_c is computed and compared with its asymptotic limit value k^2, where k is the wave number of the disturbance across the porous shell*

ϵ	Ra_c	Ra_c/k^2	k
0.05	4088.7	1.036	62.83
0.1	1067.0	1.081	31.42
0.15	500.45	1.141	20.94

In Figure 4.6 we show streamlines for the preferred mode of motion in a shell where the thickness is $\epsilon = 0.15$ and where the cylinder length is $L = \pi$. The flow domain is then a square but the flow is very much concentrated in the middle half. The flow in the upper and lower portions of the cylinder shell is an example of penetrative convection, see Nield and Bejan (1999, p. 292). Figure 4.6 shows streamlines along one half cylinder surface $(-\pi/2 < \phi < \pi/2)$, and there are equal streamlines (with opposite circulation) along the other half cylinder surface $(\pi/2 < \phi < 3\pi/2)$.

The present theory gives some qualitative insight but is quantitatively rather crude, except for very thin shells where it gives only a minor stabilizing effect compared with a strictly vertical layer.

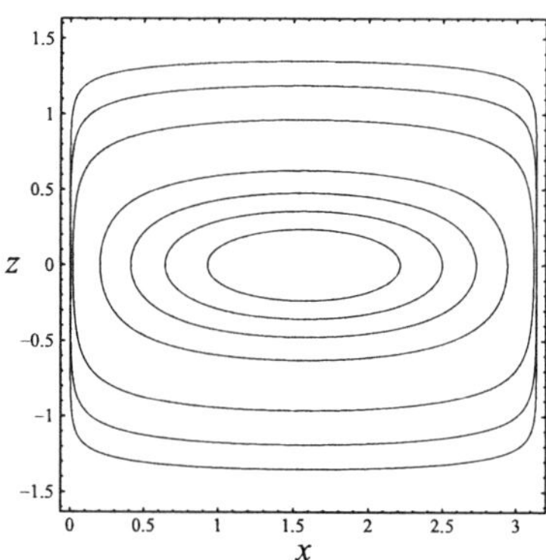

Figure 4.6 *Streamlines at the onset of convection in a horizontal cylindrical porous shell, according to thin-shell asymptotic theory. Isolines are shown for the following values of the streamfunction:* 0.0002, 0.002, 0.02, 0.2, 0.4, 0.6, 0.8

4.6 VERTICAL CYLINDERS

Here we first consider an arbitrary vertical porous cylinder which is confined by the two horizontal conducting planes $z = 0$ and $z = 1$. The vertical cylinder walls are given by vertical lines through an arbitrary closed contour in the xy plane. If we assume that all boundaries are open, then the critical Rayleigh number is given by

$$Ra_c = 12, \tag{4.83}$$

and this follows from Nield (1968). This preferred mode of convection has vertical streamlines and it is independent of the kinematic and thermal conditions valid along the vertical walls.

We now formulate the eigenvalue problem for the steady mode of the onset of convection in cylindrical coordinates (r, ϕ, z) defined by

$$(x, y, z) = (r \cos \phi, r \sin \phi, z), \tag{4.84}$$

and the velocity components are denoted by (v_r, v_ϕ, w).

After eliminating the pressure, the governing equations (4.7) – (4.9) can be written as follows:

$$\frac{\partial}{\partial r}(rv_\phi) - \frac{\partial v_r}{\partial \phi} = 0, \tag{4.85}$$

$$\frac{1}{r}\frac{\partial w}{\partial \phi} - \frac{\partial v_\phi}{\partial z} - \frac{Ra}{r}\frac{\partial \Theta}{\partial \phi} = 0, \tag{4.86}$$

$$\frac{\partial v_r}{\partial z} - \frac{\partial w}{\partial r} + Ra\frac{\partial \Theta}{\partial r} = 0, \tag{4.87}$$

$$\frac{1}{r}\frac{\partial}{\partial r}(rv_r) + \frac{1}{r}\frac{\partial v_\phi}{\partial \phi} + \frac{\partial w}{\partial z} = 0, \tag{4.88}$$

$$\frac{1}{r}\frac{\partial}{\partial r}\left(r\frac{\partial \Theta}{\partial r}\right) + \frac{1}{r^2}\frac{\partial^2 \Theta}{\partial \phi^2} + \frac{\partial^2 \Theta}{\partial z^2} + w = 0. \tag{4.89}$$

The choice of cylindrical coordinates is suitable for convection in a cylindrical porous ring restricted by $0 < z < 1$ and $a < r < b$. The thermal boundary conditions in the case of conducting boundaries (zero perturbation temperatures) are as follows:

$$\Theta = 0 \quad \text{on} \quad z = 0 \quad \text{and} \quad z = 1, \tag{4.90}$$

$$\Theta = 0 \quad \text{on} \quad r = a \quad \text{and} \quad r = b. \tag{4.91}$$

Alternatively, we may assume insulating cylinder boundaries, namely

$$\frac{\partial \Theta}{\partial r} = 0 \quad \text{on} \quad r = a \quad \text{and} \quad r = b. \tag{4.92}$$

The kinematic conditions at closed boundaries are given by

$$v_r = 0 \quad \text{on} \quad r = a \quad \text{and} \quad r = b, \tag{4.93}$$

$$w = 0 \quad \text{on} \quad z = 0 \quad \text{and} \quad z = 1. \tag{4.94}$$

Alternatively we may assume open boundaries, which are given by

$$w = 0 \quad \text{on} \quad r = a \quad \text{and} \quad r = b, \tag{4.95}$$

$$v_r = 0 \quad \text{on} \quad z = 0 \quad \text{and} \quad z = 1. \tag{4.96}$$

It should be noted that there are no boundary conditions for the azimuthal velocity v_ϕ.

4.6.1   The literature on vertical cylinders

Wooding (1959) wrote the first research paper on convection in a vertical porous cylinder and he considered a tall and slender cylinder. His result for the onset of convection can be expressed in terms of the modified critical Rayleigh number $\widetilde{Ra}$ introduced in equation (4.61), namely

$$\widetilde{Ra}_c = 13.56. \tag{4.97}$$

The length scale L is now defined as the diameter of the circular cylinder. It is interesting to compare this result for a slender cylinder with the critical Rayleigh number π^2, equation (4.61), for a box with a narrow gap between two conducting (or insulating) planes. The result for a cylinder (with the same diameter as the gap width) is that its Rayleigh number is 37.4% greater at the onset of convection. This is plausible because there is less space available for the flow. Wooding (1959) found that the preferred flow is antisymmetric with respect to a diameter in a cross-section of the cylinder.

Zebib (1978) gave a more complete solution for the onset of convection in a vertical circular cylinder with impermeable boundaries, conducting top and bottom, and insulating vertical walls. His solution for the temperature perturbation involves a Bessel function of the first kind, namely

$$\Theta = \sin \pi z \, \mathrm{J}_m \, (kr) \cos m\phi, \tag{4.98}$$

with the insulating wall condition $\mathrm{J}'_m \, (b) = 0$ where b denotes the radius of this compact porous cylinder. Zebib (1978) confirmed that the antisymmetric mode found by Wooding (1959) (here represented by $n = 1$) is preferred for tall and slender cylinders. Zebib (1978) found that the axisymmetric mode ($n = 0$) is preferred only for cylinders with dimensionless radius b in the following range:

$$1.09 < b < 1.28. \tag{4.99}$$

Bau and Torrance (1982) solved a similar onset problem as Zebib (1978), and the only difference was that the upper boundary was open instead of closed. Bau and Torrance (1982) also presented experimental observations which were in reasonable agreement with their theory.

Wang (1998) again solved a similar onset problem as Zebib (1978) with the only change being that there is a constant heat flux condition at the bottom boundary. His figure for the critical Rayleigh number as a function of the ratio of radius to height looks identical to that found by Bau and Torrance (1982). This is no surprise since we know from Nield (1968) that these two problems share the same onset criterion, namely $Ra_c = 27.1$ at the wave number 2.33 for a layer of infinite horizontal extent. Actually the criteria for the onset of convection in both these cases may be derived from the results by Zebib (1978), according to the transformation

$$\left(\frac{Ra}{4\pi^2}, \pi x, \pi y \right) \rightarrow \left(\frac{Ra}{27.1}, 2.33\,x, 2.33\,y \right). \tag{4.100}$$

This is the appropriate transformation from the solution by Zebib (1978) to produce those of Bau and Torrance (1982) and Wang (1998). The only difference in these separable eigenvalue problems is the boundary condition at the top or at the bottom. This results in a different vertical eigenfunction, which produces a different Rayleigh number and a preferred disturbance with a different radial length scale. The lateral boundary conditions are identical in the three papers by Zebib (1978), Bau and Torrance (1982) and Wang (1998). The horizontal variation of the eigenfunctions become identical after the application of the transformation (4.100).

By means of the same transformation (4.100), it is also seen that the results by Tewari and Torrance (1981) can be deduced from the original paper by Beck (1972). The transformation (4.100) states that the horizontal length scale is increased by a factor $\pi/2.33 = 1.348$ in the problems of Bau and Torrance (1982) and Wang (1998) compared with Zebib (1978). There are some computed values which may serve as a check. Wang (1998) found the radius range where he found that the preferred mode of disturbance is axisymmetric, namely

$$1.470 < b < 1.725, \tag{4.101}$$

and the corresponding values found by Zebib (1978) are given in equation (4.99). According to the transformation (4.100), Wang's values should be equal to Zebib's values multiplied by 1.348. This is in agreement with the ratio that we actually find for the upper limits $(1.725/1.28 = 1.348)$, and also for the lower limits where we find $1.470/1.09 = 1.349$.

Other papers that have been written on the Rayleigh–Bénard problem in vertical cylinders are those of Bories and Deltour (1980) and Bau and Torrance (1981). Bories and Deltour (1980) considered the effects of finite conduction in a surrounding solid medium, whilst Bau and Torrance (1981) investigated a porous medium between coaxial cylinders, where the solution is similar to equation (4.98) but the Bessel functions of the second kind Y_m enter the solution.

4.6.2 On the onset of convection in a hexagonal cylinder

One hexagonal cell is composed of three roll systems with the same wavelength being mutually rotated by an angle of $2\pi/3$. We now consider the onset of convection in a cylinder with hexagonal cross-section and impermeable boundaries, with conducting top and bottom and insulating vertical walls. The only hexagonal cell pattern that can fit into this hexagonal container is the single hexagonal cell filling the whole container.

We know that this single hexagonal cell must be the preferred mode of disturbance when the wavelength of each of its three constituting roll systems is equal to 2. This is because the critical Rayleigh number will then be $4\pi^2$, which is the global minimum for any cylindrical closed container with vertical walls and perfectly conducting top and bottom. The vertical velocity distribution in such a preferred cell can be written as follows:

$$w\left(x,y,z\right) = \left\{\cos \pi x + \cos\left[\frac{\pi}{2}\left(x - \sqrt{3}y\right)\right] + \cos\left[\frac{\pi}{2}\left(x + \sqrt{3}y\right)\right]\right\} \sin \pi z. \quad (4.102)$$

Here the maximal velocity amplitude for each of the three roll systems is chosen to be one, which means that its value at the point $(0,0,1/2)$ where these amplitudes add up, is equal to three.

The one single hexagonal cell at the onset of convection is close to axisymmetry, since there is an upwelling in the middle and a downwelling along the walls (or opposite). How much it actually deviates from axisymmetry is demonstrated in Figure 4.7. Here we show isolines for constant values of w in the downwelling outer region of a hexagonal cell, assuming (say) that the core of the cell has upwelling flow. A cut is made through the mid-plane of the cell $z = 1/2$. The inner curve connects all stagnation points given by $w = 0$, and it is close to being a circle. This implies that all the flow in the upwelling core is very close to being axisymmetric. The single hexagonal cell is clearly the closest one can get to axisymmetric flow inside a hexagonal container. One hexagonal cell fills the whole container at a critical Rayleigh number coinciding with the global minimum of $4\pi^2$ when each of the six sides of the hexagon has dimensionless length s given by

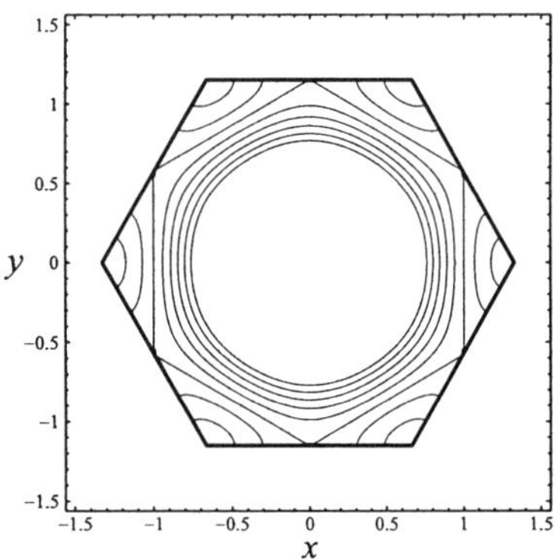

Figure 4.7   *Isolines of constant vertical velocity $w\left(x,y,1/2\right)$ in a hexagonal convection cell, given by equation (4.102). The displayed values are $w = 0$, -0.2, -0.4, -0.6, -0.8, -1.0, -1.2, -1.4*

$s = 4/3$. This is the maximal radius of the hexagon, or the distance from its centre to one of its corners. The corresponding minimal radius is the distance from the centre to the middle of each of the six sides, and it is given by $2/\sqrt{3} = 1.1547$. The effective radius-to-height ratio of the hexagonal cylinder is therefore estimated as the arithmetic mean $(1.33 + 1.15)/2 = 1.24$. This can be taken as a rough prediction of the radius for which the axisymmetric mode will be preferred in the circular cylinder. We see from the inequality (4.99) that our estimate $b = 1.24$ is indeed within the range of container radii where the axisymmetric mode is preferred. We may apply the transformation (4.100) to estimate the container radius for which we expect axisymmetric convection when one boundary is open, see Bau and Torrance (1982), or has constant heat flux, see Wang (1998). The average radius of one hexagonal cell is estimated to be $1.24 \times 1.348 = 1.67$. This value is in agreement with the parameter range for the axisymmetric mode found by Wang (1998), namely $1.470 < b < 1.725$.

The hexagonal cylinder serves to explain why an axisymmetric mode may be preferred in the circular cylinder and it gives a reasonable estimate of the radius-to-height ratio for which this axisymmetric mode can be expected to occur. To solve analytically the general onset problem for a hexagonal cylinder is very difficult but from a comparison with the circular cylinder, see Zebib (1978), we can give some qualitative information. If s is much smaller than one there is an upwelling in the left half (say) of the hexagonal container and downwelling in its right half. The hexagonal cell with all its downwelling (or upwelling) concentrated in the middle may occur only for values of s of order one. If s is considerably greater than one, there will be more than one recirculating flow cell within the container at the onset of convection. However, these cells cannot be exact hexagonal cells, because such a pattern with more than one hexagon does not fit in with the hexagonal boundary of the container.

4.7   ONSET OF CONVECTION IN SPHERICAL GEOMETRY

We formulate the general eigenvalue problem for a porous sphere. We first introduce the spherical coordinates (r, θ, ϕ) defined by:

$$(x, y, z) = (r \sin\theta \cos\phi, r \sin\theta \sin\phi, r \cos\theta). \tag{4.103}$$

Axisymmetry is defined by $\partial/\partial\phi = 0$ and also vanishing flow in the azimuthal ϕ-direction. The velocity components in the r-, θ- and ϕ-directions are denoted by v_r, v_θ and v_ϕ, respectively, and one must keep in mind the different definition of the symbol r in spherical and cylindrical coordinates.

After eliminating the pressure, the governing equations (4.7)–(4.9) reduce to the following form:

$$\frac{\partial}{\partial\theta}(v_\phi \sin\theta) - \frac{\partial v_\theta}{\partial\phi} = Ra \sin\theta \frac{\partial\Theta}{\partial\phi}, \tag{4.104}$$

$$\frac{\partial v_r}{\partial \phi} - \sin\theta \frac{\partial}{\partial r}\left(rv_\phi\right) = Ra\cos\theta \frac{\partial \Theta}{\partial \phi}, \tag{4.105}$$

$$\frac{\partial v_r}{\partial \theta} - \frac{\partial}{\partial r}\left(rv_\theta\right) = Ra\left[\sin\theta \frac{\partial}{\partial r}\left(r\Theta\right) + \frac{\partial}{\partial \theta}\left(\Theta\cos\theta\right)\right], \tag{4.106}$$

$$\sin\theta \frac{\partial}{\partial r}\left(r^2 v_r\right) + r\frac{\partial}{\partial \theta}\left(\sin\theta\, v_\theta\right) + r\frac{\partial v_\phi}{\partial \phi} = 0, \tag{4.107}$$

$$v_\theta \sin\theta - v_r \cos\theta = \frac{1}{r^2}\frac{\partial}{\partial r}\left(r^2\frac{\partial \Theta}{\partial r}\right) + \frac{1}{r^2\sin\theta}\frac{\partial}{\partial \theta}\left(\sin\theta\frac{\partial \Theta}{\partial \theta}\right) + \frac{1}{r^2\sin^2\theta}\frac{\partial^2 \Theta}{\partial \phi^2}, \tag{4.108}$$

and this choice of coordinate system is suitable for convection in a spherical porous body $0 \leqslant r < 1/2$. It should be noted that the diameter is defined as length unit according to our general conventions. The thermal boundary condition is given by

$$\Theta = 0 \quad \text{on} \quad r = \frac{1}{2}. \tag{4.109}$$

The kinematic condition in the case of a closed boundary is given by

$$v_r = 0 \quad \text{on} \quad r = \frac{1}{2}, \tag{4.110}$$

and the kinematic condition at an open boundary is given by

$$v_\theta = 0 \quad \text{on} \quad r = \frac{1}{2}. \tag{4.111}$$

4.7.1   A thin spherical shell with closed walls at given temperature

We now consider the onset of convection in the small-gap porous shell defined by

$$a < r < a + \epsilon, \tag{4.112}$$

where $\epsilon \ll 1/2$ and a is the inner radius of the shell.

We take the condition of zero perturbation temperature (4.12), namely

$$\Theta = 0 \quad \text{on} \quad r = a \quad \text{and} \quad r = a + \epsilon. \tag{4.113}$$

We assume the walls of the spherical shell to be closed (impermeable), namely

$$v_r = 0 \quad \text{on} \quad r = a \quad \text{and} \quad r = a + \epsilon, \tag{4.114}$$

and asymptotically for small ϵ we may take $v_r = 0$. The thin-shell approximation allows us to write the thermal eigenfunction as a Fourier mode in r- and ϕ-directions, i.e.,

$$\Theta = F(\theta)\sin k\,(r - a)\,e^{in\phi}. \tag{4.115}$$

Following a similar procedure as in the case of a thin horizontal porous shell, we end up with the following governing equation for the eigenfunction $F(\theta)$:

$$\frac{F''(\theta)}{a^2} + \frac{\cot\theta}{a^2}F'(\theta) + \left(Ra\sin^2\theta - k^2 - \frac{n^2}{a^2\sin^2\theta}\right)F(\theta) = 0. \qquad (4.116)$$

Only two thermal conditions are required in this eigenvalue problem and we would in principle state these conditions at the top ($\theta = 0$) and at the bottom ($\theta = \pi$) of the spherical porous shell.

According to our general definition of the Rayleigh number, see equation (4.5), we choose the outer shell diameter as length unity so that

$$a = \frac{1}{2} - \epsilon. \qquad (4.117)$$

The numerical solution procedure is the same as for equation (4.80). We utilize that the preferred mode of disturbance must be symmetrical with respect to $\theta = \pi/2$ and therefore start the integration from $\theta = \pi/2$ with the shooting conditions

$$F\left(\frac{\pi}{2}\right) = 1, \quad F'\left(\frac{\pi}{2}\right) = 0. \qquad (4.118)$$

In Table 4.3 we show computations of the critical Rayleigh number for different values of the shell thickness. It is essential to include the last term with $n = 1$ in the governing equation (4.116). Mass conservation forbids the mode $n = 0$ and $n = 1$ is the smallest possible choice for n. When $\epsilon = 0.05$, this gives the critical Rayleigh number 4096.4. This is the preferred solution since higher values of n produce higher values of Ra. The streamline pattern is omitted since in Mercator's projection (rectangular system with ϕ and θ as coordinates) it looks very similar to Figure 4.6. For all $\phi \neq 0$, the streamlines are compressed and this means less space is available for the flow than in cylindrical geometry. This is a stabilizing effect and it gives higher critical Rayleigh number for the spherical

Table 4.3 *Results for the onset of convection in the thin-shell limit of a spherical porous shell with conducting and impermeable boundaries. For various thicknesses ϵ, the critical Rayleigh number Ra_c is computed and compared with its asymptotic limit value k^2, where k is the wave number of the disturbance across the porous shell*

ϵ	Ra_c	Ra_c/k^2	k
0.05	4096.4	1.037	62.83
0.1	1077.1	1.091	31.42
0.15	514.37	1.173	20.94

shell than for the cylindrical shell (with horizontal axis) on comparing equal choice of shell thicknesses.

4.8   CONCLUDING REMARKS

The Rayleigh–Bénard instability is of broad scientific significance and it is one of the simplest known mechanisms of hydrodynamic instability in real fluids with finite viscosity, see Chandrasekhar (1961). Lord Rayleigh (1916) identified theoretically the finite threshold for the transition from pure conduction to convective flow, governed by the Rayleigh number as the order parameter. There is usually a preferred shape and a finite size of the mode of disturbance.

Rayleigh–Bénard convection is a very central phenomenon in the development of bifurcation theory for strongly nonlinear processes, with links to catastrophe theory. It is a key phenomenon in the cross-disciplinary field of synergetics, see Haken (1977). Rayleigh–Bénard cells are central in the paradigm of self-organizing dissipative structures, see Prigogine and Stengers (1984). Lorenz (1963) introduced a truncated model for nonlinear Rayleigh–Bénard convection as a first step in the modern understanding of deterministic chaos, see also Strogatz (1994). The evolution of Rayleigh–Bénard convection cells with imperfections in cell patterns, is conceptually related to crystal structures, see Manneville (1990). Rayleigh–Bénard cells have similarities with a different flow phenomenon: standing Faraday waves in oscillating containers, investigated among others by Miles (1993).

The Rayleigh–Bénard instability appears in its simplest version when the flow takes place in a porous medium. The present contribution to the topic is is to formulate a wider class of Rayleigh–Bénard problems for finite porous bodies in three dimensions. In general the eigenvalue problems for such finite bodies are considerably more complicated than the traditional case of an unlimited horizontal layer.

We have not given any new two-dimensional solutions for the onset of convection with zero perturbation temperature along an impermeable contour. Two such eigenvalue problems have been solved in the literature, see Nilsen and Storesletten (1990) who solved the problem of a rectangle and Storesletten and Tveitereid (1991) who solved the problem of a circle. In both cases there is a degeneracy in that two independent modes of disturbance, symmetric and antisymmetric modes, give the same critical Rayleigh number. In the present chapter we have found one new two-dimensional solution which is valid for the onset of convection in a rectangle with open and insulating lateral walls.

No three-dimensional problems seem to have been solved with the consistent requirement of zero perturbation temperature along the boundary of the three-dimensional porous body. Such a solution is given here for a thin spherical shell, in terms of a highly simplified thin-shell analysis. In this chapter we have focussed on three-dimensional problems in order to stimulate further research in this field.

REFERENCES

Bau, H. H. and Torrance, K. E. (1981). Onset of convection in a permeable medium between vertical coaxial cylinders. *Phys. Fluids* **24**, 382–385.

Bau, H. H. and Torrance, K. E. (1982). Low Rayleigh number convection in a vertical cylinder filled with porous materials and heated from below. *ASME J. Heat Transfer* **104**, 166–172.

Beck, J. L. (1972). Convection in a box of porous material saturated with fluid. *Phys. Fluids* **15**, 1377–1383.

Bénard, H. (1900). Les tourbillons cellulaires dans une nappe liquide. *Revue Gén. Sci. Our. Appl.* **11**, 1261–1271.

Bories, S. and Deltour, A (1980). Influence des conditions aux limites sur la convection naturelle dans une volume poreux cylindrique. *Int. J. Heat Mass Transfer* **23**, 765–771.

Chandrasekhar, S. (1961). *Hydrodynamic and Hydromagnetic Stability*. Clarendon Press, Oxford.

Chelghoum, D. E., Weidman, P. D., and Kassoy, D. R. (1987). The effect of slab width on the stability of natural convection in confined saturated porous media. *Phys. Fluids* **30**, 1941–1947.

Haken, H. (1977). *Synergetics: An Introduction*. Springer–Verlag, Berlin.

Horton, C. W. and Rogers, F. T. (1945). Convection currents in a porous medium. *J. Appl. Phys.* **16**, 367–370.

Kassoy, D. R. and Cotte, B. (1985). The effects of sidewall heat loss on convection in a saturated porous vertical slab. *J. Fluid Mech.* **152**, 361–378.

Keller, H. B. (1968). *Numerical Methods for Two-point Boundary-value Problems*. Blaisdell, Waltham, Massachusetts.

Lapwood, E. R. (1948). Convection of a fluid in a porous medium. *Proc. Camb. Phil. Soc.* **44**, 508–521.

Lorenz, E. N. (1963). Deterministic non-periodic flow. *J. Atmos. Sci.* **20**, 130–141.

Lowell, R. P. and Shyu, C. T. (1978). On the onset of convection in a water-saturated porous box: effect of conducting walls. *Lett. Heat Mass Transfer* **5**, 371–378.

Lyubimov, D. V. (1975). Convective motions in a porous medium heated from below. *J. Appl. Mech., Tech. Phys.* **16**, 257–261.

Manneville, P. (1990). *Dissipative Structures and Weak Turbulence*. Academic Press, San Diego.

McKibbin, R., Tyvand, P. A., and Palm, E. (1984). On the recirculation of fluid in a porous layer heated from below. *New Zealand J. Sci.* **27**, 1–13.

Miles, J. (1993). On Faraday waves. *J. Fluid Mech.* **248**, 671–683.

Murphy, H. D. (1979). Convective instabilities in vertical fractures and faults. *J. Geophys. Res.* **84**, 6121–6130.

Nield, D. A. (1968). Onset of thermohaline convection in a porous medium. *Water Resources Res.* **11**, 553–560.

Nield, D. A. and Bejan, A. (1999). *Convection in Porous Media* (2nd edn). Springer–Verlag, New York.

Nilsen, T. and Storesletten, L. (1990). An analytical study on natural convection in isotropic and anisotropic porous channels. *Trans. ASME J. Heat Transfer* **112**, 396–401.

Pearson, J. R. A. (1958). On convection cells induced by surface tension. *J. Fluid Mech.* **4**, 489–500.

Prigogine, I. and Stengers, I. (1984). *Order out of Chaos. Man's New Dialogue with Nature.* Bantam Books, Toronto.

Rayleigh, Lord (1916). On convection currents in a horizontal layer of fluid, when the higher temperature is on the under side. *Phil. Mag.* **32**, 529–546.

Rees, D. A. S. (1998). The instability of free convection boundary-layer flows in porous media: a critical review. In *Transport Phenomena in Porous Media* (eds D. B. Ingham and I. Pop), pp. 233–259. Pergamon, Oxford.

Rees, D. A. S. (2000). The stability of Darcy–Bénard convection. In *Handbook of Porous Media* (ed. K. Vafai), pp. 521–558. Marcel Dekker, New York.

Storesletten, L. and Tveitereid, M. (1991). Natural convection in a horizontal porous cylinder. *Int. J. Heat Mass Transfer* **34**, 1959–1968.

Strogatz, S. H. (1994). *Nonlinear Dynamics and Chaos. With Applications to Physics, Biology, Chemistry and Engineering.* Addison–Wesley, Reading, Massachusetts.

Tewari, P. K. and Torrance, K. E. (1981). Onset of convection in a box of fluid-saturated porous material with a permeable top. *Phys. Fluids* **24**, 981–983.

Wang, C. Y. (1998). Onset of natural convection in a fluid-saturated porous medium inside a cylindrical enclosure bottom heated by constant flux. *Int. Comm. Heat Mass Transfer* **25**, 593–598.

Wang, C. Y., Kassoy, D. R., and Weidman, P. D. (1987). Onset of convection in a vertical slab of saturated porous media between two impermeable conducting blocks. *Int. J. Heat Mass Transfer* **30**, 1331–1341.

Weidman, P. D. and Kassoy, D. R. (1986). The influence of sidewall heat transfer on convection in a confined saturated porous medium. *Phys. Fluids* **29**, 349–355.

Wooding, R. A. (1959). The stability of a viscous liquid in a vertical tube containing porous material. *Proc. Roy. Soc. Lond.* **A252**, 120–134.

Zebib, A. (1978). Onset of natural convection in a cylinder of water-saturated porous media. *Phys. Fluids* **21**, 699–700.

Zebib, A. and Kassoy, D. R. (1977). Onset of natural convection in a box of water-saturated porous media with large temperature variation. *Phys. Fluids* **20**, 4–9.

5 STABILITY ANALYSIS OF DOUBLE-DIFFUSIVE CONVECTION IN POROUS ENCLOSURES

M. MAMOU

Institute for Aerospace Research, National Research Council Canada, Ottawa, Ontario, K1A 0R6, Canada

http://www.meca.polymtl.ca/convection

Abstract

The stability of double-diffusive convection and finite-amplitude flows, due to opposing gradients of temperature and solute, in an inclined rectangular porous enclosure are studied analytically and numerically. The Darcy model and Boussinesq approximations are adopted to describe the double-diffusive convective flows within the enclosure. The governing equations are solved numerically using a finite element method. A closed-form analytical solution is derived on the basis of the parallel flow approximation within an infinite porous layer subject to constant fluxes of heat and solute. To study the linear stability of the diffusive state and of the fully-developed flows, a numerical technique is derived on the basis of the finite element method. The thresholds for subcritical, oscillatory, and monotonic convection are determined as functions of the governing parameters for different thermal and solutal boundary conditions and inclination angles. In a confined enclosure, the threshold for stationary convection depends on the solutal Rayleigh number, the inclination angle, and the thermal and solutal boundary conditions type. The threshold for oscillatory convection shows a strong dependence on all the governing parameters. For an infinite layer, the wavelength was found to be a function of the governing parameters. Within the overstable regime, traveling waves are also possible in horizontal and vertical enclosures, especially when the enclosure is subject to Dirichlet thermal and solutal boundary conditions. Below the threshold of overstability, subcritical convective flows were found to exist. Far from criticality, the threshold for Hopf bifurcation, which characterizes the transition from steady to oscillatory flows, is predicted. Multiple confined steady and unsteady states are found to coexist for a given set of the governing parameters. Within the parallel flow approximation, good agreement is found between the analytical results and the numerical solution of the full governing equations.

Keywords: double-diffusive convection, stability analysis, finite element, mixed boundary conditions, Hopf bifurcation, parallel flow

5.1 INTRODUCTION

Combined natural heat and mass transfer, the so-called double-diffusive or thermosolutal convection, has become increasingly an attractive field for many researchers and engineers in very diversified areas. The great interest in this phenomena stems from its rapid expansion and widespread engineering applications. The migration of moisture in fibrous insulation, contaminant transport in saturated soil, underground disposal of nuclear or non-nuclear wastes, food processing, spreading of pollutants, metallurgy, and electro-chemistry, to name but a few, are some examples where combined heat and solute transfer is encountered.

In the classical thermal natural convection within a porous enclosure, the flows are driven by the buoyancy forces resulting from density variations due to temperature gradients. For such flows, heat is transported through the fluid-saturated porous layer by both convection and diffusion. Similarly, for thermosolutal convection, for which density variations are induced by both temperature and concentration gradients, heat and solute concentration are transported by convection and diffusion. When the thermal and solutal effects are aiding each other, the convective flows remain qualitatively similar to those reported for the case of pure thermal convection. However, when these effects are opposing each other, various convective phenomena could manifest. These phenomena are characterized by the possible existence of multiple steady- and unsteady-state solutions for the same governing parameter values, subcritical flows which occur below the threshold of overstabilities, oscillatory flows (periodic or chaotic), traveling waves in relatively large aspect ratio enclosures, and asymmetric flow structures, etc., see, for example, Nield (1968), Taunton *et al.* (1972), Mamou *et al.* (1995a, 1998b), and Mamou and Vasseur (1999). These phenomena are due to the fact that the thermal and solutal diffusivities are usually different from each other. This leads essentially to different time scales for the heat and solute transfer. As a consequence, the heat and solute transfer get out of phase and give rise to the occurrence of such phenomena. Another factor which leads to similar phenomena is the porosity of the porous medium, even when the thermal and solutal diffusivities are similar or nearly equal, see, for example, Mamou *et al.* (1998c), Karimi-Fard *et al.* (1999), and Mamou and Vasseur (1999). For this situation, the heat is transferred through both the fluid and solid porous matrix. However, the solute is transported by diffusion and convection only through the fluid phase, since the porous matrix material is typically impermeable. This causes the heat and the solute transfer to have different time scales, which lead to various phenomena mentioned above.

The review of literature is focused mainly on problems related to double-diffusive convection in rectangular enclosures. The number of published papers on this topic is enormous, hence, due to space limitation, only some of the papers directly related to the present work will be referenced. The mechanism of heat and mass transfer by natural convection and the principles of convection through porous media are well documented by Bejan (1984). Recent developments and reviews of thermal convection within porous media are reported by Ingham and Pop (1998). A comprehensive review of the literature concerning double-diffusive natural convection in a fluid-saturated porous media can be found in the review article by Trevisan and Bejan (1990) and in the recent book by Nield and Bejan (1999). The present literature review is focused on two situations as follows:

5.1.1 Horizontal enclosures

This situation corresponds to a horizontal porous layer of finite/infinite extent subject to vertical thermal and solutal gradients. Double-diffusive instability in a horizontal porous layer was primarily studied by Nield (1968). On the basis of a linear stability analysis, the criteria for the onset of stationary and oscillatory convection were derived for various thermal and solutal boundary conditions. An extension of the analysis was made by Taunton *et al.* (1972), by determining the conditions for which 'salt fingers' develop in the presence of both temperature and concentration gradients. Considering a weak nonlinear stability analysis, Rudraiah *et al.* (1982) have applied the nonlinear stability analysis to the case of a porous layer with isothermal and isosolutal boundaries. The effects of Prandtl number, ratio of diffusivities, and permeability parameter on finite-amplitude convection were studied for opposing flows and the threshold for subcritical convection was obtained. Brand and Steinberg (1983) have investigated finite-amplitude convection near the threshold for both stationary and oscillatory instabilities and the temporal behavior of heat and mass transfer rates was predicted for the oscillatory regime.

The case of a sparsely packed porous medium, where viscous effects are significant, was investigated by Poulikakos (1986) on the basis of the Brinkman-extended Darcy model. The boundaries delineating the regions of monotonic and overstable regimes were obtained in terms of the governing parameters of the problem. Murray and Chen (1989) investigated experimentally and numerically double-diffusive convection in a horizontal porous layer. In the presence of stabilizing solute gradients, the onset of convection was marked by a sudden increase in heat flux at a critical temperature difference value. Furthermore, when the temperature difference was reduced to subcritical values then the heat flux curve established a hysteresis loop. Their results indicate clearly the existence of subcritical flows.

Double-diffusive fingering convection in a horizontal porous medium was considered by Chen and Chen (1993) using horizontal periodic boundary conditions and the Brinkman and Forchheimer extended Darcy model was considered. The stability boundaries which separate regions of different types of convective motion were identified in terms of the thermal and solutal Rayleigh numbers. Guo and Kaloni (1995) have studied the effect of the Darcy number and the effective viscosity on the threshold for stationary convection, when considering a Brinkman porous layer with non-slip boundaries. The effect of thermo-dependent fluid properties is also considered. Mamou *et al.* (1995c) have considered thermosolutal convection in an inclined porous layer heated and salted from the sides by uniform fluxes of heat and solute. Their analytical and numerical results, showing mutual agreement, demonstrated that subcritical steady flows, for the case of a horizontal layer, are possible. Studying the same phenomena within a horizontal porous layer, Mamou and Vasseur (1999) have reconsidered the problem using different boundary conditions for temperature and solute concentration. Multiple convective states were found to exist near criticality and general expressions for the thresholds of subcritical, oscillatory, and stationary convective flows were derived as functions of the governing parameters. In large aspect ratio enclosures, traveling waves were also observed. Amahmid *et al.* (1999a) studied double-diffusive convection in horizontal sparsely packed porous systems subject to vertical fluxes of heat and solute using the Brinkman model, and the threshold for the onset of subcritical and stationary flows have been determined. Recently, Mahidjiba *et al.* (2000) have examined the effect of mixed thermal and solutal boundary conditions (constant temperatures and mass fluxes, or *vice versa*, prescribed on the horizontal boundaries). The thresholds for oscillatory and stationary convection are obtained. It was also demonstrated that, when the thermal and solute effects are opposing each other, the flow patterns become much different from the classical Bénard convective flows. Considering the work of Mamou and Vasseur (1999), Kalla *et al.* (2001) have studied the effect of lateral heating on the bifurcation phenomena present in double-diffusive convection within a horizontal enclosure, and they found that the lateral heating acts as an imperfection brought to the bifurcation curves. Multiple steady-state solutions, with different heat and mass transfer rates, were found to coexist.

5.1.2 Vertical and tilted enclosures

Double-diffusive convection within vertical porous enclosures, subject to horizontal gradients of heat and solute, have received great interest during the last two decades. The case for aiding thermal and solutal effects, or for opposing flows (when the flow is thermally or solutably dominated), is well documented, and the flow behavior due to both temperature and concentration variations is well understood since this case compares qualitatively well with the pure thermal convection. However, when the thermal and solutal effects are opposing each other for conditions, such as the buoyancy forces are equal, or of the same order of magnitude, some interesting flow patterns may occur. Some of these flow patterns are discussed in this chapter and some new features are presented.

Starting with the work of Trevisan and Bejan (1985), heat and mass transfer by natural convection was studied numerically within a vertical enclosure subject to constant temperature and concentration on the vertical wall. Using scale analysis, the order-of-magnitude of the overall heat and mass transfer rates and their domain of validity was predicted. A good agreement was obtained between the numerical and scale analysis predictions. Furthermore, the authors have demonstrated that when the buoyancy forces are opposing each other, and are of the same magnitude ($N = -1$, where N is the buoyancy ratio), convective flow exists when $Le \neq 1$ (where Le is the Lewis number). Considering constant fluxes of heat and solute on the vertical walls, Trevisan and Bejan (1986) have investigated numerically and analytically the convective flows in enclosures of finite aspect ratio. In the boundary-layer regime, by assuming zero thermal and solutal horizontal gradients within the core region of the enclosure (outside the boundary-layer), the authors have developed an analytical solution which is valid only for $Le = 1$. On the other hand, for $Le > 1$ an analytical similarity solution was obtained, the latter was found to agree well with the numerical results. In large aspect ratio enclosures subject to constant fluxes of heat and solute, Alavyoon (1993) (in vertical enclosure) and Mamou *et al.* (1995c) (in tilted enclosure) have developed analytical solutions which are valid for a wide range of the Lewis number. In the boundary-layer regime, it was demonstrated that, for $Le \neq 1$, the horizontal thermal and solutal gradients within the core of the enclosure are not null. Alavyoon (1993) and Alavyoon and Masuda (1994) have mentioned, when $Le = 1$, that the problem does not belong to the class of problems that are described as double-diffusive flows and the diffusive regime is an exact solution of the problem when $N = -1$ (thermal and solutal buoyancy forces are equal and opposing each other) even for $Le \neq 1$. As demonstrated by Mamou *et al.* (1995b), when the Lewis number is not equal to unity, there is no reason to expect that the rest state remains unconditionally stable. An analytical solution was obtained and the threshold for the onset of convection was predicted. Using constant temperature and concentration, Nithiarasu *et al.* (1996) studied a similar problem on the basis of the generalized non-Darcy model. Their numerical results lead to the conclusion that, when $N = -1$, the convective flow disappears since the thermal and solutal buoyancy forces cancel each other. Amahmid *et al.* (1999b) have extended the case studied by Mamou *et al.* (1995b), by including the Brinkman effect. The threshold for subcritical flows was obtained as function of the Lewis and Darcy numbers. For the same problem, Mamou *et al.* (1998a) have determined the thresholds for stationary and oscillatory convection on the basis of linear stability analysis.

For opposing thermal and solutal buoyancy forces, nearly equal or of the same magnitude ($N \sim -1$), Mamou *et al.* (1995a) have predicted the existence of multiple solutions of different flow patterns within a square enclosure subject to constant fluxes of heat and solute. It was demonstrated that the transition from thermally to solutably dominated flows is characterized by the existence of multicellular flows. Using the same boundary conditions in the case of a slender vertical enclosure, Amahmid *et al.* (2000) have studied the situation where the buoyancy forces are nearly equal, i.e., $N = -1 \pm \zeta$, where $\zeta \ll 1$ is a very small positive number. As expected, multiple unicellular convective flows were predicted. Furthermore, since the system is unconditionally unstable for $\zeta \neq 0$, it was demonstrated that the addition of the quantity ζ to the buoyancy ratio is regarded as an imperfection brought to the bifurcation phenomena observed by Mamou *et al.* (1998b). The results are expected to be close to the experimented data since, in the current view, it is difficult to obtain experimentally a buoyancy ratio of $N = -1$, due to fluid property variations with temperature and concentration and to heat and mass transfer interaction.

The case of a vertical or inclined enclosure subject to opposing and equal buoyancy forces ($N = -1$) has been extensively studied during the last decade. For this situation, on the basis of both linear and nonlinear stability analysis, Charrier-Mojtabi *et al.* (1998), Mamou (1998), Mamou *et al.* (1998b, 1998c), Karimi-Fard *et al.* (1999), Marcoux *et al.* (1999), and Mojtabi and Charrier-Mojtabi (2000) have demonstrated that there exists a threshold for the onset of oscillatory and stationary convection. Different convective regimes, such as subcritical, overstable, and stationary convective modes, were delineated in terms of the governing parameters (Lewis number, enclosure aspect ratio, normalized porosity of the porous medium, inclination angle, and thermal and solutal boundary conditions). Subcritical convection was found to occur in a wide range of the Lewis number. However, the overstable regime was found to occur in a narrow range of Le (close to 1, as in the case of many gases) depending on the normalized porosity. Within an infinite layer, the wavelength at the onset of stationary convection was found to be independent of the Lewis number and this is demonstrated in the present study, but it is not the case at the onset of overstability. It was also demonstrated, when the Lewis number is close to unity, that the system remains conditionally stable, proving that the normalized porosity is lower than unity.

The present chapter is devoted to a two-dimensional study of double-diffusive convective flows within tilted porous enclosures subject to opposing thermal and solutal gradients. The situation where the thermal and solutal buoyancy forces are equal and opposing each other ($N = -1$) is considered. The case of an arbitrary buoyancy ratio is introduced for a horizontal enclosure, subject to vertical gradients of heat and solute. Similar and mixed thermal and solutal boundary conditions are considered. A reliable numerical technique is developed for determining the critical parameters for the onset of convection and, for comparison, a finite element solution of the full governing equations is obtained and the effects of the governing parameters on the convective flow behavior are studied.

5.2 PHYSICAL MODEL AND MATHEMATICAL FORMULATION

Assuming two-dimensional, laminar, and incompressible flow, double-diffusive natural convection is studied in an inclined, rectangular porous enclosure. Figure 5.1 shows the enclosure, which is of length L' and width W', tilted at an angle Φ with respect to the horizontal. The two end walls (parallel to the x-axis) are assumed impermeable and adiabatic, while Dirichlet or Neumann boundary conditions are applied on the two other side walls (parallel to the y-axis) for both the temperature and concentration. The enclosure is filled with an isotropic porous medium saturated with a binary fluid. The convective flow through the porous medium is governed by the Darcy law, upon assuming low Reynolds number flows, and the interaction between the thermal and solutal effects, known as Dufour and Soret effects, are neglected.

Applying the Boussinesq approximations, and using the stream function formulation, the dimensionless governing equations describing the conservation of momentum, energy, and constituent are as follows:

$$\xi \frac{\partial \left(\nabla^2 \Psi \right)}{\partial t} + \nabla^2 \Psi = -R_T \, \mathcal{F} \left(T + NS \right), \tag{5.1}$$

$$\frac{\partial T}{\partial t} - J \left(\Psi, T \right) = \nabla^2 T, \tag{5.2}$$

$$\varepsilon \frac{\partial S}{\partial t} - J \left(\Psi, S \right) = \frac{1}{Le} \nabla^2 S, \tag{5.3}$$

where Ψ, T, and S are the dimensionless stream function, temperature, and concentration, respectively. The operators $\mathcal{F}$ and J are defined as follows:

$$\mathcal{F} \left(\mathtt{f} \right) = \sin \Phi \frac{\partial \mathtt{f}}{\partial x} + \cos \Phi \frac{\partial \mathtt{f}}{\partial y}, \quad J \left(\mathtt{f}, \mathtt{g} \right) = \frac{\partial \mathtt{f}}{\partial x} \frac{\partial \mathtt{g}}{\partial y} - \frac{\partial \mathtt{f}}{\partial y} \frac{\partial \mathtt{g}}{\partial x}. \tag{5.4}$$

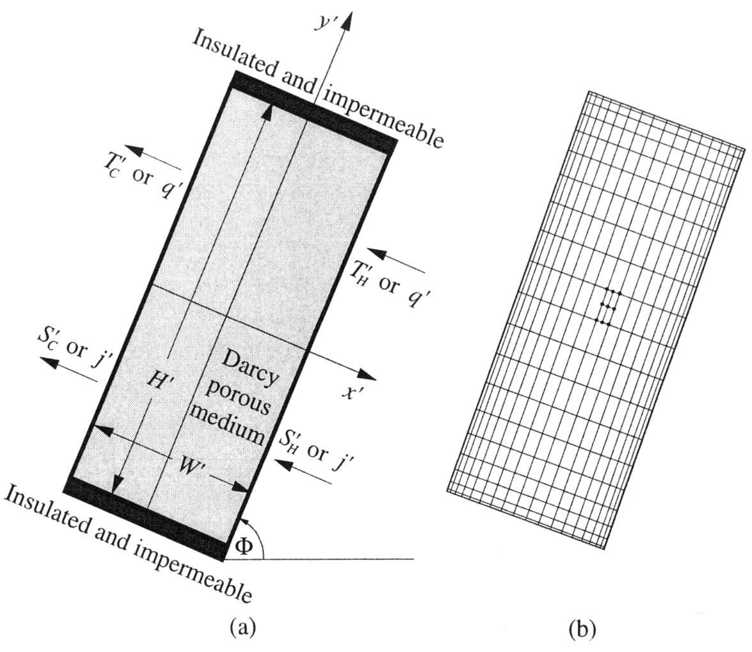

Figure 5.1 *(a) Physical model and coordinate systems; (b) computational domain*

The above governing equations were obtained by using the following scales:

$$x = \frac{x'}{W'}, \quad y = \frac{y'}{W'}, \quad t = \frac{t'\alpha_T}{\sigma W'^2}, \quad \Psi = \frac{\Psi'}{\alpha_T}, \quad T = \frac{T' - T'_0}{\Delta T^*}, \quad S = \frac{S' - S'_0}{\Delta S^*},$$

(5.5)

where t' is the time, α_T is the thermal diffusivity of the saturated porous medium, and σ is the saturated porous medium to fluid heat capacity ratio. Other definitions are as follows:

$$T'_0 = a_T T'_{(0,0)} + (1 - a_T) \frac{T'_H + T'_C}{2}, \quad \Delta T^* = a_T \frac{q'H'}{\kappa} + (1 - a_T)\left(T'_H - T'_C\right),$$

$$S'_0 = a_S S'_{(0,0)} + (1 - a_S) \frac{S'_H + S'_C}{2}, \quad \Delta S^* = a_S \frac{j'H'}{\alpha_S} + (1 - a_S)\left(S'_H - S'_C\right),$$

(5.6)

where the subscript $(0,0)$ refers to the origin of the coordinate system (center of the enclosure), and H and C refer to the hot and the cold boundaries, respectively. The parameter α_S is the mass-averaged diffusivity through the fluid saturated porous medium, κ is the effective thermal conductivity of the saturated porous medium, and the quantities q' and j' are the constant fluxes of heat and solute, respectively, applied on the active walls.

The dimensionless boundary conditions are stated as follows:

$$\left.\begin{array}{l} \Psi = 0 \\ a_T\dfrac{\partial T}{\partial x} \pm (1 - a_T)\,T = \dfrac{a_T + 1}{2} \\ a_S\dfrac{\partial S}{\partial x} \pm (1 - a_S)\,S = \dfrac{a_S + 1}{2} \end{array}\right\} \quad \text{on} \quad x = \pm\frac{1}{2}, \qquad \left.\begin{array}{l} \Psi = 0 \\ \dfrac{\partial T}{\partial y} = 0 \\ \dfrac{\partial S}{\partial y} = 0 \end{array}\right\} \quad \text{on} \quad y = \pm\frac{A}{2}. \tag{5.7}$$

The parameters a_T and a_S are set to zero for Dirichlet boundary conditions and to 1 for Neumann, and they can have different values, i.e., mixed boundary conditions with $(a_T, a_S) - (1, 0)$ or $(0, 1)$.

For the present problem, six dimensionless parameters appear in the governing equations (5.1) – (5.3) and (5.7). They are the thermal Darcy–Rayleigh number, R_T, the solutal to thermal buoyancy ratio, N, the Lewis number, Le, the enclosure aspect ratio, A, the normalized porosity, ε, and the inclination angle of the enclosure, Φ. They are defined as follows:

$$R_T = \frac{g\beta_T K \Delta T^* W'}{\alpha_T \nu}, \quad N = \frac{\beta_S \Delta S^*}{\beta_T \Delta T^*}, \quad Le = \frac{\alpha_T}{\alpha_S},$$

$$\varepsilon = \frac{\epsilon}{\sigma}, \quad \xi = \frac{Da}{\epsilon\sigma Pr}, \quad A = \frac{H'}{W'}, \tag{5.8}$$

where K is the permeability of the porous medium, and β_T and β_S are the thermal and concentration expansion coefficients, respectively. The parameter ξ is regarded as an acceleration coefficient of the porous medium, $Da = K/W'^2$ is the Darcy number, and $Pr = \nu/\alpha_T$ is the Prandtl number. For Darcy porous media $Da \ll 1$, and for fluids having moderate or large values of Pr, the parameter ξ is very small, so that the temporal term in the momentum equation can be neglected. However, this term is retained in the present work for generality, so that the foregoing stability analysis can be extended easily to sparsely porous media, where inertial and viscous forces may have significant effects on the flow behavior, i.e., when using the Forchheimer and Brinkman models.

The normalized porosity, ε, expressed in terms of the porosity of the porous medium, ϵ, and the solid to fluid heat capacity ratio, $r = (\rho C)_s / (\rho C)_f$, is given by $\varepsilon = \epsilon/\left[\epsilon + (1 - \epsilon)\,r\right]$. Since $0 < \epsilon < 1$, it is clear that $0 < \varepsilon < 1$. As it will be demonstrated later, ε has a strong effect on the oscillatory behavior of the convective flows and on the stability of the steady fully-developed flows.

The local heat and mass transfer rates, which are of interest in engineering applications, are expressed in terms of the local Nusselt and Sherwood numbers as follows:

$$Nu = \frac{a_T}{T_{(1/2,y)} - T_{(-1/2,y)}} + (1 - a_T)\left.\frac{\partial T}{\partial x}\right|_{x=\pm 1/2},$$

$$Sh = \frac{a_S}{S_{(1/2,y)} - S_{(-1/2,y)}} + (1 - a_S)\left.\frac{\partial S}{\partial x}\right|_{x=\pm 1/2}, \tag{5.9}$$

and the averaged values of Nu and Sh along the active walls can be computed as follows:

$$Nu_m = \frac{1}{A} \int_{-A/2}^{A/2} Nu \, dy \quad \text{and} \quad Sh_m = \frac{1}{A} \int_{-A/2}^{A/2} Sh \, dy. \tag{5.10}$$

5.3 FINITE-AMPLITUDE CONVECTION

The solution of the present problem is obtained by solving numerically and analytically the full governing equations. The numerical solution is obtained on the basis of the finite element method, while the analytical solution is derived on the basis of the parallel flow approximation within slender enclosures subject to Neumann boundary conditions.

5.3.1 Numerical solution

For finite-amplitude convection, the full governing equations (5.1) – (5.3), with the associated boundary conditions, equation (5.7), are solved using a finite element method. The calculus domain is discretized into nine-noded Lagrange cubic elements, as shown in Figure 5.1. Within each element the stream function, temperature, and concentration fields are quadratic. The number of elements in the x- and y-directions are given, respectively, by N_{ex} and N_{ey}, and the number of nodes in the x- and y-directions are $2N_{ex} + 1$ and $2N_{ey} + 1$, respectively. The Galerkin weak formulation is first obtained and then the Bubnov–Galerkin procedure, using the implicit scheme, is performed to discretize the governing equations. Global matrix systems of linear equations are derived and the solution is obtained by the iterative procedure described in Mamou *et al.* (1998b, 1998c). Further details regarding the finite element method and the validation of the numerical simulation can be found in Mamou *et al.* (1995c, 1998b, 1998c) and Mamou (1998).

5.3.2 Analytical solution

For a slender enclosure (large aspect ratio), Alavyoon (1993) and Mamou *et al.* (1995c) have demonstrated that when the enclosure is subject to uniform fluxes of heat and solute from the sides, an analytical solution can be derived on the basis of the parallel flow approximation. Using the parallel flow approximation, the governing equations (5.1) – (5.3) can be reduced to a set of ordinary differential equations that can be easily solved to yield a closed-form solution for the stream function, temperature and concentration distributions. In this section, the analytical solution is developed only for the cases of vertical and horizontal enclosures.

Table 5.2 *Critical Rayleigh number, R_{TC}^{sup} (positive and negative values), and grid size effects for the onset of stationary convection within a square enclosure with $Le = 2$, $N = -1$, $\Phi = 0°$ and $90°$, and different thermal and solutal boundary conditions*

		$N_{ex} \times N_{ey}$			
Φ	(a_T, a_S)	4×4	8×8	12×12	16×16
	$(0,0)$	$186.597, -186.597$	$184.221, -184.221$	$184.091, -184.091$	$184.069, -184.069$
$90°$	$(1,1)$	$214.790, -214.790$	$210.175, -210.175$	$209.904, -209.904$	$209.858, -209.858$
	$(0,1)$	$270.785, -270.785$	$259.898, -259.898$	$259.288, -259.288$	$259.185, -259.185$
	$(1,0)$	$235.412, -235.412$	$235.412, -235.412$	$230.072, -230.072$	$230.013, -230.013$
	$(0,0)$	$-39.518, \quad \infty$	$-39.481, \quad \infty$	$-39.479, \quad \infty$	$-39.479, \quad \infty$
$0°$	$(1,1)$	$-22.975, \quad \infty$	$-22.948, \quad \infty$	$-22.946, \quad \infty$	$-22.946, \quad \infty$
	$(0,1)$	$-16.060, \quad \infty$	$-16.033, \quad \infty$	$-16.032, \quad \infty$	$-16.031, \quad \infty$
	$(1,0)$	$325.718, -70.449$	$308.025, -69.695$	$307.103, -69.650$	$306.948, -69.642$

bottom or from the top, the threshold for stationary convection is finite. For $a_T = a_S = 0$, the critical value is seen to agree well with the classical value for pure thermal convection, namely $4\pi^2$.

The effect of the boundary conditions and the Lewis number on the threshold for stationary convection ($R_{TC}^{\mathrm{sup}} |1 - Le| = R^{\mathrm{sup}}$) is depicted in Figure 5.4(a) for $\Phi = 90°$, and the results are presented only for the positive eigenvalue. It is observed that, for similar boundary conditions, the parameter R^{sup} is independent of the Lewis number. However, R^{sup} becomes a function of Le for mixed boundary conditions. For large or small Lewis numbers, the constant R^{sup} is seen to tend towards that corresponding to similar boundary conditions and it becomes independent of Le. For this situation, the flow structures are similar to those observed for $a_T = a_S$, but for relatively small Lewis number ($Le \sim 1$) the flow structure becomes different and more complex, see Figures 5.4(b) and (c).

Onset of oscillatory convection

The threshold corresponds to the marginal overstable regime. For this situation, the perturbation growth rate parameter ($p = p_r + \mathrm{i}p_i$) is a pure complex number (i.e., $p_r = 0$ and $p_i \neq 0$). The linear discretized equations (5.31) – (5.33) can be rearranged to the following linear eigenvalue problem:

$$\begin{bmatrix} [K_\psi] & [B_{\psi\theta}] & [B_{\psi\phi}] \\ [B_\theta] & [K_\theta] & 0 \\ [B_\phi] & 0 & [K_\phi] \end{bmatrix} \begin{Bmatrix} \{F\} \\ \{G\} \\ \{H\} \end{Bmatrix} = p \begin{bmatrix} [M_\psi] & 0 & 0 \\ 0 & [M_\theta] & 0 \\ 0 & 0 & [M_\phi] \end{bmatrix} \begin{Bmatrix} \{F\} \\ \{G\} \\ \{H\} \end{Bmatrix}. \quad (5.43)$$

The matrices are defined by $[B_{\psi\theta}] = -R_T [B_\psi]$, $[B_{\psi\phi}] = -R_T N [B_\psi]$, $[B_\theta] = -[B_\theta]$, $[B_\phi] = -[B_\phi]$, $[M_\psi] = -\xi [M_\psi]$, $[M_\theta] = -[M_\theta]$, $[M_\phi] = -\varepsilon [M_\phi]$, $[K_\theta] = [K_\theta]$, $[K_\phi] = (1/Le) [K_\phi]$, and $[K_\psi] = [K_\psi]$, where $[B_\psi]$, $[K_\psi]$, $[K_\theta]$, $[K_\phi]$, $[M_\psi]$, $[M_\theta]$,

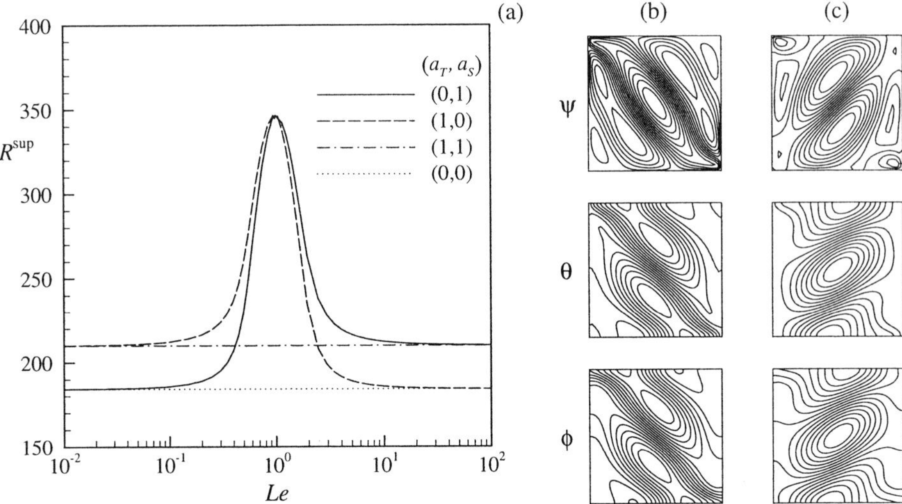

Figure 5.4 *(a) Threshold for stationary convection within a square enclosure for $N = -1$ and $\Phi = 90°$: effect of the Lewis number and boundary conditions. Perturbation profiles (ψ, θ, and ϕ) at the threshold for stationary convection for $N = -1$, $A = 1$, $\Phi = 90°$, and $(a_T, a_S) = (0,1)$: (b) $Le = 0.9$, $R_{TC}^{\mathrm{sup}} = \pm 3402.97$ and (c) $Le = 1.1$, $R_{TC}^{\mathrm{sup}} = \pm 3436.80$*

and $[M_\phi]$ are defined by equation (5.34) for confined enclosures and by equation (5.36) for an infinite layer. It should be noted that the matrices of the above system are real for confined enclosures and complex for an infinite layer, but in both cases the eigenvalues and eigenfunctions are complex.

The above eigenvalue system is solved by using the subroutines *DGVCRG* and *DGVCCG* of the IMSL library for general real and complex matrices, respectively. For a given set of the governing parameters, R_S, Le, A, Φ, ε, ξ, a_T, and a_S, the eigenvalues, p, and eigenfunctions, $\{F\}$, $\{G\}$, and $\{H\}$, are computed for different thermal Rayleigh numbers, R_T, (for finite aspect ratio enclosure, A) and different wavelengths (for an infinite layer, A_C). For a given aspect ratio or wavelength, the onset of the oscillatory convection is obtained by only one eigenvalue when its real part changes from negative to positive value (when $p_r < 0$ the perturbation is decaying, and when $p_r > 0$ the perturbation is growing). If the imaginary part of the eigenvalue is different from zero, then the instability is oscillatory. For an infinite layer, the minimum (critical) Rayleigh number is obtained for different wavelengths. So, the minimum Rayleigh number for all wavelengths represents then the threshold for oscillatory convection and the corresponding aspect ratio represents the critical wavelength. At the onset of overstabilities, two complex conjugate eigenvalues having zero real parts were found to exist. The flow patterns of these two solutions are mirror images of each other and the resulting convective flows could oscillate between the two solutions. In large aspect ratio enclosures, the two solutions may be superposed to lead to traveling wave flows along the porous layer.

Discussion

To assess the efficiency and the validity of the present numerical technique, two typical cases are considered. The first one corresponds to the horizontal enclosure subject to Dirichlet boundary conditions, for which an analytical exact solution exists, see Nield (1968). The second corresponds to a vertical square enclosure for the same boundary conditions. For the second one, recent numerical results have been reported by Mamou *et al.* (1998c) and Karimi-Fard *et al.* (1999). Results obtained for different grid sizes with $N = -1, \xi = 0, \varepsilon = 0.2, A = 1$, and $a_T = a_S = 0$ are depicted in Table 5.3. The results are presented in terms of the threshold for the onset of oscillatory convection, R_{TC}^{over}, and the oscillation frequency, fr. It is observed that when the number of grid points increases the results converge to those reported in the past, see Table 5.3. As can be seen, even with a grid size of 2×2, the precision is within 2%. It is worth noting that the threshold for overstability is obtained by only one eigenvalue when its real part, p_r, changes from a negative to a positive value. Numerically, it is assumed that $p_r = 0$ when the numerical value of $|p_r|$ is less than 10^{-6}. The numerical technique presented in this section can also be used to find the critical Rayleigh number for the onset of stationary convection ($p_r = p_i = 0$) and for this situation the oscillatory frequency is null. Another efficient way to determine the critical point without excessive computations is to interpolate the critical value between negative and positive values of p_r. The variation of p_r with R_T is nearly linear and thus a first-order interpolation is largely sufficient without loss of accuracy.

It is well known in the literature that the overstable regime occurs below the threshold of stationary convection. Above the threshold, $R_T > R_{TC}^{\text{sup}}$, overstable solutions are possible but they are not dominant since, in this regime, the strongest perturbation is described by the one having highest growth rate p_r. Usually (right above R_{TC}^{sup}), the perturbation that is growing in a monotonic manner (no oscillation) has the highest growth rate. Thus, it becomes dominant and overcomes the overstable perturbations. Karimi-Fard *et al.* (1999)

Table 5.3 *Critical Rayleigh number, R_{TC}^{over}, and oscillation frequency, $fr = |p_i|/2\pi$, for the onset of overstabilities in a vertical and horizontal square enclosure with $N = -1$, $Le = 2$, $\varepsilon = 0.2$, $\xi = 0$, and $a_T = a_S = 0$. The superscripts N and K refer to Nield (1968) and Karimi-Fard et al. (1999), respectively*

Φ	Grid size	2×2	4×4	8×8	10×10	Reported values
$0°$	R_{TC}^{over}	34.55925	34.54403	34.54363	34.54362	$3.5\,\pi^2 = 34.5436^N$
	fr	1.75666	1.75621	1.75620	1.75620	1.75620^N
$90°$	R_{TC}^{over}	173.48301	170.76035	170.68308	170.68222	170.7^K
	fr	2.22405	2.24016	2.24053	2.24121	2.24^K

have considered the case of a vertical square enclosure with the following parameters: $N = -1$, $A = 1$, $\varepsilon = 0.5$, $\xi = 0$, $\Phi = 90°$, and different Lewis numbers $Le = 0.5$, 1, $\sqrt{2}$, and 2. Mamou *et al.* (1998c) demonstrated that, for $Le > 1$, the transition from oscillatory to monotonic convection occurs exactly at $Le = 1/\sqrt{\varepsilon}$. In other words, for $Le \geqslant \sqrt{2}$, the instability is stationary and the oscillation frequency vanishes. The supercritical Rayleigh number is given by $R_{TC}^{\text{sup}} = \pm 184.06/(1 - Le)$. For $Le < \sqrt{2}$, the overstable regime may exist; for $Le = 1$, the critical parameters are given by $R_{TC}^{\text{sup}} = \infty$, $R_{TC}^{\text{over}} = 774.2$, and $p_i = \pm 69.1$, which are slightly different from those reported by Karimi-Fard *et al.* (1999); for $Le = 0.5$, it is found that the overstable regime does not exist (since $R_{TC}^{\text{over}} > R_{TC}^{\text{sup}}$). It was found by Karimi-Fard *et al.* (1999) that the threshold for overstabilities is 1591 and the pulsation $p_i = 259.3$. In the present analysis, at Rayleigh number value of 1591, there are eight types of growing perturbations. Seven are monotonic, for which the maximum growth rate is given by $p = 67.0 + 0i$, and the eighth one is oscillatory with $p \simeq 0 \pm 259.26i$. It appears from these results that the most dominant perturbation is the stationary one. Therefore, for $Le = 0.5$, overstabilities are dominated by the stationary one. The present study concerns a two-dimensional stability analysis, such that the perturbation profiles are supposed to be independent of the depth of the enclosure (i.e., they are localized in the x–y plane). This supposition is supported by the three-dimensional stability analysis conducted by Karimi-Fard *et al.* (1999), who have demonstrated that the most dangerous perturbations (that give lower critical Rayleigh number) occur in the x–y plane. Maybe it is not the case for finite-amplitude convection within cubic enclosures, as reported by Sezai and Mohamad (1999). The authors found that three-dimensional flows are possible near $N = -1$. Their results are obtained by solving the steady governing equations and for Dirichlet boundary conditions. For this situation, oscillatory flows may be possible.

Within an infinite vertical layer, the effects of the Lewis number on the thresholds of oscillatory and stationary convection are illustrated in Figure 5.5 for $\varepsilon = 0.2$, $\xi = 0$, and for different boundary conditions (a_T, a_S). It should be noted that the critical wavelength, $A_C = 2.504$, at the onset of stationary convection is independent of Le. For oscillatory convection, the wavelength was found to depend on Le, ε, ξ, and (a_T, a_S). As depicted in Figure 5.5, the critical parameters do not depend on the boundary conditions (for example, the results are the same for $(a_T, a_S) = (0, 0)$ or $(1, 1)$ and for $(a_T, a_S) = (1, 0)$ or $(0, 1)$). The most interesting observation in the results presented in Figure 5.5 is the effect of Le on the transition from oscillatory to stationary convection. For $Le > 1$, the transition occurs at nearly zero oscillation frequency. In Figure 5.5(c), the frequency is observed to decrease towards zero when approaching the transition boundary depicted by symbols in the figure. However, the frequency is maximum at the transition boundary for $Le < 1$. It is also noted that there is a discontinuity of the wavelength at the threshold of transition. The flow structure at the onset of overstability is non-symmetric and there exists two conjugate solutions, which may be combined to lead to traveling wave flows along the porous layer.

For inclined enclosures and similar thermal and solutal boundary conditions, closed-form expressions for the thresholds were derived by Mamou *et al.* (1998b, 1998c) for the onset of stationary and oscillatory convection. For the threshold of the overstable regime,

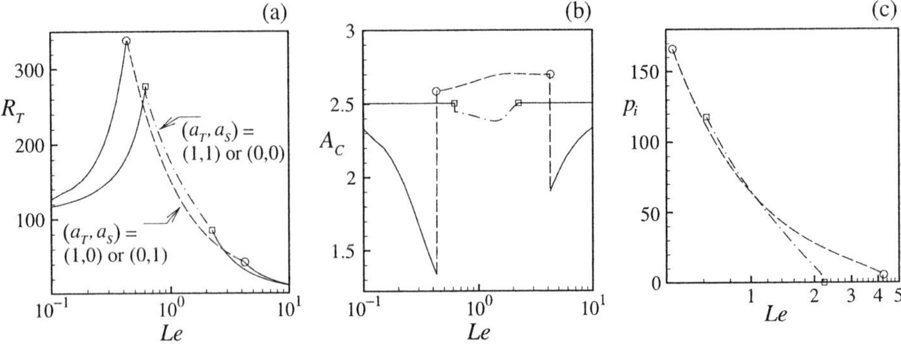

Figure 5.5 *Effect of Le on the thresholds of oscillatory and stationary convection within a vertical infinite layer for $\varepsilon = 0.2$, $N = -1$, and $\xi = 0$. The threshold of stationary convection is depicted by solid lines and that for oscillatory convection by dashed lines*

their results are approximate since they have supposed that the perturbation profiles are qualitatively the same at the onset of oscillatory and stationary convection. In the present analysis, it is observed that, at the onset of overstabilities, the flow structure could be non-symmetric (the convective cells are seen to be shifted to the left vertical wall) and the temperature and concentration profiles are not identical, as can be seen from Figure 5.6. This difference could contribute to the unsteadiness of the flow.

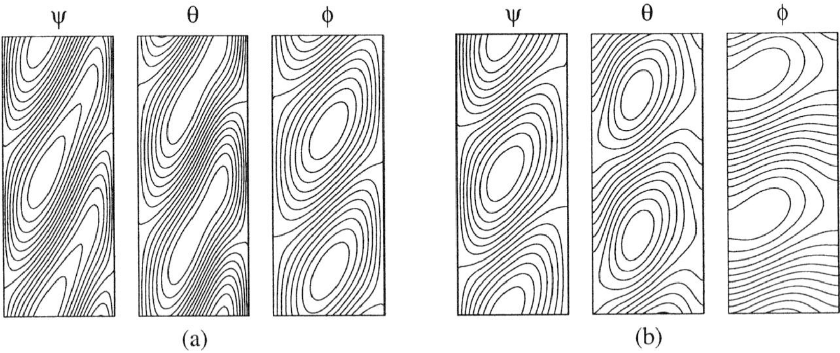

Figure 5.6 *Perturbation profiles (ψ, θ, and ϕ) at the onset of overstabilities in an infinite vertical layer for $N = -1$, $Le = 1$, $\Phi = 90°$, $\varepsilon = 0.1$, and (a) $a_T = a_S = 0$ and (b) $a_T = a_S = 1$. Here $R_{TC}^{over} = 138.57$, $A_C = 2.44$, and $p_i = \pm 92.18$*

It can be demonstrated that, for $Le = 1$, the only possible convective flows are the oscillatory ones, see for instance Mamou *et al.* (1998b, 1998c) and Karimi-Fard *et al.* (1999). The effect of the normalized porosity, ε, on the threshold of oscillatory convection is displayed in Figure 5.7 for equal thermal and solutal diffusivities ($Le = 1$). The results correspond to an infinite vertical layer subject to different boundary conditions. For similar boundary conditions, the threshold for oscillatory convection tends towards infinity when ε approaches unity. However, it remains finite for mixed boundary conditions. The critical wavelength is observed to be much affected by ε variations for the case of mixed boundary conditions. The pulsation p_i is seen to decrease with increasing ε, and the same trend was reported by Karimi-Fard *et al.* (1999) within a square enclosure.

For an horizontal infinite porous layer with similar thermal and solutal boundary conditions ($a_T = a_S$), the threshold for oscillatory convection for stress-free boundaries were determined by Nield (1968). Taunton *et al.* (1972) have extended the work of Nield by introducing the acceleration effects. The threshold for oscillatory convection has been determined in terms of the thermal Rayleigh number as a function of the governing parameters. Mamou and Vasseur (1999) have reconsidered the problem by examining the effect of the enclosure confinement on the onset of oscillatory convection, without the acceleration parameter. General expressions for the thresholds were derived for the Dirichlet and Neumann boundary conditions. Following the same analysis described in Taunton *et al.*

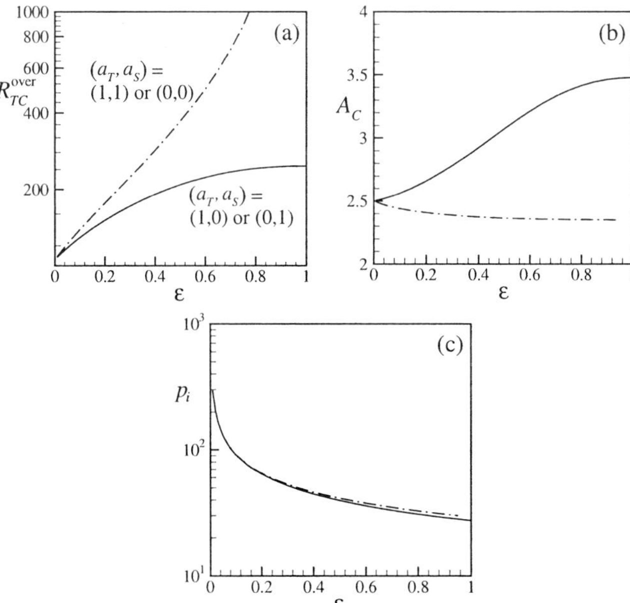

Figure 5.7 *Threshold of oscillatory convection in an infinite vertical layer for* $N = -1$, $Le = 1$, *and* $\xi = 0$

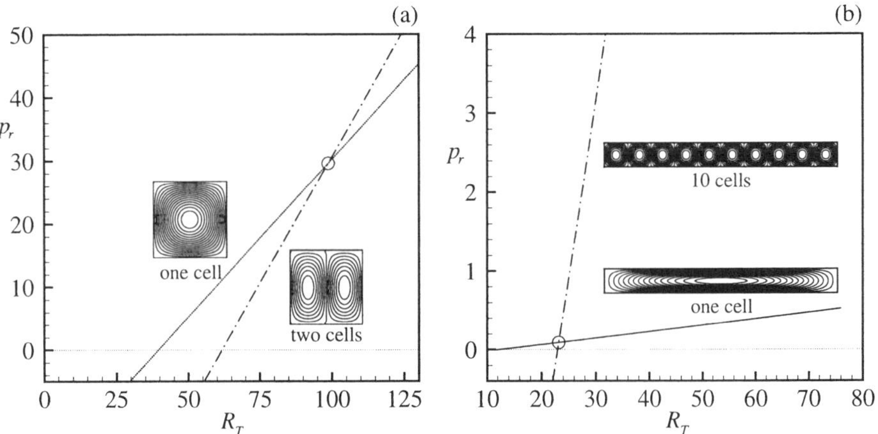

Figure 5.9 *Growth rate of different convective modes for pure thermal convection in a horizontal enclosure:* $R_S = 0$, $\xi = 0$, *and (a)* $A = 1$, *with* $a_T = 0$, *and (b)* $A = 10$, *with* $a_T = 1$

For $A = 1$ with $a_T = 0$, $p_r = 0$ corresponds to the threshold of the onset of stationary convection. For one cell flow $R^{\mathrm{sup}} = 4\pi^2$, and for two cell flow $R^{\mathrm{sup}} = 25\pi^2/4$. The growth rate of these two flow types is seen to increase, at different rates, as R_T increases and they intersect at $R_T^{\mathrm{int}} = 10\pi^2$. Below the intersection point, $R_T = R_T^{\mathrm{int}}$, the growth rate of the one cell mode is higher than that of the two cell mode. As a result, the one cell mode is the preferable one for $R_T < R_T^{\mathrm{int}}$. However, for $R_T > R_T^{\mathrm{int}}$, the growth rate of the two cell mode is the highest one and, therefore, this mode is now the favorable one. It is expected that, when solving the governing equations with the pure diffusive state as initial conditions, the flow solution will be monocellular for $R_T < R_T^{\mathrm{int}}$ and bicellular for $R_T > R_T^{\mathrm{int}}$.

For Neumann boundary conditions (Figure 5.9b), the intersection point between the one cell flow and the 10 cell flow growth rates is obtained for $R_T^{\mathrm{int}} = 23.15$ (the flow with $2, 3, \ldots, 9$ cells are not presented here). The thresholds for stationary convection are $R_{TC}^{\mathrm{sup}} = 12.017$ (one cell flow) and $R_{TC}^{\mathrm{sup}} = 22.94$ for square convective cells (10 cell flow). As discussed above, when $R_T > R_T^{\mathrm{int}}$ the Bénard convective flow is favorable and the finite-amplitude flow may be stable. Also, it is worth mentioning that, for $A = 10$, the flows with $1, 2, \ldots, 10$ cell modes may be possible, providing that the Rayleigh number is sufficiently high.

For mixed boundary conditions, the stability of the diffusive state is discussed for a horizontal square enclosure. The thresholds for oscillatory and stationary convection are presented in the stability diagram in Figure 5.10 for positive thermal Rayleigh number (heating from below). The threshold for stationary convection is first discussed. Let us recall that, for pure thermal convection with Dirichlet ($a_T = 0$) and Neumann ($a_T = 1$)

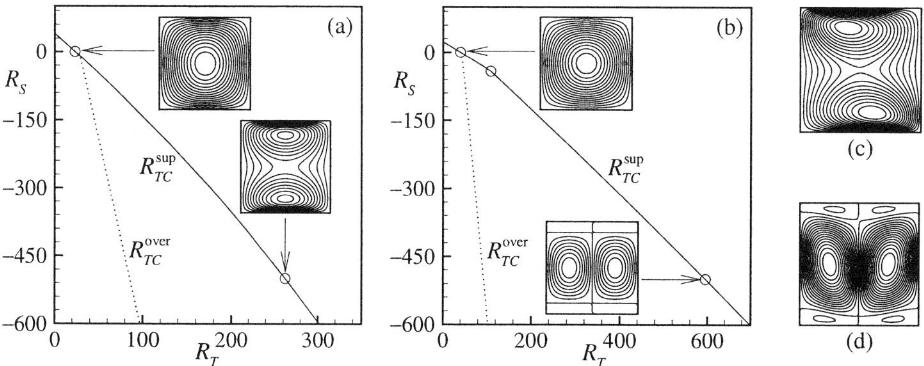

Figure 5.10 *Stability diagram for mixed boundary conditions within a horizontal square enclosure for $Le = 10$, $\varepsilon = 1$, $\xi = 0$, and $\Phi = 0°$, with (a) $(a_T, a_S) = (1, 0)$ and (b) $(a_T, a_S) = (0, 1)$. Finite-amplitude convection for $R_S = -500$ and $Le = 0.5$: streamlines for (c) $R_T = 300$ and $(a_T, a_S) = (1, 0)$, with $R_{TC}^{\text{sup}} = 262.18$, and (d) $R_T = 650$ and $(a_T, a_S) = (0, 1)$, with $R_{TC}^{\text{sup}} = 595.20$*

boundary conditions, the supercritical Rayleigh numbers, R^{sup}, are given by $R_{0}^{\text{sup}} = 4\pi^2$ and $R_{1}^{\text{sup}} = 22.95$, respectively. According to the linear stability results, the flow structure consists of a square single cell for relatively small solutal Rayleigh number $|R_S|$. From the stability diagram, it was found that the threshold of stationary convection curves could be fitted by the following relationship:

$$R_{TC,(a_T,a_S)}^{\text{sup}} = -\frac{R_{a_T}^{\text{sup}}}{R_{a_S}^{\text{sup}}} R_S + R_{a_T}^{\text{sup}}, \tag{5.50}$$

such that, for $(a_T, a_S) = (1, 0)$, we have

$$R_{TC,(1,0)}^{\text{sup}} = -\left(4\pi^2/22.95\right) R_S + 22.95 = -1.72\, R_S + 22.95, \tag{5.51a}$$

and, for $(a_T, a_S) = (0, 1)$, we have

$$R_{TC,(0,1)}^{\text{sup}} = -\left(22.95/4\pi^2\right) R_S + 4\pi^2 = -0.58\, R_S + 39.85. \tag{5.51b}$$

The above relation is valid as long as the incipient flow remains monocellular and the streamlines circular. As shown in Figure 5.10(a), for $(a_T, a_S) = (1, 0)$, it is observed that the flow patterns are strongly affected by the increase of $-R_S$. For relatively high solutal Rayleigh number $-R_S$, the circular original cell observed for $R_S \geqslant 0$ is significantly distorted and broken into two co-rotating cells. One cell is near the lower boundary and the other near the upper boundary. For large $-R_S$, a weak recirculation flow is observed between the two cells. Similar results have been obtained by Mahidjiba *et al.* (2000).

The authors have also studied the possible occurrence of overstability when the solute is stabilizing. The curves of the threshold for stationary convection are slightly curved due to the change in the flow structure.

For $(a_T, a_S) = (0, 1)$, it is observed in Figure 5.10(b) that, at the onset of stationary convection, the flow pattern is monocellular for $R_S > -41.9$. For $R_S < -41.9$, the flow structure that gives the minimum critical Rayleigh number is characterized by the existence of six convective cells. Two main vertical contour-rotating cells, see Figure 5.10(b), occupy the central region of the enclosure, and four small cells located near the top and the lower boundaries (i.e., two on the top and two on the bottom). For such boundary conditions, Mahidjiba *et al.* (2000) have found that, within an infinite layer, the wavelength decreases to 0.93. For the onset of overstabilities, the flow pattern was found to be always monocellular, and, as shown in Figure 5.10, the threshold of oscillatory convection, obtained for $Le = 10$, is well below the threshold for stationary convection. Finite-amplitude flow patterns near the threshold for stationary convection are depicted in Figure 5.10(c) and (d) for $(a_T, a_S) = (1, 0)$ and $(a_T, a_S) = (0, 1)$, respectively. It is observed that the flow patterns are similar to the incipient flow patterns predicted by the linear stability analysis. It worth mentioning that, for the case with $(a_T, a_S) = (0, 1)$, the flow was found to be oscillating without any noticeable change in the flow patterns.

5.4.2 Convective state

The convective solution is obtained analytically for the case of a horizontal or a vertical infinite layer subject to constant fluxes of heat and solute ($a_T = a_S = 1$). The steady convective solution is expected to become unstable far from criticality (i.e., for high Rayleigh number). It was discussed in the past by Kimura *et al.* (1995), for pure thermal convection, that, when the thermal Rayleigh number is increased progressively, there is a certain value above which the convective flows become oscillatory. The transition from steady to oscillatory flow is termed a Hopf bifurcation and these authors predicted its threshold.

Onset of subcritical and stationary convection

The analytical solution has served as a guide to understand the steady double-diffusive mechanism and it is regarded as a useful nonlinear stability analysis to explain the conditions under which subcritical flows exist. According to the solution given by equation (5.22), there exist five possible steady-state solutions. The first one corresponds to the pure diffusive regime ($\psi_0 = 0$) and the others to convective flow regimes. In equation (5.22), the $\pm$ sign outside the brackets indicates the existence of convective solutions of the same amplitude, one solution corresponding to a clockwise circulation and the other to a counterclockwise one. On the other hand, the $\pm$ sign within the brackets indicates that two convective solutions of different amplitudes are possible.

From equation (5.22), it can be demonstrated that there are one or two types of bifurcations, depending on the governing parameter values. The first one is called a supercritical bifurcation, since the transition from diffusive to convective states occurs through zero amplitude at supercritical Rayleigh number $R_T = R_{TC}^{\mathrm{sup}} = 12$, which corresponds to the threshold of stationary convection. The second bifurcation is subcritical; two branches of solutions bifurcate from the rest-state solutions and these two branches are connected to each other at a saddle-node point, $R_T = R_{TC}^{\mathrm{sub}}$, corresponding to the subcritical Rayleigh number, which characterizes the onset of finite-amplitude convective flows.

In general, supercritical bifurcation occurs for aiding flows ($R_S > 0$) or for opposing flows ($R_S < 0$) when $Le < 1$. For this situation, the supercritical Rayleigh number, R_{TC}^{sup}, corresponding to the onset of stationary convection, is obtained from the conditions $d_1 < 0$ and $d_2 = 0$ (i.e., $\psi_0 = 0$) as

$$R_{TC}^{\mathrm{sup}} = -R_S + R^{\mathrm{sup}} \quad \text{or} \quad R_{TC}^{\mathrm{sup}} = \frac{R^{\mathrm{sup}}}{1 + NLe}, \tag{5.52}$$

which is the same result as that predicted by the linear stability analysis, see equation (5.41).

On the other hand, subcritical bifurcation is possible only for the case of opposing flows ($R_S < 0$) and when the stabilizing agent is the slower diffusive component ($Le > 1$). The subcritical Rayleigh number, R_{TC}^{sub}, for the onset of subcritical convection can be obtained, from the conditions $d_1 > 0$ and $d_1^2 + d_2 = 0$, as follows:

$$R_{TC}^{\mathrm{sub}} = Le^{-2} \left(\sqrt{Le^2 - 1} + \sqrt{-R_S^0} \right)^2 R^{\mathrm{sup}}. \tag{5.53}$$

At the threshold, R_{TC}^{sub}, the flow intensity is given by $\psi_0 = \pm \sqrt{bd_1}/Le$ and, for the existence of subcritical convection, the conditions $R_S < 0$ and $Le > [(R_S - R^{\mathrm{sup}})/R_S]^{1/2}$ must be satisfied. For Dirichlet boundary conditions and an infinite layer, Rudraiah *et al.* (1982) have determined the threshold of subcritical flows using a weak nonlinear stability analysis. The threshold expression is similar to that given in equation (5.53) but with $R^{\mathrm{sup}} = 4\pi^2$. Mamou and Vasseur (1999) have extended the work of Rudraiah *et al.* (1982) by considering general boundary conditions and finite aspect ratio enclosures, and they have obtained the same expression with R^{sup} as a function of the boundary conditions and the enclosure aspect ratio.

Onset of oscillatory convection: Hopf bifurcation

The procedure to determine the threshold of Hopf bifurcation is the same as that used to compute the threshold of overstabilities in the previous sections for an infinite layer. For this situation, the basic solution Ψ_b, T_b, and S_b is given by equation (5.18). To validate the present numerical procedure, the results of Kimura *et al.* (1995) for thermal convection are considered. To simulate their case with the present governing equations, equation (5.29), the solutal buoyancy force is nullified ($R_S = 0$). Using 4 to 16 finite elements in the vertical direction, the critical Rayleigh number for Hopf bifurcation is presented in Table 5.4, and the corresponding critical wavenumber $k_c = 2\pi/A_C$ and frequency $fr = |p_i|/2\pi$ are also displayed. It can be seen that the present results agree

Table 5.4 *Critical Rayleigh number, R_{TC}^{Hopf}, critical wavelength, A_C, and oscillation frequency, fr, for Hopf bifurcation in an infinite horizontal layer*

	Number of elements				
	4	8	12	16	Kimura *et al.* (1995)
R_{TC}^{Hopf}	510.0195	506.1841	506.0815	506.0742	506.07
$k_c = 2\pi/A_C$	4.7696	4.8239	4.8257	4.8251	4.825
$fr = |p_i|/2\pi$	22.0287	22.1070	22.1122	22.1095	22.11

well with those obtained by Kimura *et al.* (1995) and a good precision can be obtained by using only a few finite elements in the x-direction.

In a horizontal porous layer with Neumann boundary conditions, the effects of the Lewis number and the normalized porosity on the threshold of Hopf bifurcation are studied in Figure 5.11. The results are presented for $R_S = -100$, $\Phi = 0°$, $\xi = 0$, and $a_T = a_S = 1$. A strong effect of ε and Le on the critical parameters is observed, and when Le becomes

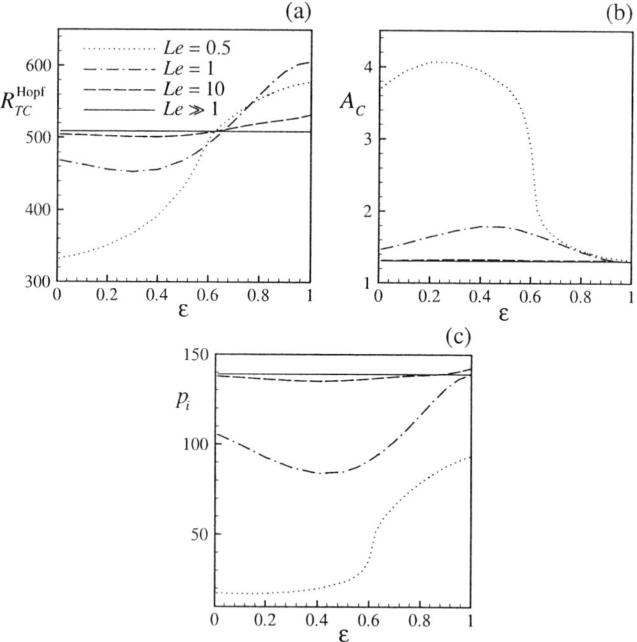

Figure 5.11 *Critical parameters at the onset of Hopf bifurcation for an infinite layer. Effects of Le and ε for $R_S = -100$, $\Phi = 0°$, $a_T = a_S = 1$, and $\xi = 0$*

large enough then the results become independent of this parameter. This trend could be explained by the fact that the solute concentration becomes uniform within a large part of the enclosure, and so the solutal buoyancy force is weakened. For this situation, the critical values tend towards those corresponding to pure thermal convection. The perturbation profiles at the threshold of Hopf bifurcation are depicted in Figure 5.12 and they consist of small vortices aligned near the horizontal enclosure boundaries. Since the boundaries of the vortices are not vertical, traveling waves could be initiated above the threshold of Hopf bifurcation.

To examine the oscillatory behavior of the convective flows above the threshold of Hopf bifurcation, some typical results are presented in Figure 5.13 for $R_S = -100$, $Le = 10$, $\varepsilon = 1$, $\xi = 0$, and $(a_T, a_S) = (1, 1)$, and linear stability analysis gives the value of the threshold to be $R_{TC}^{\text{Hopf}} = 532$ and the critical wavelength is $A_C = 1.3$. To simulate a flow with an infinite layer, an aspect ratio of $A = 10$ is chosen. It is observed numerically that the amplitude of the oscillations is very small near criticality and the flow pattern evolution is indistinguishable. For this reason, a thermal Rayleigh number of $R_T = 600$ is chosen (a little far from criticality). For this situation, the flow is periodically oscillating and the flow pattern remains unicellular, but the parallelism of the streamlines is broken, especially in the mid-height of the enclosure. The unsteadiness of the flow is characterized by a series of secondary circulations traveling from the center to the ends of the enclosure. Time evolution of the perturbation demonstrates that the vortices were observed to move along the enclosure walls like a closed belt, see Figure 5.13(c) for the ψ_p profile. The shape and the wavelength of the formed vortices are approximately the same as those predicted by the stability analysis when superposing the two conjugate solutions. Similar time evolution of the concentration field is observed, see Figure 5.13(c).

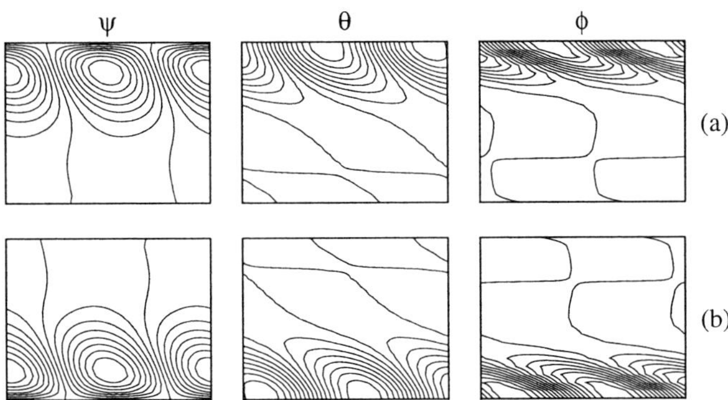

Figure 5.12 *Perturbation profiles (ψ, θ, and ϕ) at the threshold of Hopf bifurcation for an infinite layer with $R_S = -100$, $Le = 10$, $\Phi = 0°$, $\varepsilon = 1$, $\xi = 0$, and $a_T = a_S = 1$: $R_{TC}^{\text{Hopf}} = 532.12$, $A_C = 1.297$, and (a) $p_i = 142.46$ and (b) $p_i = -142.46$*

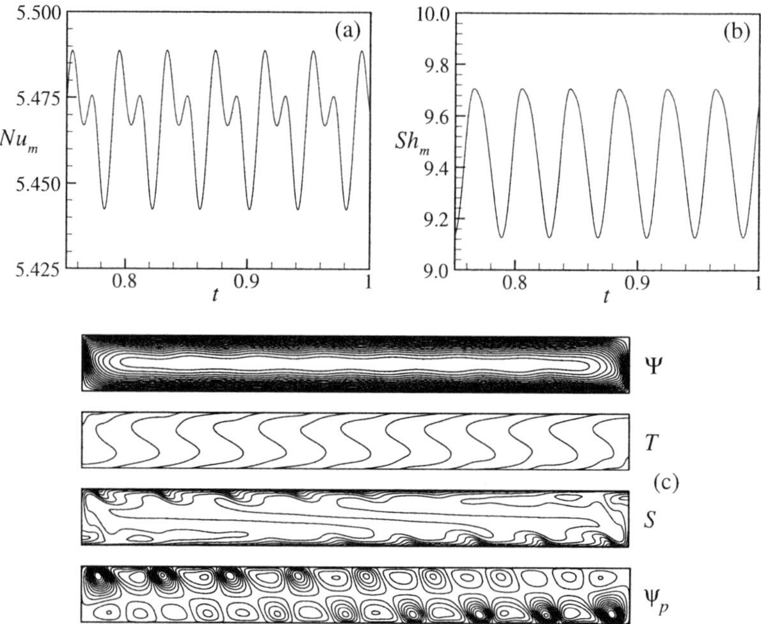

Figure 5.13 *(a,b) The time history of the heat and mass transfer rates. (c) Streamlines, temperature, solute, and perturbed stream function contours for $R_T = 600$, $R_S = -100$, $Le = 10$, $A = 10$, $\varepsilon = 1$, $\xi = 0$, $a_T = a_S = 1$, and $\Phi = 0°$ at $t = 0.785$*

For the same problem, the effect of the acceleration parameter, ξ, on the threshold of the Hopf bifurcation is studied. Some typical results are depicted in Table 5.5 for $R_S = -100$, $Le = 10$, $\varepsilon = 1$, and $(a_T, a_S) = (1, 1)$, and it is observed that, when ξ increases progressively, the threshold for transition is delayed and the perturbation wavelength becomes very large. The oscillation frequency (i.e., pulsation) is also observed to decrease. As a result, the oscillatory flows could be stabilized by increasing ξ. This effect could be obtained for the fluid having a low Prandtl number or relatively high Darcy number.

The bifurcation diagram is illustrated in Figure 5.14 for $R_T = 100$, $R_S = -100$, $Le = 10$, $A = 10$, $a_T = a_S = 1$, and $\Phi = 0°$. The results are presented in terms of the local heat and mass transfer rates (Nu and Sh) as a function of the thermal Rayleigh number, and five modes are delineated in the graph. The first one, region I, corresponds to the stable diffusive regime in which all perturbations decay; region II corresponds to subcritical flows, where the diffusive state is unstable to finite-amplitude perturbations, and it should be noted that the heat and mass transfer rates are finite at $R_T = R_{TC}^{\text{sub}}$; region III illustrates the overstable regime in which perturbations grow in an oscillatory manner; region IV represents the stationary convection regime; and region V delineates the oscillatory finite-amplitude convection which occurs right above the threshold of Hopf bifurcation.

Table 5.5 *Effect of the acceleration parameter, ξ, on the threshold of Hopf bifurcation in an infinite horizontal layer:* $Le = 10$, $R_S = -100$, $\varepsilon = 1$, $\Phi = 0°$, *and* $(a_T, a_S) = (1, 1)$

ξ	R_{TC}^{Hopf}	A_C	p_i
0	532.12	1.30	± 142.46
10^{-3}	521.19	1.38	± 128.10
2×10^{-3}	526.25	1.49	± 115.37
5×10^{-3}	583.69	1.91	± 91.48
10^{-2}	748.39	2.60	± 74.19
2×10^{-2}	1360.71	4.02	± 62.68

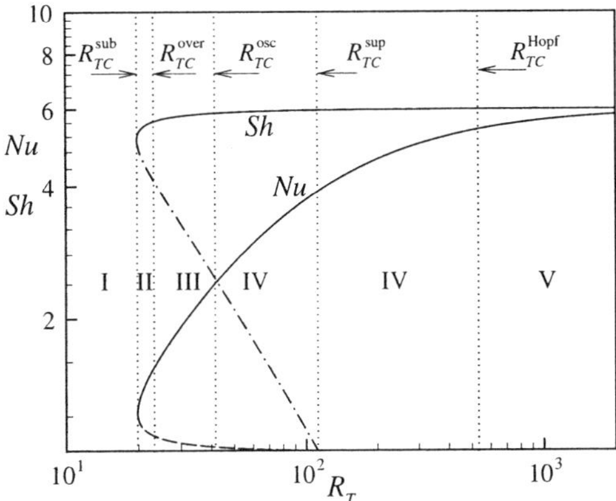

Figure 5.14 *Bifurcation diagram (Nu and Sh versus R_T) for $R_T = 100$, $R_S = -100$, $Le = 10$, $A = 10$, $a_T = a_S = 1$, and $\Phi = 0°$. The critical parameters are obtained for $\xi = 0$ and $\varepsilon = 1$ as $R_{TC}^{\text{sub}} = 19.8$, $R_{TC}^{\text{over}} = 23.2$, $R_{TC}^{\text{osc}} = 41.6$, $R_{TC}^{\text{sup}} = 112.0$, and $R_{TC}^{\text{Hopf}} = 532.1$*

The stability of the basic convective solution is also studied within a vertical porous layer. As for the case of a horizontal porous layer, the basic solution becomes unstable above a certain threshold, R_{TC}^{Hopf}. It was found that the normalized porosity, ε, and the acceleration parameter, ξ, have a strong effect on the onset of Hopf bifurcation. The results are illustrated in Table 5.6 for $Le = 10$, $N = -1$, $\Phi = 90°$, and $(a_T, a_S) = (1, 1)$. When the acceleration parameter is cancelled, the critical Rayleigh number is observed to decrease significantly with decreasing ε. This results are in agreement with those reported by Mamou *et al.* (1998b) when considering an aspect ratio $A = 4$. However, when increasing the acceleration parameter from 0 to 1, for $\varepsilon = 1$, then a reverse trend is observed. The threshold and the wavelength are seen to increase with ξ, i.e., the effect

Table 5.6 *Effect of normalized porosity, ε, and the acceleration parameter, ξ, on the threshold of Hopf bifurcation in an infinite vertical layer for $Le = 10$ and $N = -1$ when (a) $\xi = 0$ and (b) $\varepsilon = 1$*

(a)

ε	R_{TC}^{Hopf}	A_C	p_i
0.2	20.22	8.09	± 13.32
0.4	37.20	4.52	± 15.96
0.6	53.00	3.54	± 17.72
0.8	67.95	2.99	± 15.32
1	79.92	2.65	± 14.60

(b)

ξ	R_{TC}^{Hopf}	A_C	p_i
0	79.92	2.65	± 14.60
10^{-2}	88.88	2.97	± 13.40
5×10^{-2}	153.88	5.11	± 9.31
10^{-1}	274.26	10.38	± 5.60
2×10^{-1}	520.82	22.68	± 3.17

of the acceleration parameter is to delay the transition to oscillatory flows. Similar results have been observed by Amahmid *et al.* (2000) when reconsidering the case studied by Alavyoon and Masuda (1994). The oscillatory flows obtained by Alavyoon and Masuda (1994) are stabilized by Amahmid *et al.* (2000) by increasing the parameter ξ.

The perturbation profiles are illustrated in Figure 5.15 and are qualitatively similar to those observed in a horizontal layer, see Figure 5.12. The threshold of Hopf bifurcation is $R_{TC}^{\mathrm{Hopf}} = 80$. Typical finite-amplitude results are presented in Figure 5.16 right above the threshold, i.e., $R_T = 90$. The convective flow was found to be oscillatory, as can be seen from the time history of Nu_m and Sh_m. The flow remains unicellular but the streamlines are slightly distorted, indicating the presence of small vortices travelling along the vertical wall of the layer, see Figure 5.16(d).

Figure 5.17 illustrates the influence of the Rayleigh number, R_T, on the maximum and the minimum values of the stream function ($\Psi_{\max}$ and $\Psi_{\min}$, respectively) for the case $A = 1$, $\Phi = 90°$, $Le = 10$, and $\varepsilon = 1$. Starting the computations from the purely

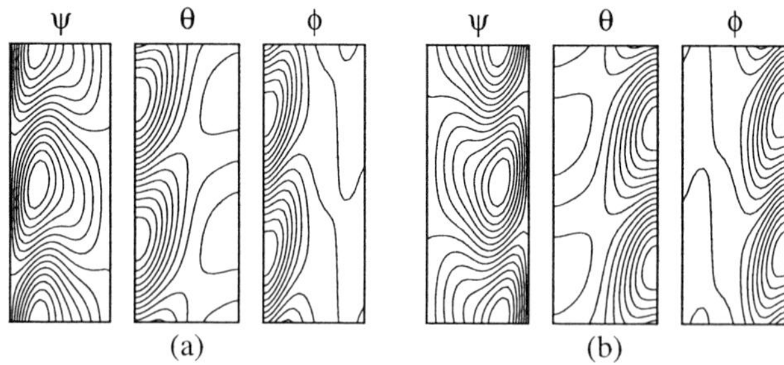

Figure 5.15 *Perturbation profiles (ψ, θ, and ϕ) at the threshold of Hopf bifurcation for an infinite layer with $Le = 10$, $\Phi = 90°$, $\varepsilon = 1$ and $a_T = a_S = 1$: $R_{TC}^{\mathrm{Hopf}} = 79.92$, $A_C = 2.65$, and (a) $p_i = 14.60$ and (b) $p_i = -14.60$*

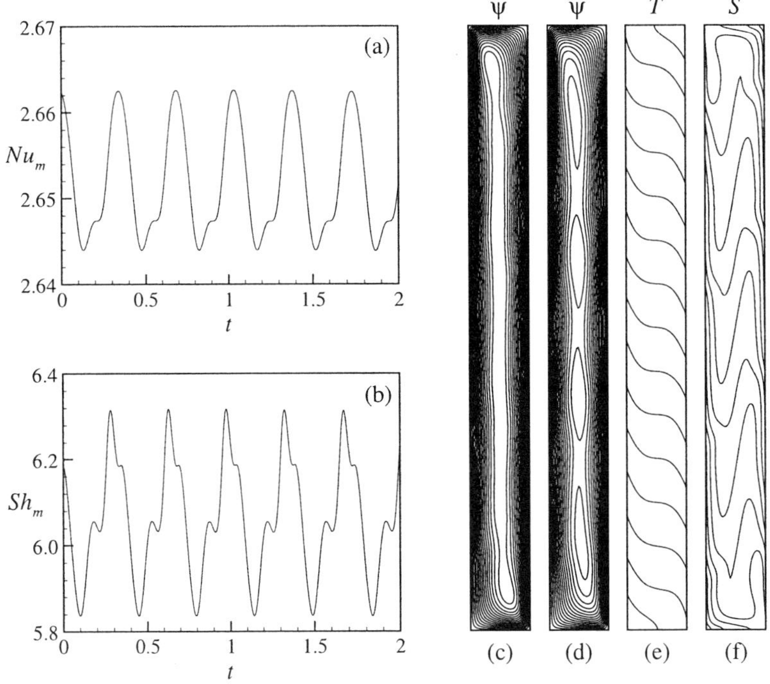

Figure 5.16 *(a,b) The time history of the heat and mass transfer rates, (c,d) streamlines at t = 1.664 and 1.728, and (e,f) temperature and solute contours at t = 1.664, for $R_T = 90$, $N = -1$, $Le = 10$, $A = 10$, $\varepsilon = 1$, and $\Phi = 90°$*

diffusive solution, it was found that this solution can be maintained when R_T is below the critical value $R_{TC}^{\mathrm{sup}} = 20.45$. However, for $R_T > R_{TC}^{\mathrm{sup}}$, this rest state is unstable, although it continues to be the solution of the governing equations. A stable convective regime bifurcates from the rest state at $R_T = R_{TC}^{\mathrm{sup}}$. The resulting supercritical convective regime is characterized, as expected, by symmetrical solutions, as exemplified by Figures 5.17(c) and (d) for $R_T = 25$ and 100, respectively. However, upon using a finite-amplitude flow as an initial condition, another branch of solutions was found to exist in the range $13 \leqslant R_T \leqslant 35$. This second set of solutions corresponds to non-symmetrical flow patterns, as illustrated by Figure 5.17(b) for $R_T = 25$. It is noted that this non-symmetrical branch is maintained down to $R_T = 13$, i.e., below $R_{TC}^{\mathrm{sup}} = 20.45$, the critical Rayleigh number for the onset of supercritical convection. Thus, for $13 \leqslant R_T \leqslant 20.45$, two different types of solution are possible, a purely diffusive regime and a subcritical, finite-amplitude convective regime. The coexistence of the two finite-amplitude solutions are also observed to occur in the range $20.45 \leqslant R_T \leqslant 25.5$. Similar results were reported by Charrier-Mojtabi *et al.* (1998).

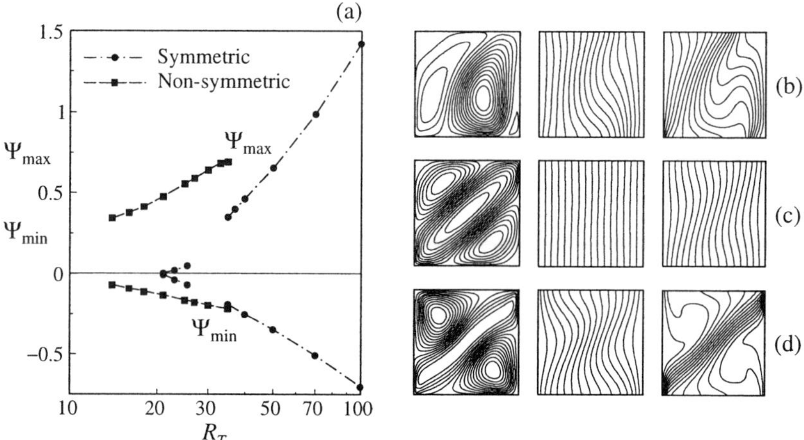

Figure 5.17 R_T *effect on (a) flow intensity versus* R_T, *for* $N = -1$, $Le = 10$, $\varepsilon = 1$, $A = 1$, $\Phi = 90°$, *and* $a_T = a_S = 0$, *and on the streamlines, isotherms, and isoconcentrations, for (b)* $R_T = 25$, $Nu_m = 1.015$, *and* $Sh_m = 1.480$, *(c)* $R_T = 25$, $Nu_m = 1.001$, *and* $Sh_m = 1.047$, *and (d)* $R_T = 100$, $Nu_m = 1.164$, *and* $Sh_m = 4.595$

For Dirichlet boundary conditions ($a_T = a_S = 0$), it has been reported in the past that, within the overstable regime, the double-diffusive convection may be oscillatory. Especially in horizontal, large aspect ratio enclosures, the flow is characterized by traveling waves along the porous layer, see, for example, Predtechensky *et al.* (1994) and Mamou and Vasseur (1999). Similar phenomena are observed in a vertical porous layer. It is found that, for a relatively large aspect ratio with $R_T = 90$, $Le = 2$, $\varepsilon = 0.1$, and $\xi = 0$, the oscillatory flow consists of multiple cells traveling vertically along the porous layer. For $A = 10$ (the results are not presented here), the numerical results show that the oscillatory flow is aperiodic. However, for $A = 15$, the oscillations becomes periodic and the flow cells are seen to travel from bottom to top, as indicated in Figure 5.18, at a nearly constant velocity. The cells are generated in the bottom region of the enclosure and move upward, without loss of intensity, until they reach the top of the layer, where they collapse. This phenomenon is not observed for a vertical slot with Neumann boundary conditions. The preferred mode for this case is the unicellular flow. The convenient way to study this type of phenomenon is to consider a vertical slot of a finite aspect ratio, with the use of periodic boundary conditions. However, since the flow is nonlinear, it is very difficult to predict the wavelength of the convective rolls.

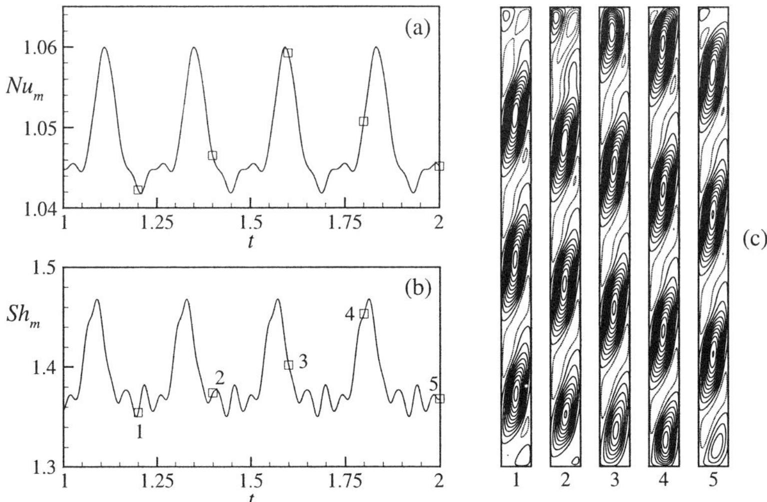

Figure 5.18 *(a,b) Time history of the heat and mass transfer rates and (c) flow structure at different instants of time for* $R_T = 90$, $N = -1$, $Le = 2$, $A = 15$, $\varepsilon = 0.1$, $a_T = a_S = 0$, and $\Phi = 90°$

5.5 CONCLUSIONS

The present work is devoted to numerical and analytical stability analyses of double-diffusion convection within tilted porous enclosures of arbitrary aspect ratio, subject to opposing thermal and solutal gradients. Interest in this type of problems is motivated by its importance in many practical situations, such as in chemical engineering, metallurgy and underground disposal of pollutants, where convection in multi-component fluids is involved.

The governing equations are solved numerically using the finite element method and analytically using the parallel flow approximation in an infinite layer, subject to constant fluxes of heat and solute. Reliable numerical techniques are developed, on the basis of Galerkin and finite element methods, to perform a linear stability analysis of the pure diffusive state, when it exists, and of the fully-developed flows within the enclosure. The effect of the governing parameters on the thresholds of stationary or oscillatory convective flows are studied for many different situations and boundary conditions. The stability of the fully-developed solution in slender vertical or horizontal enclosures is studied and the threshold which characterizes the transition from steady to oscillatory convection is determined. The porosity of the porous medium and the acceleration parameter are found to have a strong effect on the thresholds of overstability and Hopf bifurcation. Multiple convective states were found to exist for the same governing parameters.

Up to the present day, most of the studies of double-diffusive convection are theoretical and based on numerous approximations and assumptions. In addition, most of the studies are limited to two-dimensional situations, in order to understand the fundamentals of double-diffusive convection on one hand, and to avoid excessive computational time to obtain a solution on the other hand. It is well known that three-dimensional effects could have a significant impact on the flow behavior and on the heat and mass transfer rates. In addition, turbulence in double-diffusive convection, which describes the real physics of the flows, remains, as yet, unexploited. Also, it is observed that there is a big lack of extensive experimental studies to confirm the theoretical findings. Therefore, it is recommended that much effort is required to obtain experimental confirmation of the available theoretical results, in order to have a better understanding of the combined heat and mass transfer within porous media.

Acknowledgements

The author is grateful to Professors D. B. Ingham and I. Pop for reviewing this chapter, and for their fruitful suggestions which have considerably improved the general quality of this chapter. The author would like to thank Dr A. Laoudi, Dr M. Khalid, and Professor P. Vasseur for their comments and constructive criticisms. The author would also like to thank Dr S. D. Harris for his assistance given during the preparation of this chapter.

REFERENCES

Alavyoon, F. (1993). On natural convection in vertical porous enclosures due to prescribed fluxes of heat and mass at the vertical boundaries. *Int. J. Heat Mass Transfer* **36**, 2479–2498.

Alavyoon, F. and Masuda, Y. (1994). On natural convection in vertical porous enclosures due to opposing fluxes of heat and mass prescribed at the vertical walls. *Int. J. Heat Mass Transfer* **37**, 195–206.

Amahmid, A., Hasnaoui, M., Mamou, M., and Vasseur, P. (1999a). Double-diffusive parallel flow induced in a horizontal Brinkman porous layer subjected to constant heat and mass fluxes: Analytical and numerical studies. *Heat Mass Transfer* **35**, 409–421.

Amahmid, A., Hasnaoui, M., Mamou, M., and Vasseur, P. (2000). On the transition between aiding and opposing double-diffusive flows in a vertical porous cavity. *J. Porous Media* **3**, 123–137.

Amahmid, A., Hasnaoui, M., and Vasseur, P. (1999b). Etude analytique et numérique de la convection naturelle dans une couche poreuse de Brinkman doublement diffusive. *Int. J. Heat Mass Transfer* **42**, 2991–3005.

Bejan, A. (1984). *Convection Heat Transfer*. Wiley, New York.

Brand, H. and Steinberg, V. (1983). Nonlinear effects in the convective instability of a binary mixture in a porous medium near threshold. *Phys. Lett.* **93A**, 333–336.

Charrier-Mojtabi, M. C., Karimi-Fard, M., Mejdi, A., and Mojtabi, A. (1998). Onset of a double-diffusive convective regime in a rectangular porous cavity. *J. Porous Media* **1**, 107–121.

Chen, F. and Chen, C. F. (1993). Double-diffusive fingering convection in a porous medium. *Int. J. Heat Mass Transfer* **36**, 793–807.

Guo, J. and Kaloni, P. N. (1995). Double-diffusive convection in a porous medium, nonlinear stability, and the Brinkman effect. *Studies Appl. Math.* **94**, 341–358.

Ingham, D. B. and Pop, I. (eds) (1998). *Transport Phenomena in Porous Media.* Pergamon, Oxford.

Kalla, L., Mamou, M., Vasseur, P., and Robillard, L. (2001). Multiple solutions for double-diffusive convection in a shallow porous cavity with vertical fluxes of heat and mass. *Int. J. Heat Mass Transfer.* In press.

Karimi-Fard, M., Charrier-Mojtabi, M. C., and Mojtabi, A. (1999). Onset of stationary and oscillatory convection in a tilted porous cavity saturated with a binary fluid: Linear stability analysis. *Phys. Fluids* **11**, 1346–1358.

Kimura, S., Vynnycky, M., and Alavyoon, F. (1995). Unicellular natural circulation in a shallow horizontal porous layer heated from below by a constant flux. *J. Fluid Mech.* **294**, 231–257.

Mahidjiba, A., Mamou, M., and Vasseur, P. (2000). Onset of double-diffusive convection in a rectangular porous cavity subject to mixed boundary conditions. *Int. J. Heat Mass Transfer* **43**, 1505–1522.

Mamou, M. (1998). *Convection Thermosolutale dans des Milieux Poreux et Fluides Confinés.* Ph.D. thesis. Ecole Polytechnique of Montreal, University of Montreal, Quebec, Canada.

Mamou, M. and Vasseur, P. (1999). Thermosolutal bifurcation phenomena in porous enclosures subject to vertical temperature and concentration gradients. *J. Fluid Mech.* **395**, 61–87.

Mamou, M., Hasnaoui, M., Amahmid, A., and Vasseur, P. (1998a). Stability analysis of double diffusive convection in a vertical Brinkman porous enclosure. *Int. Comm. Heat Mass Transfer* **25**, 491–500.

Mamou, M., Vasseur, P., and Bilgen, E. (1995a). Multiple solutions for double-diffusive convection in a vertical porous enclosure. *Int. J. Heat Mass Transfer* **38**, 1787–1798.

Mamou, M., Vasseur, P., and Bilgen, E. (1995b). Thermosolutal convection instability in a vertical porous layer. In *Proceedings of 2nd International Thermal Energy Congress*, Agadir Maroco, pp. 463–467.

Mamou, M., Vasseur, P., and Bilgen, E. (1998b). Double diffusive convection instability problem in a vertical porous enclosure. *J. Fluid Mech.* **368**, 263–289.

Mamou, M., Vasseur, P., and Bilgen, E. (1998c). A Galerkin finite-element study of the onset of double-diffusive convection in an inclined porous enclosure. *Int. J. Heat Mass Transfer* **41**, 1513–1529.

Mamou, M., Vasseur, P., Bilgen, E., and Gobin, D. (1995c). Double-diffusive convection in an inclined slot filled with porous medium. *Eur. J. Mech. B – Fluids* **14**, 629–652.

Mamou, M., Vasseur, P., and Hasnaoui, M. (2001). On numerical stability analysis of double-diffusive convection in confined enclosures. *J. Fluid Mech.* **433**, 209–250.

Marcoux, M., Karimi-Fard, M., and Charrier-Mojtabi, M. C. (1999). Naissance de la convection thermosolutale dans une cellule rectangulaire poreuse soumise á des flux de chaleur et de masse. *Int. J. Therm. Sci.* **38**, 258–266.

Mojtabi, A. and Charrier-Mojtabi, M. C. (2000). Double-diffusive convection in porous media. In *Handbook of Porous Media* (ed. K. Vafai), pp. 559–603. Marcel Dekker, New York.

Murray, B. T. and Chen, C. F. (1989). Double-diffusive convection in a porous medium. *J. Fluid Mech.* **201**, 147–166.

Nield, D. A. (1968). Onset of thermohaline convection in porous medium. *Water Resources Res.* **4**, 553–560.

Nield, D. A. and Bejan, A. (1999). *Convection in Porous Media* (2nd edn). Springer–Verlag, New York.

Nithiarasu, P., Seetharamu, K. N., and Sundararajan, T. (1996). Double-diffusive natural convection in an enclosure filled with fluid-saturated porous medium: A generalized non-Darcy approach. *Numer. Heat Transfer, Part A* **30**, 413–426.

Poulikakos, D. (1986). Double-diffusive convection in a horizontal sparsely packed porous layer. *Int. Comm. Heat Mass Transfer* **13**, 587–598.

Predtechensky, A. A., McCormick, W. D., Swiff, J. B., Rossberg, A. G., and Swinney, H. L. (1994). Traveling wave instability in sustained double-diffusive convection. *Phys. Fluids* **6**, 3923–3935.

Rudraiah, N., Shrimani, P. K., and Friedrich, R. (1982). Finite amplitude convection in a two-component fluid saturated porous layer. *Int. J. Heat Mass Transfer* **25**, 715–722.

Sezai, I. and Mohamad, A. A. (1999). Three-dimensional double-diffusive convection in a porous cubic enclosure due to opposing gradients of temperature and concentration. *J. Fluid Mech.* **400**, 333–353.

Taunton, J. W., Lightfoot, E. N., and Green, T. (1972). Thermohaline instability and salt fingers in a porous medium. *Phys. Fluids* **15**, 748–753.

Trevisan, O. V. and Bejan, A. (1985). Natural convection with combined heat and mass transfer buoyancy effects in a porous medium. *Int. J. Heat Mass Transfer* **28**, 1597–1611.

Trevisan, O. V. and Bejan, A. (1986). Mass and heat transfer by natural convection in a vertical slot filled with porous medium. *Int. J. Heat Mass Transfer* **29**, 403–415.

Trevisan, O. V. and Bejan, A. (1990). Combined heat and mass transfer by natural convection in a porous medium. *Adv. Heat Transfer* **20**, 315–352.

6 CONVECTION IN ORDERED AND DISORDERED POROUS LAYERS

L. E. HOWLE

Department of Mechanical Engineering and Materials Science, Center for Nonlinear and Complex Systems, Duke University, Durham, NC 27708-0300, USA

email: laurens.howle@duke.edu

Abstract

In this chapter we investigate buoyancy-driven convection in a horizontal fluid-saturated porous medium. Our particular focus is on the discrepancy that can exist between theoretical results and the results from laboratory experiments. When one conducts laboratory experiments then there exist a number of competing criteria that make the design of the *ideal* laboratory experiment difficult. The compromises one must make in building the media result in media that are spatially inhomogeneous except in cases of specially constructed systems. These heterogeneities create the rounding of the heat transfer curve in the neighborhood of convection onset, convection wave-pattern pinning, and in some cases, media in which no-flow (conductive) state cannot exist for any nonzero destabilizing temperature gradient.

Keywords: ordered media, disordered media, convection, instability, pattern formation, defect pinning

6.1 INTRODUCTION

One need only search the scientific, mathematical, and engineering literature to gauge the importance of the study of fluid flow in porous media. Practical applications such as building insulation, packed-bed catalytic reactors, electronic cooling, geothermal systems, and groundwater contaminant transport motivate many of these investigations. Many of these systems are complicated and involve two-phase flow, phase change, property variations, localized effects such as inclusions, etc.

One of the more easily understood and widely studied areas of scientific research in porous media flow is the canonical Horton–Rogers–Lapwood convection (HRLC) system. This problem derives its name from the independent works of Horton and Rogers (1945) and Lapwood (1948). We direct the reader to the recent text by Nield and Bejan (1999) for a detailed review of the HRLC problem. In the idealized HRLC problem, a fluid-saturated porous medium is bounded above and below by two impermeable solid planes. The porous layer is oriented so that the direction of gravity is perpendicular to the upper and lower solid planes. The isothermal bounding planes differentially heat the fluid-saturated medium with the lower plane being the warmer of the two. A slight modification of the problem places a spatially uniform heat flux at the lower plate and maintains the upper plate at a constant temperature. In this system, one uses the vertical temperature difference, or the vertical heat flux, as a control parameter. Below a critical value of the control parameter, there is no motion of the fluid that saturates the medium and heat transport is due solely to the conduction of heat through both the fluid and medium (and between the two). After one exceeds the critical value of the control parameter, there is a spontaneous initiation of fluid motion though the interstitial spaces of the stationary solid medium. The motion of the fluid coincides with the initiation of certain flow patterns that may or may not be stationary in time (after some initial transients). HRLC is one of the simplest nonlinear pattern forming systems and therefore it is studied by researchers not only interested in the applications listed previously but by researchers interested in pattern formation and nonlinear and complex systems.

In this chapter, we focus on HRLC in media which are completely saturated with a single fluid. In particular, we investigate the pattern formation near the onset of convection and the associated heat transfer. We also investigate the influence of the variability of the medium on the formation of convection patterns and on heat transfer. Relying on some of our previous experimental results and the results of others (both experimental and theoretical) we are led to the conclusion that the construction of the medium can have a considerable influence on the convective wave patterns and upon the heat transfer.

6.2 HORTON–ROGERS–LAPWOOD EXPERIMENTS

Of the many experimental investigations on buoyancy-driven convection in porous media with heating from below, the majority report heat transfer data. A few report pattern formation at the onset of convection and pattern evolution at higher control parameter values. In Figure 6.1, we show heat transfer data from nine experimental investigations. These heat transfer data are reported in terms of the porous Rayleigh–Darcy number, Ra_m, defined by

$$Ra_m = \frac{g\beta K H \Delta T}{\nu \alpha_m}. \tag{6.1}$$

Here, g is the magnitude of the gravitational acceleration, β is the isobaric volumetric expansion coefficient of the fluid, K is the medium permeability, H is the spacing between the horizontal bounding plates, ΔT is the temperature difference, ν is the fluid viscosity, and α_m is the thermal diffusivity of the fluid-saturated porous medium. The medium thermal diffusivity can be expressed in terms of the fluid thermal diffusivity as

$$\alpha_m = \frac{k_m}{k_f}\alpha_f, \tag{6.2}$$

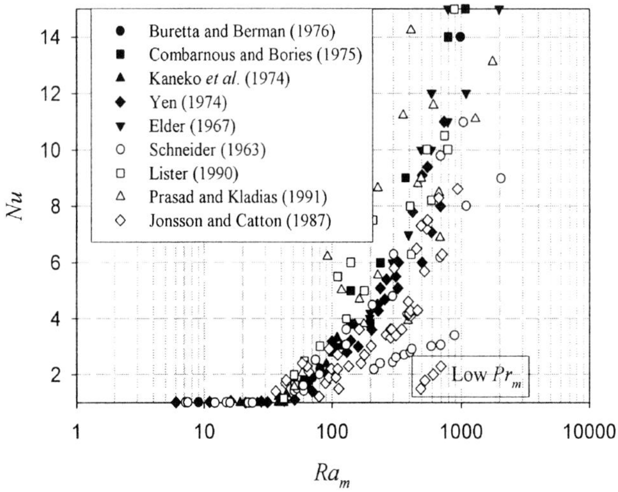

Figure 6.1 *Experimental heat transport data. Reprinted from Howle (1993), with the permission of the author*

where α_f is the thermal diffusivity of the fluid, k_f is the fluid thermal conductivity, and k_m is the effective thermal conductivity of the fluid-saturated medium. For media with fluid and solid whose conductivities are similar, we can approximate the medium thermal conductivity with the expression

$$k_m = \phi k_f + (1 - \phi)\, k_s. \tag{6.3}$$

In equation (6.3), ϕ is the porosity (void fraction) and k_s is the thermal conductivity of the solid. The Nusselt number, Nu, used in Figure 6.1, is the total heat transfer normalized by the conductive heat transport at the same temperature difference. Accordingly, Nu departs from 1 as convection begins and the convective motion begins to augment the total heat transport. From Figure 6.1 we note the scatter in the data, particularly at larger values of Ra_m. This scatter can be reduced somewhat by a different choice of the dimensionless numbers, see, for example, Wang and Bejan (1987). However, most of the experimental results available in the literature report the heat transfer data in terms of Nu versus Ra_m so we do the same here.

Elder's (1967) experiments included glass spheres of diameters 3, 5, 8 and 18 mm and styropor beads of 6 mm diameter that he glued into a porous matrix. Additionally, Elder performed experiments in a Hele–Shaw cell. He focused on cellular convection in systems with heating from below, on the role of end effects and on the role of mass discharge. Kaneko *et al.* (1974) constructed porous layers with approximately spherical double-screened silica sand as the solid medium. They used heptane and ethanol as saturating fluids. Additionally, Kaneko *et al.* used 55 thermocouples embedded within the medium to gather convection pattern information. Although their experiments studied various angles of inclination to the horizontal, only the zero angle results are shown in Figure 6.1. Yen (1974) investigated the effect of density inversion on convection in a water–glass porous medium. For convection in the non-density-inverted system, Yen found that convection begins at the predicted critical Rayleigh–Darcy number of $4\pi^2$. Combarnous and Bories (1975) report the data from experiments using a Hele–Shaw qualitative analogy to porous convection, convection in a layer with a free upper surface and in layers bounded by isothermal planes. The experiments of Buretta and Berman (1976) utilized porous layers of glass beads with diameters of 3, 6 and 14.3 mm saturated with demineralized water. They varied the bed thickness and permeability while heating from below with Rayleigh–Darcy numbers from 10 to 10 000. Buretta and Berman estimated the critical Rayleigh–Darcy number to be 38, which is in excellent agreement with theoretical predictions. Jonsson and Catton (1987) studied the effect of Prandtl number on heat transfer in HRLC experiments using glass, stainless steel and lead particles. They used silicon oil, water and mercury as working fluids. In Figure 6.1, we include the results of Jonsson and Catton's experiments with low Prandtl number lead–mercury experiments. These results are significant because they show a substantial departure of the critical Rayleigh–Darcy number from the higher Prandtl number experiments shown in the figure. We should note that lead is somewhat soluble in mercury and therefore Jonsson and Catton's experiment might have exhibited binary porous medium behavior. Lister (1990) performed HRLC experiments in a 3 m diameter 30 cm high hexagonal container with two different types of media. The first medium was rubberized curled coconut fibers, while the second medium was clear polymethylmethacrylate beads. Convection pattern visualization revealed the preferred wave pattern began as hexagons near the onset of convection. At higher Rayleigh–Darcy numbers, Lister observed complicated, irregular and three-dimensional patterns. Lister reported that the heat transfer curve of the bead medium jumped upward immediately after the onset of convection and then settled to a slope of about 0.52. At higher Rayleigh numbers the slope increased to a value greater than 1. He attributed this behavior to lateral thermal dispersion. Prasad and Kladias (1991) critically compared data gathered from experiments in water–glass, oil–glass, water–steel, heptane–glass, and glycol–glass systems with the various extensions of the Darcy flow model such as inertia effects, boundary viscous effects, variable porosity–wall channeling and dispersion effects. Among their findings, Prasad and Kladias found that wall channeling and variable porosity near the wall region had a significant influence on heat transfer and convective flows. Also included in Figure 6.1 are the results of Schneider (1963).

6.3   ONSET OF CONVECTION IN A HOMOGENEOUS ISOTROPIC MEDIUM

In the section, we derive the critical Rayleigh–Darcy number for which a horizontally infinite, isotropic, homogeneous porous layer loses stability to infinitesimal disturbances. This serves as a basis against which we can compare similar analysis from more complicated media. We begin by writing the transport equations in dimensional form for flow of an incompressible fluid through a porous medium. We write the respective equations for the mass, momentum, and energy conservation as follows:

$$\nabla \cdot \boldsymbol{v}' = 0, \tag{6.4}$$

$$\rho_f c_a \partial_{t'} \boldsymbol{v}' = \rho_f \boldsymbol{g} - \nabla p' - \frac{\mu}{K} \boldsymbol{v}', \tag{6.5}$$

$$\sigma \partial_{t'} T' + \boldsymbol{v}' \cdot \nabla T' = \alpha_m \nabla^2 T', \tag{6.6}$$

where the prime denotes the dimensional variable. The symbols in equations (6.4) – (6.6) are fluid density, ρ_f, acceleration coefficient, c_a, dynamic fluid viscosity, μ, and heat capacity ratio,

$$\sigma = \frac{(\rho c_p)_m}{(\rho c_p)_f}, \tag{6.7}$$

where the subscripts m and f refer to medium and fluid, respectively. In order to non-dimensionalize these equations, we choose as scales

$$\boldsymbol{v} = \frac{H}{\alpha_m} \boldsymbol{v}', \quad T = \frac{T'}{\Delta T}, \quad p = \frac{H^2}{\alpha_m \nu \rho_0} p', \quad x = \frac{x'}{H}, \quad t = \frac{\alpha_m}{H^2} t' \tag{6.8}$$

and rewrite the scaled system of equations as follows:

$$\nabla \cdot \boldsymbol{v} = 0, \tag{6.9}$$

$$c_a Pr_m^{-1} \partial_t \boldsymbol{v} = Ra T \boldsymbol{e}_z - \nabla p - Da^{-1} \boldsymbol{v}, \tag{6.10}$$

$$\sigma \partial_t T + \boldsymbol{v} \cdot \nabla T = \nabla^2 T. \tag{6.11}$$

In writing equations (6.9) – (6.11), we have made use of the Boussinesq approximation, which allows us to neglect fluid property dependence on temperature, except in the body force term—the first term on the right-hand side of equation (6.10)—where we assume the temperature-dependent fluid density has the form $\rho_f = \rho_0 \left[1 - \beta \left(T - T_0 \right) \right]$ with the subscript 0 representing the reference state. In the dimensionless equations, Prandtl, bulk–fluid Rayleigh and Darcy numbers are respectively defined as follows:

$$Pr = \frac{\nu}{\alpha_m}, \quad Ra = \frac{g \beta \Delta T H^3}{\nu \alpha}, \quad Da = \frac{K}{H^2}. \tag{6.12}$$

For boundary conditions, we take no-penetration, slip hydrodynamic conditions on the horizontal surfaces and fixed temperature on these same surfaces. By taking the double curl of equation (6.11), we eliminate the pressure gradient and use the mass conservation equation to remove the resulting velocity divergence terms. The perturbation equations for the conduction state are given by

$$T = (1 - z) + \Theta \tag{6.13}$$

Now assume that the density dependence on the temperature is well approximated by a linear function and rewrite this equation as follows:

$$\rho_0 \left[1 - \beta \left(T - T_0\right)\right] \boldsymbol{g} = \nabla p. \tag{6.35}$$

By taking the curl of equation (6.35) we obtain

$$\nabla T \times \boldsymbol{g} = 0. \tag{6.36}$$

This result clearly indicates that there can be no horizontal temperature gradient in order for the conduction solution to exist (for a nonzero driving temperature difference). This result holds for either homogeneous or heterogeneous media.

Chen and Hsu (1991) performed a stability analysis on the convection of a thin fluid layer overlaying and saturating a porous layer. The permeability and thermal conductivity were anisotropic and heterogeneous. The heterogeneity was restricted to the vertical direction, thus ensuring the existence of the no-motion state. They found that vertically increasing permeability decreased stability, while vertically decreasing permeability increased stability. Braester and Vadasz (1993) showed that non-isothermal heterogeneous media, with conductivity functions that are not of the form $k_m\left(x, y, z\right) = f\left(z\right) h\left(y, z\right)$—where z is the vertical direction—must always have convection. Neel (1992) studied the situation in which there is a small non-uniformity in the boundary data and found a smooth transition towards convection. Braester and Vadasz (1993) examined the implications of the motionless state on a medium with spatially varying thermal conductivity. They considered weak heterogeneity of the medium, resulting in spatial variability of permeability and thermal conductivity. They found that the existence of horizontal thermal gradients was generally sufficient for the existence of natural convection. Braester and Vadasz also found that variability in the effective thermal conductivity had a more pronounced impact on the flow pattern than variability in permeability. Their analysis on heterogeneous systems with perfect boundaries showed a smooth transition from conduction to convection at the critical Rayleigh–Darcy number. In a later section of this chapter, we discuss one experiment in a disordered porous medium that shows this type of smooth transition from conduction to convection.

6.6   CONSTRUCTION OF LABORATORY EXPERIMENTS

In the construction of a porous medium for laboratory heat transfer or convection wave pattern measurements, there are a number of competing factors that affect the uniformity of the medium. These factors, as elucidated by Shattuck *et al.* (1997), include the thermal relaxation time of the medium, the value of the Darcy number, and the extent to which the Boussinesq approximation is valid. The first restriction is that the medium particle size must be smaller than the layer height. This restricts the particle size, d, to reside above a line of slope 1 on a plot of d as a function of H as shown in Figure 6.2. Ideally, we would like short thermal relaxation times so that the experiments avoid long transient periods, a

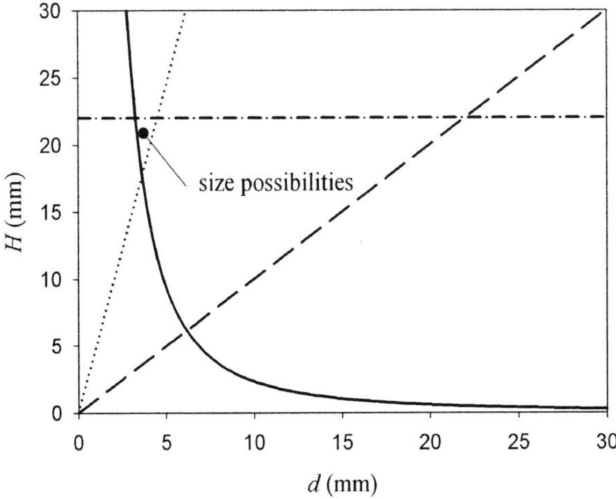

Figure 6.2 *Size possibilities for the design of a porous medium experiment. Here, d is the representative medium (bead or bar) size and H is the medium layer height. The combination of maximum thermal relaxation time, maximum Darcy number and Boussinesq limits the possible layer height and particle size to the triangle shown. Note that these results produce a ratio $d/H \approx 0.25$, not a spatially refined medium. Reprinted from Shattuck et al. (1997), with the permission of the author*

small Darcy number and an approximately linear dependence of the density on temperature with negligible dependence of other fluid properties on temperature. First, consider the vertical thermal relaxation time $\tau \equiv H^2/\alpha_m$. For a given maximum acceptable relaxation time and medium diffusivity, the layer height is given by

$$H = (\alpha_m \tau)^{1/2} . \tag{6.37}$$

If, for example, we impose the a maximum allowable thermal relaxation time of 16 min. and take the medium thermal diffusivity to be that of water, we get the dash–dot line in Figure 6.2. Our experiment must have a layer height below this line. Next, consider the desire to maintain the Boussinesq approximation. Suppose we wish to keep the temperature difference at the onset of convection before $\Delta T_c = 10\,^{\circ}\mathrm{C}$, where the subscript denotes the critical value. The porous Rayleigh–Darcy number relates the critical temperature difference to the layer height but not the particle size. We can, however, relate the critical Rayleigh number to the particle size through the use of a correlation between the particle size and permeability, such as the Ergun (1952) relationship

$$K = \frac{d^2 \phi^3}{150\,(1 - \phi)^2} \tag{6.38}$$

for beads of diameter d in a medium of porosity ϕ. If we use equation (6.38), together with equation (6.1), and take the critical value of porous Rayleigh number to be $Ra_c = 4\pi^2 \approx 40$, then we arrive at the expression (at critical conditions)

$$d^2 H = \frac{6 \times 10^3 \alpha_m \nu \left(1 - \phi\right)^2}{g\beta\Delta T_c \phi^3}. \tag{6.39}$$

Now, use the selected critical temperature difference, take properties of water at $20\,^\circ$C and set $\phi = 0.4$, which is a reasonable value for media constructed from packed beads and one gets the solid curve in Figure 6.2. Our porous medium must fall to the right of this line. For a spatially homogeneous medium, we need $H/d \gg 1$, which also implies $Da \ll 1$, and thus yields the constraint

$$H = d \left[\frac{150 Da \left(1 - \phi\right)^2}{\phi^3}\right]^{-1/2}. \tag{6.40}$$

If one chooses the rather large upper limit of $Da = 5 \times 10^{-5}$, we get the right-limiting dotted line in Figure 6.2. Given the constraints of this example, Figure 6.2 shows that there is a limited region of possible sizes for which we can construct a medium. It should be noted that this allows at most $4d/H$. This is not a homogeneous medium!

By relaxing these constraints even slightly, we can quickly gain unacceptably large thermal relaxation times, or large critical temperature differences. Shattuck et $al.$ give the following example. Suppose we are interested in an experiment with water at $40\,^\circ$C, 3.175 mm plastic beads and a layer height of $H = 1$ cm. In this case, $\Delta T_c = 10\,^\circ$C. If, because of concerns about heterogeneity, we decrease the bead size to 1 mm, this changes the critical temperature difference to $\Delta T_c = 100\,^\circ$C. This is clearly non-Boussinesq. We can reduce this critical temperature difference by increasing the layer height but this will lead to unacceptably large thermal diffusion times. Using Shattuck et $al.$'s example, by increasing the layer height by a factor of 3, we increase the horizontal relaxation time from 26 hrs to 3.6 mths. This trend is true even for small aspect ratio experiments.

6.7 EXPERIMENTAL MEASUREMENTS

In this section we describe some of our previous experiments using specially constructed ordered media and disordered media originally reported in Howle (1993) and Howle et $al.$ (1993, 1997). These media are made so that a modified shadowgraph can collect wave pattern information while we simultaneously gather precision heat transport data. Two different experiments, with the same apparatus, demonstrate that

(i) ordered media produce parallel rolls,

(ii) side-wall forcing may not determine pattern orientation,

(iii) ordered media have sharp bifurcations from conduction to convection,

(iv) disordered media have irregular polygonal wave patterns, and

(v) disordered media generally have a rounded bifurcation from conduction to convection.

Many experiments report wave pattern evolution for convection in bulk fluid (Rayleigh–Bénard convection). There are relatively few experiments which have pattern information for HRLC because of the difficulty associated with signal transmission through the medium. Methods that have been successfully used for pattern visualization include Hele–Shaw analogy to porous media particle tracers on a thin overlaying clear fluid layer, thermal or velocity probes embedded in the medium, Christiansen effect, pH variation and electrolysis, magnetic resonance imaging, and modified shadowgraphy.

Elder (1967) and Hartline and Lister (1977) used a two-dimensional Hele–Shaw cell to approximate convection in a porous medium. With this technique, fluid saturates the small gap between two transparent vertical surfaces. Because there is a significant amount of viscous drag on the fluid, caused by the close spacing between the vertical surfaces, this is a reasonable analogy for flow in a porous medium where viscous drag, due to flow though the pore network, is also significant. Two-dimensional flow pattern visualization with the Hele–Shaw analogy is gathered with interferometry or shadowgraphy. Three-dimensional polygonal patterns approximating hexagons were observed by Bories and Thirriot (1969). In their experiments, a thin clear fluid layer over the porous layer allowed floating aluminium flakes to gather and disperse according to the convection pattern. The flakes gathered in regions over down flow. Combarnous and Bories (1975) and Murray and Chen (1989) used probes embedded within the medium to gather convection pattern information. With this method thermocouples or thermistors are placed in the medium, usually on the horizontal mid-plane. The resulting thermal map allows the reconstruction of the flow pattern. Bories *et al.* (1991) used the Christiansen effect to visualize convection patterns in a vertical annulus and in a rectangular layer. The Christiansen effect uses the differing dependence of the index of refraction on the temperature of the fluid and solid phases. By correctly choosing the materials, the refractive index will be the same in both phases at a certain temperature and the image is undistorted on this isotherm. Away from this reference isotherm, the temperature is inferred by the amount of image distortion. Chellaiah and Viskanta (1987) used the pH variation and electrolysis for the pattern visualization. Lister (1990) used electrolysis of a bromothymol blue solution that was acidified with acetic and hydrochloric acids. Lister was then able to observe dye transport through a clear upper boundary. The regions of dye collection corresponded with the down flows. Shattuck *et al.* (1995, 1997) used the powerful imaging method of magnetic resonance imaging to gather the temperature in the velocity information on convection porous layers. This method allows velocity and thermal structures as small as $10 \, \mu m$ to be imaged. Howle (1993) and Howle *et al.* (1993, 1997) used the modified shadowgraphic technique (MST) to non-invasively gather convection pattern in media that are constructed so that the fluid–solid interfaces are parallel and perpendicular to the direction of the light travel. Below, we give a detailed discussion of the method and of results gathered with MST.

In the experiments discussed in Howle *et al.* (1993, 1997), we place the fluid-saturated porous medium between the upper boundary consisting of a transparent sapphire window and a lower boundary consisting of a large, oxygen free high conductivity copper plate. The working fluid we use for all of the experiments is deionized degassed water. The upper surface of the copper plate is diamond polished to a flatness of one half of a wavelength of the visualizing light (632.8 nm) per inch and gold plated to provide a highly reflective surface for shadowgraphic visualization. A thin foil heater is affixed to the lower surface of the copper plate. By coupling the resistive heating to the large thermal capacity of the lower boundary we gain a lower boundary condition that is approximately at a constant temperature, rather than a constant heat flux.

In order to provide a transparent upper boundary for shadowgraphic visualization, we use a 1 cm thick optically-flat sapphire window. Sapphire has the advantage that the high thermal conductivity ($0.40\,\mathrm{W\,cm^{-1}K^{-1}}$) compared to the thermal conductivity of the medium reduces the horizontal thermal gradients at the upper boundary of the medium. Cooling water flows between the sapphire upper boundary and a crown glass optical flat. A Neslab cooler–circulator and specially constructed thermal regulation system maintains the upper boundary at a temperature of $25 \pm 0.002\,^{\circ}\mathrm{C}$.

A 20 mW He–Ne laser provides light for shadowgraphy. The light passes through a pin hole spatial filter and columnator before reflecting off of a beam splitter. The light then passes through the crown glass optical flat, cooling water, sapphire upper boundary, porous medium and reflects off of the gold plated lower boundary. The light then passes a second time through the porous medium, sapphire, cooling water, and optical flat before passing though a beam splitter and projecting onto a screen. The image is then captured by a CCD camera and digitized by a frame-grabber card in a computer.

We make thermal measurements with ten YSI 44006 precision thermistors eight of which are in thermal contact with the upper boundary and two of which are in thermal contact with the lower boundary. We gather the thermal data with a computer controlled scanning multimeter. The resolution of our thermal measurement is $\pm 0.0003\,^{\circ}\mathrm{C}$.

6.7.1   Ordered medium

In this subsection, we discuss the construction of the ordered medium and present results from our laboratory experiments. We construct the ordered medium by stacking layers of bars cut from 0.159 cm clear polycarbonate sheet. Slots of width 0.159 cm are machined in the polycarbonate sheet with a Bridgeport milling that maintains tolerances of ± 0.005 cm. We construct the second bar layer similar to the first but rotate the layer by 90°. The third layer is identical to the first, except that we offset the bars by one half of the grid period. This prevents the fluid from having a straight path when traveling vertically through the medium. We continue the rotation, offset and stacking until we have the desired number of layers. For the results discussed there, we use either six or seven layers of bars and for the medium with six layers of bars the properties are shown in Table 6.1. We show a drawing of the ordered medium with a rectangular planform in Figure 6.3.

Table 6.1 *Properties of ordered medium. * denotes property estimated from available transport data. Prandtl and Darcy numbers are calculated with vertical property values*

Property	Horizontal	Vertical
Permeability, K (cm^2)	3.9×10^{-4}	2.6×10^{-4}
Conductivity, k_m (W cm^{-1}K^{-1})	Not measured	3.6×10^{-3} *
Size (cm)	5.56, 2.70	0.902
Prandtl number, Pr_m		3.35
Darcy number, Da		3.2×10^{-4}
Porosity		0.5

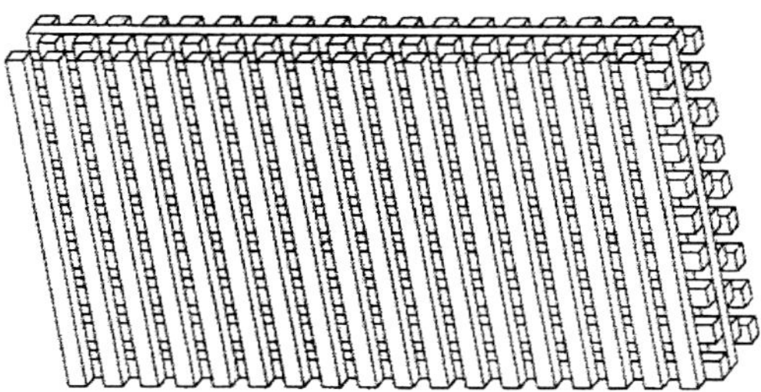

Figure 6.3 *Ordered rectangular grid porous medium. The side walls are not shown for clarity. In most of the experiments discussed here, we use six layers of bars. Reprinted from Howle (1993), with the permission of the author*

The onset of convection in the ordered medium shows a sharp, well defined discontinuity in the slope of Nu as a function of Ra_m, see Figure 6.4. The onset of convection in this medium occurs at the critical Rayleigh–Darcy number of $Ra_m^c = 50 \pm 6$. In reporting the heat transfer data in Figure 6.4, we normalize the Rayleigh–Darcy number with the critical value but there is no rounding of the heat transfer curve near the onset as is typical with heterogeneous media to within our experimental resolution of 0.1%. An interesting feature of Figure 6.4 is the slope of the convective portion of the heat transfer curve, namely

$$S \equiv \frac{\mathrm{d}Nu}{\mathrm{d}Ra_m}. \tag{6.41}$$

Joseph (1976) predicts $S = 2$, while our experiments show $S = 0.53 \pm 0.17$. A possible cause for the difference is the rather large value of the ratio of the bar height to the layer height in these experiments. Stated another way, this medium may not be refined enough to be considered homogeneous. Further, we note that the experiments of Close *et al.* (1985) also found a dependence of the heat transfer slope on the refinement of the porous

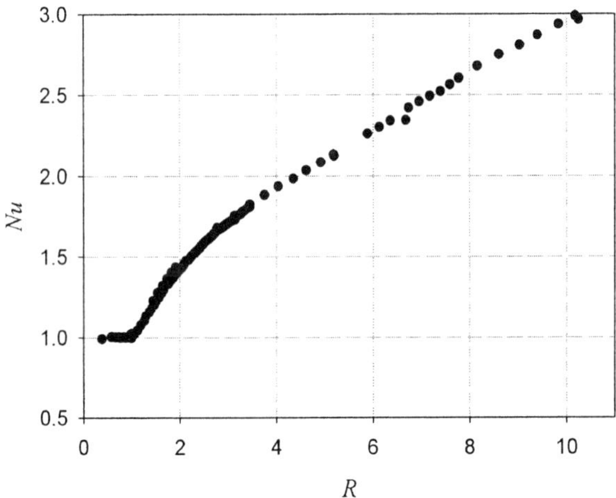

Figure 6.4 *Heat transport data for five consecutive experiments using the ordered medium. Onset of convection occurs at $Ra^c_m = 50 \pm 6$. The reduced Rayleigh number reported here, R, is the Rayleigh–Darcy number normalized by the critical value. Reprinted from Howle (1993), with the permission of the author*

layer. Other experimentally observed values are $S \approx 1$ by Elder (1967) and $S = 0.78$ by Shattuck *et al.* (1995, 1997).

The slope discontinuity in the heat transfer curve in Figure 6.4 corresponds to the initiation of parallel convection rolls. To our knowledge, these experiments and the experiments of Shattuck *et al.* (1995, 1997) are the first observation of parallel convection rolls in HRLC. In Figure 6.5, we show an example of convection rolls for $R = 1.93$. We observe parallel rolls from the smallest values of R at which they can be observed, $R = 1.05$, to approximately $R = 3$. Above this second value a secondary instability occurs and rolls are rarely observed.

For the pattern shown in Figure 6.5, the wave number is $q = 3.6 \pm 0.3$. An interesting feature of this convection pattern is its $45°$ orientation with respect to the side wall and the grid alignment. In Howle *et al.* (1997) we show that this is a result of the grid orientation and not of the side-wall forcing.

6.7.2   Disordered medium

In this subsection we consider another specially constructed medium that allows shadowgraphic visualization. For this medium, however, we created a system with some amount of disorder. We construct this medium with the same 0.159 cm thick polycarbonate sheet. We drill 362 randomly placed non-overlapping densely packed holes with 0.159 cm di-

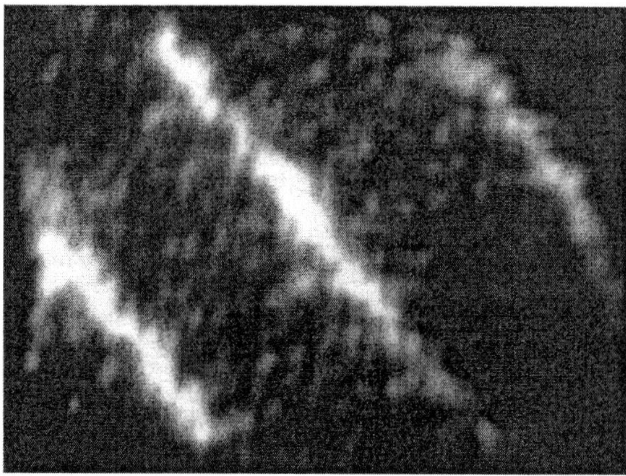

Figure 6.5 *Shadowgraphic image of the parallel convection rolls at $R = 1.93$. The bright regions correspond to cool, down flow while the dark regions correspond to warm up flow. Reprinted from Howle (1993), with the permission of the author*

ameter. We use five disks stacked vertically with 0.051 cm annular spacers between each disk. Additionally, we place spacers between the top disk and the upper boundary and the lower disk and the lower boundary. Figure 6.6 shows a drawing of the medium and Table 6.2 lists its thermal properties.

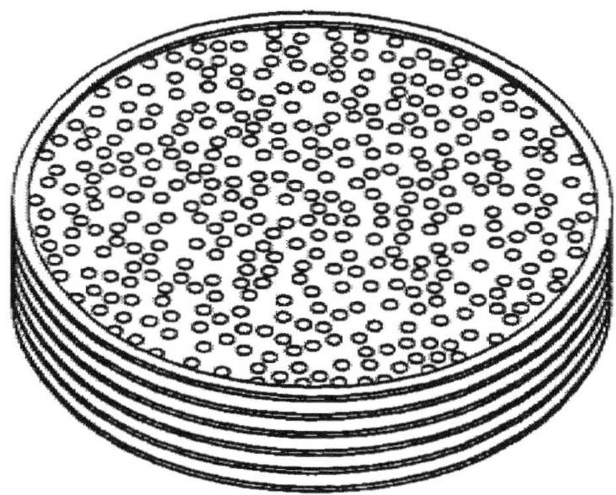

Figure 6.6 *Schematic drawing of the disordered porous medium. Each of the five disk layers contains 362 randomly placed non-overlapping holes. Reprinted from Howle (1993), with the permission of the author*

Table 6.2 *Properties of the disordered medium. Prandtl and Darcy numbers
are calculated with the vertical property values*

Property	Horizontal	Vertical
Permeability, K (cm^2)	2.3×10^{-4}	1.8×10^{-4}
Conductivity, k_m (W cm^{-1}K^{-1})	0.010	0.008
Size (cm)	3.016 (radius)	1.056
Prandtl number, Pr_m	4.19	
Darcy number, Da	1.6×10^{-4}	
Porosity	0.314	

In contrast to the ordered medium that shows a sharp primary bifurcation, the disordered
medium shows rounding of the heat transfer curve at the primary bifurcation. This is
shown in Figure 6.7 which displays heat transfer data for five experiments. Onset of
convection in the disordered medium occurs at $Ra_m^c = 37 \pm 4$ which is in good agreement
with the expected critical value of $Ra = 41.1$ (including the effects of anisotropy, see
equation (6.32)). For defining the critical point, we use the point of intersection of the
fully conductive and fully convective portions of the heat transfer curves. The rounding
noted at criticality corresponds to localized convection in regions where the properties
combine to create a local Rayleigh–Darcy number greater than the mean value. As the
temperature difference across the medium increases, the convection spreads out from these
seed areas and eventually fills the entire layer. This corresponds to the point where the heat
transfer curve first has a constant positive slope. As in the case of our ordered medium,
the disordered medium also exhibits a Nu versus Ra_m slope that is less than the predicted
value. For the disordered medium $S = 1.35 \pm 0.15$.

An additional interesting feature shown in Figure 6.7 is the different slopes for the fully
convective portion of Nu as a function of Ra_m. With each cycling through the onset of
convection, we generally obtain a different convection pattern. Each convection pattern,
although frozen once initiated, produces a slightly different Nu versus Ra_m curve.

A typical wave pattern is show in Figure 6.8. This medium always produces irregular
polygons as the preferred pattern with each cycling through criticality producing a different
pattern. Due to the irregularity of the pattern, the critical wave number, $q = 3 \pm 0.4$, has
a relatively large uncertainty. The irregular nature of the pattern suggests that the local
variability in the Rayleigh–Darcy number plays an important role in selecting the critical
convection pattern. It should be noted that these experiments and the experiments in the
ordered media from the previous section using the same apparatus are subject to the same
thermal regulation. Therefore, we can reasonably assume that the rounding of the primary
bifurcation is solely due to the local variability of the disordered medium.

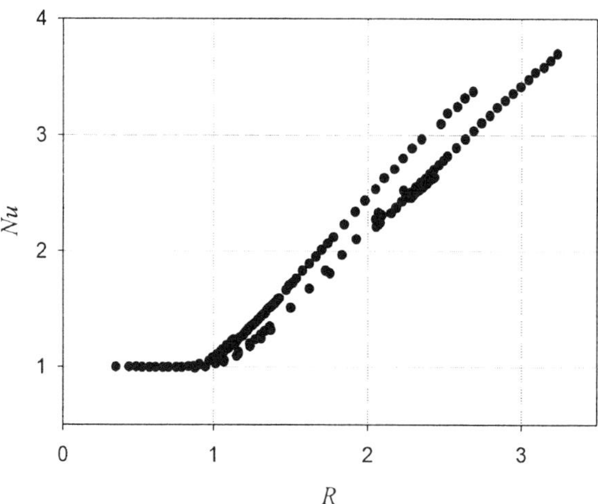

Figure 6.7 *Experimental heat transfer data for five experiments using the disordered medium. The onset of convection occurs at $Ra_m^c = 37 \pm 4$. Reprinted from Howle (1993), with the permission of the author*

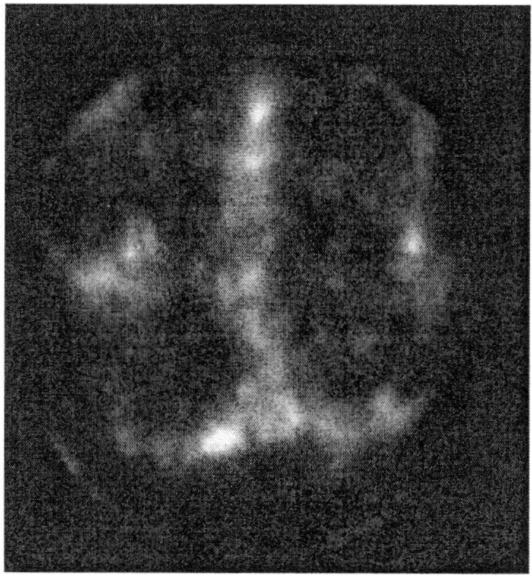

Figure 6.8 *Typical irregular polygon-like convection pattern in the disordered medium at $R = 1.27$. Reprinted from Howle (1993), with the permission of the author*

6.8 CONCLUSIONS

In this chapter, we have discussed two different experiments using the same apparatus. We have found that order or disorder within the medium profoundly influenced the convective wave patterns near the onset of convection. For the ordered medium, consisting of a few layers of rectangular bars, we have found the preferred wave pattern at the onset of convection to be parallel rolls. In the disordered medium, consisting of randomly drilled disks, we have found irregular polygons as the preferred wave pattern. Further, with each cycling through the onset of convection, a new pattern has emerged. Because we have carried out both sets of experiments in the same apparatus, we can reasonably conclude that the character of the medium (ordered versus disordered) caused these differences in the pattern preference.

Our experimental results have also shown that the wave pattern influences the slope of the heat transfer curve. For the ordered medium, where most experiments have produced the same pattern and wave number, the heat transfer curve was repeatable. The disordered medium produced different heat transfer curve slopes. In this medium, we have found that the slope is dependent on the convection pattern. Since each cycling through convection onset produced a unique convection pattern, we can expect to see different heat transfer curves.

The majority of porous convection experiments are produced with a random structure and therefore most experiments cannot be considered to be homogeneous. This is because of the competing needs to have short thermal relaxation times, small-scale structure, and Boussinesq thermal properties ultimately limit the spatial resolution of the medium. The resulting inhomogeneous media produce complicated, irregular convection patterns. We find that most laboratory experiments, ours included, have a mesostructure that is neither much smaller than the size of the experiment nor much larger than the size of the representative elementary volume used to derive the porous flow equations.

Acknowledgements

I would like to thank Professors Derek B. Ingham and Ioan Pop for the opportunity to contribute to this book.

REFERENCES

Bories, S. and Thirriot, C. (1969). Echanges thermiques et tourbillons dans une couche poreuse horizontale. *La Houille Blanche* **24**, 237–245.

Bories, S., Charrier-Mojtabi, M. C., Houi, D., and Raynaud, P. G. (1991). Non-invasive measurement techniques in porous media. In *Convective Heat and Mass Transfer in Porous Media* (eds S. Kakaç, B. Kilkis, F. A. Kulacki, and F. Arinç), pp. 173–224. NATO ASI Series E **196**. Kluwer, Holland.

Braester, C. and Vadasz, P. (1993). The effect of a weak heterogeneity of a porous medium on natural convection. *J. Fluid Mech.* **254**, 345–362.

Buretta, R. J. and Berman, A. S. (1976). Convective heat transfer on a liquid saturated porous layer. *J. Appl. Mech.* **43**, 249–253.

Castinel, G. and Combarnous, M. (1975). Natural convection in an anisotropic porous layer. *Int. Chem. Eng.* **17**, 605–614.

Chellaiah, S. and Viskanta, R. (1987). Natural convection flow visualization in porous media. *Int. Comm. Heat Mass Transfer* **14**, 607–616.

Chen, F. L. and Hsu, L. H. (1991). Onset of thermal convection in an anisotropic and inhomogeneous porous layer underlying a fluid layer. *J. Appl. Phys.* **69**, 6289–6301.

Close, D. J., Symmons, J. G., and White, R. F. (1985). Convective heat transfer in shallow, gas-filled porous media. *Int. J. Heat Mass Transfer* **28**, 2371–2378.

Combarnous, M. A. and Bories, S. (1975). Hydrothermal convection in saturated porous media. *Adv. Hydrosci.* **10**, 231–307.

Elder, J. W. (1967). Steady free convection in a porous medium heated from below. *J. Fluid Mech.* **27**, 29–48.

Epherre, J. F. (1975). Criterion for the appearance of natural convection in an anisotropic porous layer. *Int. Chem. Eng.* **17**, 615–616.

Ergun, S. (1952). Fluid flow through packed columns. *Chem. Eng. Progress* **48**, 89–94.

Gustafson, M. R. and Howle, L. E. (1999). Effects of anisotropy and boundary plates in the critical values of a porous medium heated from below. *Int. J. Heat Mass Transfer* **42**, 3419–3430.

Hartline, B. K. and Lister, C. R. B. (1977). Thermal convection in a Hele–Shaw cell. *J. Fluid Mech.* **79**, 379–389.

Horton, C. W. and Rogers, F. T. (1945). Convection currents in a porous medium. *J. Appl. Phys.* **16**, 367–370.

Howle, L. E. (1993). *Pattern formation at the onset of convection in porous media*. Ph.D. thesis. Duke University, Durham, NC.

Howle, L. E., Behringer, R. P., and Georgiadis, J. G. (1993). Visualization of convective fluid flow in a porous medium. *Nature* **362:6417**, 230–232.

Howle, L. E., Behringer, R. P., and Georgiadis, J. G. (1997). Convection and flow in porous media. 2. Visualization by shadowgraph. *J. Fluid Mech.* **332**, 247–262.

Jonsson, T. and Catton, I. (1987). Prandtl number dependence of natural convection in porous media. *J. Heat Transfer* **109**, 371–377.

Joseph, D. D. (1976). *Stability of Fluid Motions II*. Springer–Verlag, Berlin.

Kaneko, T., Mohtadi, M. F., and Aziz, K. (1974). An experimental study of natural convection in inclined porous media. *Int. J. Heat Mass Transfer* **17**, 485–496.

Kvernvold, O. and Tyvand, P. A. (1979). Nonlinear thermal convection in anisotropic porous media. *J. Fluid Mech.* **90**, 609–624.

Lapwood, E. R. (1948). Convection of a fluid in a porous medium. *Proc. Camb. Phil. Soc.* **44**, 508–521.

Lister, C. R. B. (1990). An explanation for the multivalued heat transport found experimentally for convection in a porous medium. *J. Fluid Mech.* **214**, 287–320.

Murray, B. T. and Chen, C. F. (1989). Double-diffusive convection in a porous medium. *J. Fluid Mech.* **201**, 147–166.

Neel, M. C. (1992). Inhomogeneous boundary conditions and the choice of the convective patterns in a porous layer. *Int. J. Eng. Sci.* **30**, 507–521.

Nield, D. A. and Bejan, A. (1999). *Convection in Porous Media* (2nd edn). Springer–Verlag, New York.

Prasad, V. and Kladias, N. (1991). Non-Darcy natural convection in saturated porous media. In *Convective Heat and Mass Transfer in Porous Media* (eds S. Kakaç, B. Kilkis, F. A. Kulacki, and F. Arinç), pp. 89–119. NATO ASI Series E **196**. Kluwer, Holland.

Schneider, K. J. (1963). Investigation of the influence of free convection on heat transport through granular material. In *Proceedings of 11th International Congress of Refrigeration,* Paper II-4, pp. 247–254. Pergamon, Oxford.

Shattuck, M. D., Behringer, R. P., Johnson, G. A., and Georgiadis, J. G. (1995). Onset and stability of convection in porous media—visualization by magnetic resonance imaging. *Phys. Rev. Lett.* **75**, 1934–1937.

Shattuck, M. D., Behringer, R. P., Johnson, G. A., and Georgiadis, J. G. (1997). Convection and flow in porous media. 1. Visualization by magnetic resonance imaging. *J. Fluid Mech.* **332**, 215–245.

Strauss, J. M. (1974). Large amplitude convection in porous media. *J. Fluid Mech.* **64**, 51–63.

Wang, M. and Bejan, A. (1987). Heat transfer correlation for Bénard convection in a fluid saturated porous layer. *Int. Comm. Heat Mass Transfer* **14**, 617–626.

Yen, Y.-C. (1974). Effects of density inversion on free convective heat transfer in porous layer heated from below. *Int. J. Heat Mass Transfer* **17**, 1349–1356.

7 MICROMECHANICS OF ORDERED, UNIDIRECTIONAL HETEROGENEOUS MATERIALS

C. Y. WANG[†]

Department of Mathematics, Michigan State University, East Lansing, MI 48824, USA

email: cywang@mth.msu.edu

Abstract

The pore-scale mechanics of thermal conduction and creeping flow in ordered, unidirectional heterogeneous media is considered and only exact methods which yield detailed fields are included. The effective conductivity and the effective permeability depend on the shape, orientation, and relative location of the parallel inclusions. Of particular interest is the property of anisotropy, which is absent for the frequently studied circular cylinders arranged in square or equilateral triangular arrays. The singular case of zero-thickness strip inclusions, for which all theories based on the volume fraction would fail, is also discussed.

Keywords: parallel fibers, micromechanics, effective conductivity, effective permeability, anisotropy

7.1 INTRODUCTION

Heterogeneity is prevalent in numerous natural and artificial materials occurring in, but not limited to, chemical, civil, and mining engineering, physical and earth sciences, and biology (e.g., Happel and Brenner, 1973). The theoretical prediction of the effective properties of heterogeneous media has always been difficult. There exists a variety of semi-empirical and approximate methods, including the rule of mixtures, bounding techniques, self-consistent schemes, homogenization, local averaging, etc., see for example Aboudi (1991), Adler (1992), Kaviany (1995), and Storesletten (1998). Most approximate methods do not take into account the shapes and relative positions of the phases, such that not only local field properties but also some global properties (such as anisotropy) cannot

[†]Visiting Professor, Department of Mechanical Engineering, National University of Singapore, Singapore 119260

be determined. For example, the averaging method needs closure assumptions which can only be obtained from micromechanics (e.g., Whitaker, 1999). Given the properties of the individual phases and their geometric arrangements, the most accurate exact methods solve the governing differential equations and the exact boundary conditions micromechanically. In particular, this is facilitated for ordered heterogeneous materials where a representative cell can be identified. The exact methods also give a detailed field, which would be useful in the identification of local extrema, such as temperature hot spots. On the other hand, a detailed flow field is essential in the computation of convective heat or mass transfer.

In what follows we shall discuss only the exact methods (as described above) applied to a heterogeneous matrix containing ordered unidirectional (2D) fibers and the phases are individually homogeneous and are assumed to be in perfect contact with each other. Two related steady-state problems will be considered. The first problem is the prediction of the effective thermal conductivity of a composite containing two solid phases or a porous material with a stagnant fluid phase. The second problem deals with the prediction of the effective creeping flow permeability through a solid porous array.

7.2 EFFECTIVE CONDUCTIVITY

Thermal conduction of a composite or a stagnant porous medium is governed by Laplace's equation, and is thus analogous to electrical conduction, mass diffusion, potential flow, Darcy flow in porous media (of an even smaller pore scale), longitudinal elastic shear, etc. Therefore the solutions to such problems may appear in a variety of different fields. The governing equation for thermal conduction is given by

$$\nabla^2 T = 0, \tag{7.1}$$

where T is the temperature. Between the phase boundaries, the temperature and the flux are to be continuous. For a given mean temperature gradient $\nabla \bar{T}$, the mean flux per area $\bar{q}$ is obtained micromechanically. Then from Fourier's law

$$\bar{q} = \boldsymbol{K}\left(-\nabla\bar{T}\right) \tag{7.2}$$

one can deduce the theoretical effective conductivity tensor $\boldsymbol{K}$.

Consider a material containing an ordered array of unidirectional fibers. The longitudinal effective conductivity may be obtained exactly by the rule of mixtures (an average weighted by volume fractions). This is because the flux vector is parallel to the longitudinal direction everywhere and the phases become independent of each other. In this chapter, we shall concentrate on the more difficult transverse effective conductivity.

7.2.1 Some properties of the effective conductivity

It can be shown that the effective conductivity (and effective permeability of creeping flow) is a symmetric matrix, and this can be diagonalized if the coordinate axes coincide with the principal axes of an orthotropic medium (Adler, 1992; Dullien, 1992). We now show that if the effective conductivity is the same for two independent transverse directions then the material is isotropic in all transverse directions.

Let K_{ij} be the conductivity tensor in a coordinate system which is aligned with the principal axes of an orthotropic medium. Thus K_{ij} is diagonal. If K'_{pq} denotes the conductivity tensor in the new primed axes which are rotated through an angle δ about the longitudinal (third) direction, the relations are given by

$$K'_{11} = K_{11} \cos^2 \delta + K_{22} \sin^2 \delta, \quad K'_{22} = K_{11} \sin^2 \delta + K_{22} \cos^2 \delta, \tag{7.3}$$

$$K'_{12} = K'_{21} = (K_{22} - K_{11}) \cos \delta \sin \delta. \tag{7.4}$$

For a square array of circular cylinders, the conductivities along the transverse principal axes are equal, i.e., $K_{11} = K_{22} = K$. Thus from equations (7.3) and (7.4), $K'_{11} = K'_{22} = K$ and $K'_{12} = K'_{21} = 0$ for all rotations δ and the medium is transversely isotropic. For an equilateral triangular array (sometimes misidentified as a hexagonal array) of circular cylinders, the conductivities along the two transverse perpendicular principal axes (parallel and perpendicular to one side of the triangle) may not be equal. However, due to symmetry, the conductivities in a coordinate system which is rotated through an angle $\delta = 2\pi/3$ are the same as those of the original system. Since in this new system the conductivity tensor must also be diagonal then from equation (7.4) we have $K_{11} = K_{22}$. Applying the previous arguments for the square array shows that the equilateral triangular array is transversely isotropic. Similarly, the regular hexagonal array (a tiling of hexagons with circular cylinders at the corners) is transversely isotropic.

Keller (1964), using potential fields, proved that the electric effective conductivity of a composite containing a rectangular array of circular cylinders has the following property. The effective conductivity ratio (effective conductivity over the matrix conductivity) in one transverse principal direction is the reciprocal of the effective conductivity ratio in the other transverse principal direction, with the conductivities of the fiber and the matrix interchanged. Keller's theorem has been verified for the square and triangular arrays (Lu, 1995). If Keller's theorem is true, we only need to consider half of the number of cases, say, the fibers are less conductive than the matrix, since the more conductive cases can be deduced from the same data using Keller's theorem.

7.2.2 Fibers of circular cross-sections

Basically three different types of methods are used in the micromechanical analysis. The earliest method, used by Rayleigh (1892), summed the infinite singularities due to the fibers. Rayleigh's method was improved and extended by Perrins *et al.* (1979) who published tables of effective conductivities for various relative fiber conductivities. The second method, by Han and Cosner (1981) and Lu (1995), used eigenfunction expansions in a representative cell and collocated on the boundary. Lu's results agree very well with those of Perrins *et al.* Finally, finite-difference (Keller and Sachs, 1964; Adams and Doner, 1967) and finite element (Grove, 1990; Rolfes and Hammerschmidt, 1995; Islam and Pramila, 1999) methods have also been used. The results of the latter cannot be compared due to the poor quality of the graphs. All the above sources considered ordered arrays of circular cylindrical fibers in square or triangular arrays, except two specific rectangular arrays computed by Han and Cosner (1981). The related case of elliptic fibers in a rectangular array was studied by Adams and Doner (1967) using finite differences. Lu (1994) transformed the ellipse to a circle and then used collocation. Nicorovici and McPhedran (1996) extended Rayleigh's method to elliptic cylinders.

In what follows we shall briefly discuss the anisotropic properties of a material with a rectangular array of circular cylinders. We shall use a modified method of eigenfunction expansion and collocation and Figure 7.1(a) shows the dimensions. The representative cell is partitioned into three regions: Region III is a quarter of a cylinder, Region II is a square region with a corner deleted by Region III, and a complementary rectangular Region I. The normalized boundary conditions for Case A, where the transverse heat flux is through the larger cylinder gaps, are shown in Figure 7.1(b). The general solution, T_{I}, for Region I is given by

$$T_{\mathrm{I}}(x,y) = x + \sum_{n=1}^{\infty} A_n \sin(\alpha_n x)\left(\mathrm{e}^{\alpha_n(y-a)} + \mathrm{e}^{-\alpha_n(y+a)}\right), \qquad (7.5)$$

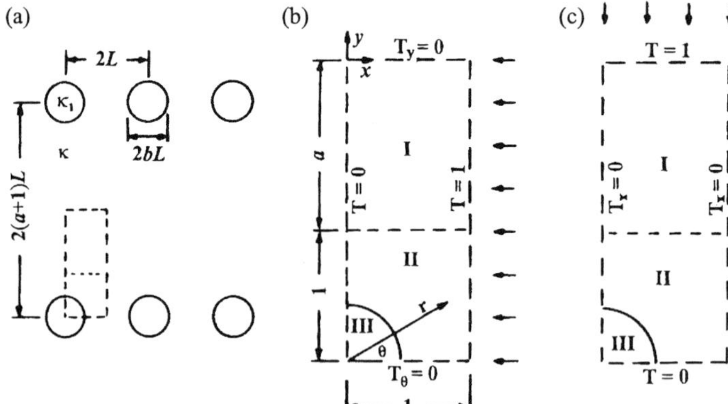

Figure 7.1   *(a) Cross-sectional geometry, (b) Case A, flux through the larger gaps, and (c) Case B, flux through the smaller gaps*

where $\alpha_n \equiv n\pi$ and A_n are coefficients to be determined. The general solution, $T_{\rm II}$, for Region II is given by

$$T_{\rm II}\,(r,\theta) = \sum_{n=1}^{\infty} \cos\left(\beta_n\theta\right)\left(B_n r^{\beta_n} + C_n r^{-\beta_n}\right),\tag{7.6}$$

where $\beta_n \equiv 2n - 1$. Similarly, the solution, $T_{\rm III}$, for Region III is given by

$$T_{\rm III}\,(r,\theta) = \sum_{n=1}^{\infty} D_n \cos\left(\beta_n\theta\right) r^{\beta_n}.\tag{7.7}$$

The temperature and heat flux are then matched at their common boundaries. We find

$$T_{\rm II}\,(b\theta) = T_{\rm III}\,(b\theta),\tag{7.8}$$

$$\frac{\partial T_{\rm II}}{\partial r}\,(b\theta) = \lambda\frac{\partial T_{\rm III}}{\partial r}\,(b\theta),\tag{7.9}$$

where $\lambda \equiv \kappa_1/\kappa$ is the ratio of the conductivities of the two phases. Equations (7.8) and (7.9) yield

$$C_n = \mu b^{2\beta_n}\,B_n, \quad D_n = (1+\mu)\,B_n,\tag{7.10}$$

where $\mu \equiv (1 - \lambda)\,/\,(1 + \lambda)$. Between Regions I and II the matching conditions are as follows:

$$T_{\rm I}\,(x, -a) = T_{\rm II}\,(\csc\theta, \theta),\tag{7.11}$$

$$\frac{\partial T_{\rm I}}{\partial y}\,(x, -a) = \mathbb{K}T_{\rm II}\,(\csc\theta, \theta),\tag{7.12}$$

where

$$\mathbb{K} \equiv \frac{\partial}{\partial y} = \sin\theta\frac{\partial}{\partial r} + \frac{\cos\theta}{r}\frac{\partial}{\partial\theta}.\tag{7.13}$$

The remaining condition is for $T_{\rm II}$ to be unity on the right-hand side, i.e.,

$$T_{\rm II} = (\sec\theta, \theta) = 1.\tag{7.14}$$

Equations (7.11) – (7.14) are to be satisfied by collocation.

Next consider Case B where the mean temperature gradient is applied through the smaller gaps. The general solutions for the three regions are as follows:

$$T_{\rm I} = 1 + A_0 y + \sum_{n=1}^{\infty} A_n \cos\left(\alpha_n x\right)\left(e^{\alpha_n (y-a)} + e^{-\alpha_n (y+a)}\right),\tag{7.15}$$

$$T_{\rm II}\,(r,\theta) = \sum_{n=1}^{\infty} B_n \sin\left(\beta_n\theta\right)\left(r^{\beta_n} + \mu b^{2\beta_n} r^{-\beta_n}\right),\tag{7.16}$$

$$T_{\mathrm{III}}(r,\theta) = \sum_{n=1}^{\infty} (1+\mu)\, B_n \sin(\beta_n\theta)\, r^{\beta_n}. \tag{7.17}$$

The matching conditions are equations (7.11) and (7.12) and the symmetry condition

$$\mathbb{L}T_{\mathrm{II}}(\sec\theta,\theta) = 0, \tag{7.18}$$

where

$$\mathbb{L} \equiv \frac{\partial}{\partial x} = \cos\theta\frac{\partial}{\partial r} - \frac{\sin\theta}{r}\frac{\partial}{\partial\theta}. \tag{7.19}$$

After the temperature distribution is found, the mean heat flux can be integrated analytically and one can obtain the effective conductivity. Let σ denote the ratio of the effective conductivity over the conductivity of the matrix and the aspect ratio be the distance of two adjacent cylinders normal to the mean flux over that parallel to the mean flux. We find Keller's theorem holds and only the results for λ or σ less than unity need to be presented. Figure 7.2 shows the effective conductivity ratio as a function of the volume fraction ϕ for insulated cylinders ($\lambda = 0$). The rectangular arrangements are complementary when the product of the two aspect ratios (AR) is unity and prominent anisotropy effects are

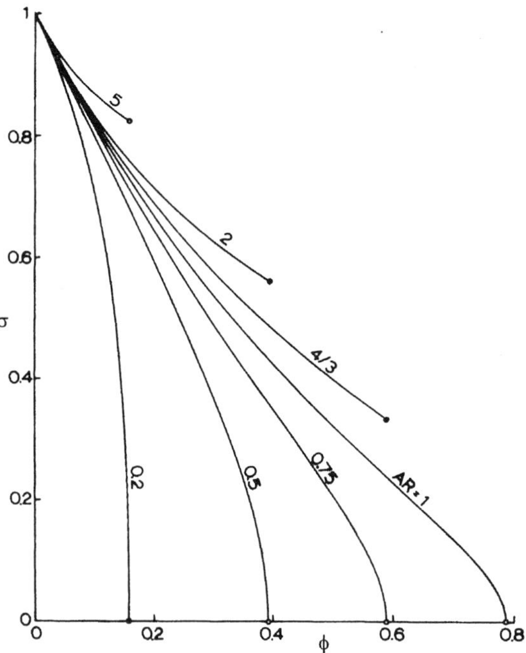

Figure 7.2 *Effective conductivity ratio as a function of the volume fraction for insulated cylinders*

evident. Further, we observe that the anisotropy becomes less if the conductivity ratio λ is closer to one and Figure 7.3 shows the variation of the effective conductivity ratio as a function of the volume fraction ϕ for partially conducting cylinders with $\lambda \equiv \kappa_1/\kappa = 0.2$. It should be noted that the small circles in Figures 7.2 and 7.3 are for the situation when the cylinders touch each other.

7.2.3 Fibers of rectangular cross-sections

Studies of square cylinders in a (staggered) checkerboard array have been performed by Milton *et al.* (1981), Fogelholm and Grimvall (1983), Gu and Tao (1988), Bao *et al.* (1990), and Lu and Bao (1992) using a variety of methods including numerical, integral equation, eigenfunction expansion, and collocation. For square cylinders with touching corners then an exact solution is possible (Bedichevskii, 1985) but it should be noted that Rayleigh's method does not apply. Square cylinders in a square array was solved by Lu (1995) using cylindrical coordinates and collocation. Since the effective conductivities in the two perpendicular directions are the same then the material is isotropic.

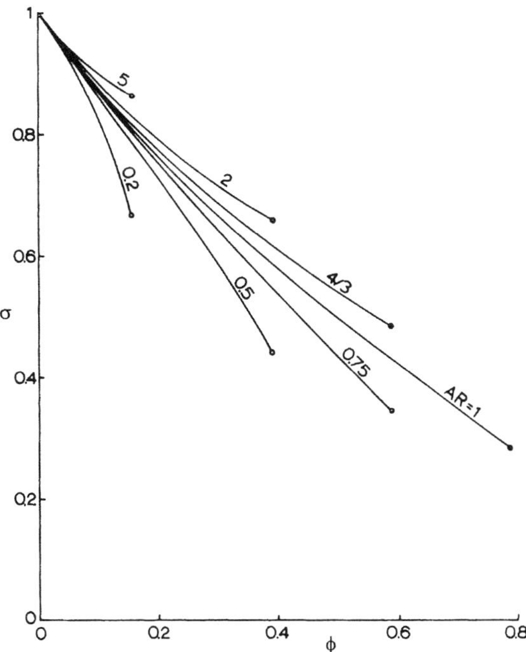

Figure 7.3 *Effective conductivity ratio for partially conducting cylinders, namely* $\lambda = \kappa_1/\kappa = 0.2$

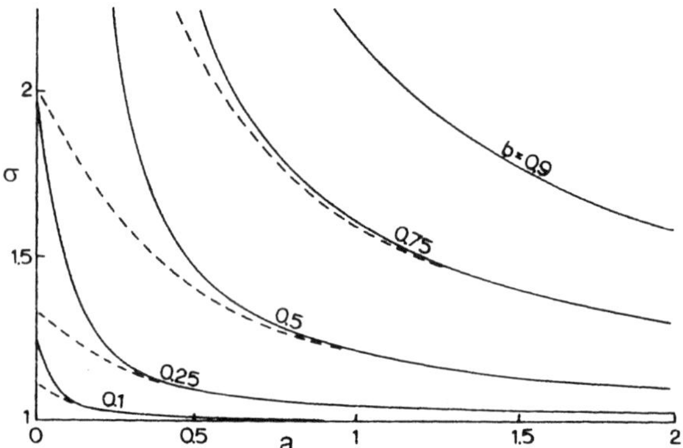

Figure 7.6 *Effective transverse tangential conductivity as a function of a for perfectly conducting strips (Wang, 1999a). Dashed lines are for a rectangular array (Wang, 1994)*

7.3   EFFECTIVE PERMEABILITY

The slow viscous flow through an array of cylinders is important in the flow across a filter and biological fibers, catalytic converters, drying, manufacturing of composite materials, and seepage through porous media. The governing Stokes equation for steady, incompressible creeping flow is given by

$$\mu \nabla^2 \boldsymbol{u} - \nabla p = 0,$$
$$\nabla \cdot \boldsymbol{u} = 0,$$

(7.22)

where μ is the viscosity of the fluid, $\boldsymbol{u}$ is the velocity vector, and p is the pressure. Due to linearity, the Stokes equation leads to Darcy's law, e.g., Scheidegger (1974), Adler (1992), and Nield and Bejan (1999), i.e.,

$$\boldsymbol{U} = \frac{1}{\mu} \boldsymbol{K} \left(-\nabla p \right),$$

(7.23)

where $\boldsymbol{U}$ is the mean velocity vector and $\boldsymbol{K}$ is the effective permeability tensor, to be found micromechanically by solving equation (7.22) with the no-slip boundary conditions on all solid surfaces. As noted before, the effective permeability is symmetrical, and is diagonal if the axes are aligned with the principal axes of the medium. The previous isotropy proof still holds, but Keller's theorem is inapplicable. For unidirectional fibers, due to linearity, one can separate the flow into longitudinal and transverse directions.

For longitudinal flow, equation (7.22) reduces to the Poisson equation

$$\nabla^2 w = 1,$$

(7.24)

where ∇^2 is the two-dimensional Laplacian normalized by a length scale L and w is the longitudinal velocity normalized by $L^2 \Delta p / \mu$. It is of interest to note that equation (7.24) also governs the torsion of elastic bars and the deflection of membranes. The longitudinal flow problem is not as trivial as the longitudinal conduction problem.

For the transverse flow, equation (7.22) reduces to the biharmonic equation

$$\nabla^4 \psi = 0, \tag{7.25}$$

where ψ is the stream function obtained from the continuity equation. Further, the biharmonic equation also governs plane elasticity problems.

7.3.1 Fibers of circular cross-sections

Longitudinal flow between parallel circular cylinders was solved by Sparrow and Loeffler (1959) using an eigenfunction expansion and collocation, Banerjee and Hadaller (1973) using a variational technique, and Larson and Higdon (1986) using a boundary integral method. However, the transverse flow, governed by a fourth-order differential equation, is more difficult to solve. Sangani and Acrivos (1982) used eigenfunction expansion and collocation, Drummond and Tahir (1984) used singularity summations, and Larson and Higdon (1987) used boundary integrals. Numerical methods, such as finite differences (Gebart, 1992; Skartsis *et al.*, 1992; Alcocer *et al.*, 1999) and smooth particle hydrodynamics (Zhu *et al.*, 1999), have also been used. The above sources mostly considered circular cylinders in a square or triangular array.

We shall present the non-isotropic rectangular array, see Figure 7.7(a), considered by Wang (2001), and Figure 7.7(b) shows the domain decomposition for longitudinal flow. The solution to Region I, satisfying equation (7.24) and the reflective conditions at $y = 0$, $x = 0$, and $x = 1$, is given by

$$w_{\mathrm{I}}(x, y) = -\frac{y^2}{2} + A_0 + \sum_{n=1}^{\infty} A_n \left[e^{\alpha_n (y-a)} + e^{-\alpha_n (y+a)} \right] \cos(\alpha_n x), \tag{7.26}$$

where $\alpha_n = n\pi$. The general solution of Region II satisfying equation (7.24), the no-slip condition at $r = b$, and the reflective conditions at $\theta = 0$ and $\theta = \pi/2$ is given by

$$w_{\mathrm{II}}(r, \theta) = \frac{b^2 - r^2}{4} + B_0 \ln\left(\frac{r}{b}\right) + \sum_{n=1}^{\infty} B_n \left(r^{2n} - b^{4n} r^{-2n}\right) \cos(2n\theta). \tag{7.27}$$

The remaining conditions are the reflective condition at $x = 1$, namely

$$\mathbb{L}\, w_{\mathrm{II}}(\sec\theta, \theta) = 0, \tag{7.28}$$

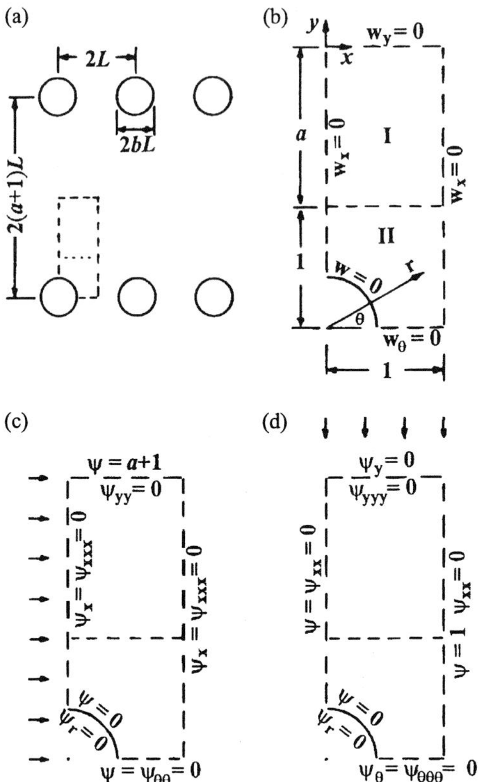

Figure 7.7 *(a) The cross-section, (b) boundary conditions for longitudinal flow, (c) boundary conditions for transverse flow through the larger gaps, and (d) boundary conditions for transverse flow through the smaller gaps*

and the matching conditions at $y = -a$, i.e.,

$$w_{\mathrm{I}}\left(\cot\theta, -a\right) = w_{\mathrm{II}}\left(\csc\theta, \theta\right), \tag{7.29}$$

$$\frac{\partial w_{\mathrm{I}}}{\partial y}\left(\cot\theta, -a\right) = \mathbb{K}w_{\mathrm{II}}\left(\csc\theta, \theta\right). \tag{7.30}$$

The infinite series are then truncated and equations (7.28) – (7.30) are collocated on θ angles spaced between 0 and $\pi/2$. The coefficients A and B, and thus the flow field, are determined. Since the solution is analytic, the mean velocity and the effective permeability can be integrated easily and Figure 7.8 shows the normalized permeability $K^* = K/L^2$ as a function of b.

The boundary conditions for transverse flow parallel to the cylinder rows are shown in Figure 7.7(c), where ψ is the stream function. The general solution of Region I that

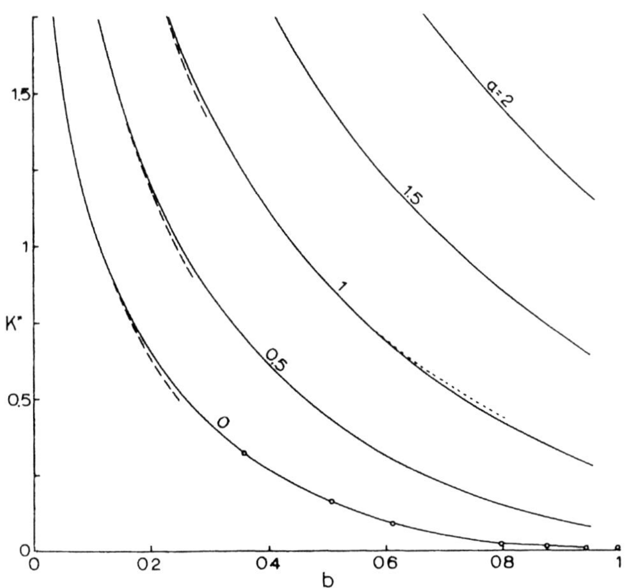

Figure 7.8 *Permeability K^* as a function of b for longitudinal flow. Small circles are from Sparrow and Loeffler (1959), the dashed lines are from equation (7.42) and the dotted lines are from Drummond and Tahir (1984)*

satisfies equation (7.25) and the reflective conditions at $y = 0$, $x = 0$, and $x = 1$ is given by

$$\psi_{\mathrm{I}}(x, y) = 1 + a + C_0 y + D_0 y^3 + \sum_{n=1}^{\infty}\left\{ C_n \left[e^{\alpha_n(y-a)} - e^{-\alpha_n(y+a)} \right] \right.$$
$$\left. + D_n \left[e^{\alpha_n(y-a)} + e^{-\alpha_n(y+a)} \right] \right\} \cos(\alpha_n x).$$
$$(7.31)$$

The general solution of Region II, which satisfies equation (7.25), the no-slip on the cylinder, and the reflective conditions at $\theta = 0$ and $\theta = \pi/2$, is given by

$$\psi_{\mathrm{II}}(r, \theta) = \left\{ \left(r^3 - 2b^2 r + \frac{b^4}{r} \right) E_1 + \left[r \ln\left(\frac{r}{b}\right) - \frac{r}{2} + \frac{b^2}{2r} \right] F_1 \right\} \sin\theta$$
$$+ \sum_{n=2}^{\infty}\left\{ \left[r^{2+\beta_n} - \left(1 + \frac{1}{\beta_n} \right) b^2 r^{\beta_n} + \frac{b^{2+2\beta_n}}{\beta_n r^{\beta_n}} \right] E_n \right.$$
$$\left. + \left[r^{2-\beta_n} - \frac{b^{2-2\beta_n}}{\beta_n} r^{\beta_n} + \left(\frac{1}{\beta_n} - 1 \right) \frac{b^2}{r^{\beta_n}} \right] F_n \right\} \sin(\beta_n \theta),$$
$$(7.32)$$

8 MODELING TURBULENCE IN POROUS MEDIA

J. L. LAGE*, M. J. S. DE LEMOS[†] and D. A. NIELD[‡]

*Department of Mechanical Engineering, Southern Methodist University, Dallas, TX 75275-0337, USA

email: jll@engr.smu.edu

[†]Departamento de Energia, Instituto Tecnológico de Aeronáutica, 12228-900 São José dos Campos, SP, Brazil

email: delemos@mec.ita.br

[‡]Department of Engineering Science, University of Auckland, Private Bag 92019, Auckland, New Zealand

email: d.nield@auckland.ac.nz

Abstract

Four available methodologies for developing macroscopic turbulence models for incompressible single-phase flow in rigid, fully saturated porous media are reviewed. The first method, known as the Antohe–Lage (A–L) method, starts with the closed volume-averaged equations, which are then averaged in time to produce the turbulence equations. The second, known as the Nakayama–Kuwahara (N–K) method, makes use, first, of the closed time-averaged equations, and then proceeds with volume-averaging for deriving the turbulence equations. These two methodologies lead, in general, to distinct sets of turbulence equations because of the different averaging order, i.e., space–time and time–space, respectively. A third, and probably the most consistent method, based on double-decomposition, is the Pedras–de Lemos (P–dL) method. In this method, the momentum equation is closed by using the Hazen–Dupuit–Darcy model for the total drag effect only after the space–time averaging (or time–space averaging) is performed. Although for the P–dL method the averaging order is immaterial when deriving the turbulence momentum equation, the difference between space–time and time–space averaging remains in the k–ε equations. Unfortunately, detailed experimental model validation, which remains to be seen, is tremendously challenging because of the need to obtain time-averaged and volume-averaged quantities simultaneously in order to compare experimental and analytical (numerical) results directly. A fourth method, the Travkin–Catton (T–C) morphology method, is discussed only briefly because it follows the N–K method (time–space integrating order) and no closure to the final equations is yet available.

Keywords: turbulence, modeling, transport, porous media, averaging

8.1 INTRODUCTION

Modeling turbulent transport in porous media can impact several critical and practical engineering areas. For instance, the accurate simulation of turbulent air flow permeating through forests, where the vegetation is seen as a porous structure, is extremely important for predicting bio-diversity (spreading of seeds) and mitigation of fire propagation. The transport and dispersion of smog through heavily built cities can also benefit from the accurate modeling of turbulent flow through porous media.

Efficient and realistic overall pressure-drop along oil extraction porous wells is also very important. In this application, the flow of oil and gas along a radial-inward path is accelerated by getting near a more permeable region (the well) becoming turbulent. Proper mathematical characterization of the flow is necessary in order to reduce uncertainties on the well lifetime performance.

Processes of solidification and fusion of certain alloys are characterized by the presence of three distinct domains, namely, a fluid, a mushy and a solid zone. When the flow in the fluid zone is turbulent, the accurate prediction of the final product (the metal) depends on the proper characterization of the turbulent transport process inside the mushy zone. A similar process is the manufacturing of optical fiber and glass, which involves the turbulent flow of a doping gas during the melting process.

These engineering and environmental processes are only a few examples establishing the variety and importance of applications that can benefit from a proper mathematical analysis of turbulent flow in porous media. In a broader sense, the study of turbulence in porous media embraces fluid and thermal sciences, materials, chemical, geothermal, bio, petroleum and combustion engineering.

So far, the term turbulence has been used here to denote turbulence anywhere within the pores of the porous medium. The turbulent transport within the pore network of a porous medium can be studied, in principle, through direct numerical simulation. However, the direct numerical simulation at the pore (microscopic) level is impractical not only because of the tremendous computational effort required to resolve all the different turbulence scales, but also because of the additional effort required to access, map and resolve the complicated internal morphology of the porous medium.

Modeling is a natural alternative to direct numerical simulation. The objective of good modeling is to reduce the complexity of the mathematical formulation for studying the phenomenon. *Averaging* is a powerful modeling tool, which must be used carefully not to compromise the fundamental mathematical information ruling the phenomenon. When studying turbulent flow in individual fluid conduits, for instance, the Navier–Stokes equation can be time-averaged and closed with the stress–strain relations for the Reynolds stress, leading to the well-known k–ε turbulence model. This model reduces the complexity of the problem by eliminating the need to follow the rapid fluctuations in time of fluid velocity and pressure, so characteristic of turbulence. Instead, the final model equation deals only with time-averaged quantities. Although information is invariably lost when time averaging the equations, it is hoped that the major characteristics of the transport process be retained by the model.

> Antohe and Lage (1997) examined their model equations for the turbulence kinetic
> energy and its dissipation rate, assuming a unidirectional fully-developed flow through
> an isotropic porous medium. Their model demonstrates that the only possible steady
> state solution for the case is 'zero' macroscopic turbulence kinetic energy. This
> solution should be re-examined, since the macroscopic turbulence kinetic energy in
> a forced flow through a porous medium must stay at a certain level, as long as the
> presence of porous matrix keeps generating it. (The situation is analogous to that of
> turbulent fully-developed flow in a conduit.) Also, it should be noted that the small
> eddies must be modeled first, as in the case of LES (Large Eddy Simulation). Thus
> we must start with the Reynolds averaged set of the governing equations and integrate
> them over a representative control volume, to obtain the set of macroscopic turbulence
> model equations. Therefore, the procedure based on the Reynolds averaging of the
> spatially averaged continuity and momentum equations is questionable, since the
> eddies larger than the scale of the porous structure are not likely to survive long
> enough to be detected.

Evidently, Nakayama and Kuwahara (1999) neither considered the possibility of a pore-network with morphology very different from that of a single conduit (damping turbulence, instead of producing it), nor realized the special meaning of the turbulence kinetic energy defined by Antohe and Lage (1997). The Antohe–Lage result says nothing about the existence or otherwise of microscopic turbulence, and its failure to do so should not be used as negative criticism of the model.

Nakayama and Kuwahara (1999) went on to describe their own work:

> The macroscopic turbulence kinetic energy and its dissipation rate are derived by
> spatially averaging the Reynolds-averaged transport equations along with the k–ε
> turbulence model. For the closure problem, the unknown terms describing the pro-
> duction and dissipation rates inherent in porous matrix are modified collectively.
> In order to establish the unknown model constants, we conduct an exhaustive nu-
> merical experiment for turbulent flows though a periodic array, directly solving the
> microscopic governing equations, namely, the Reynolds-averaged set of continuity,
> Navier–Stokes, turbulence kinetic energy and dissipation rate equations. The micro-
> scopic results obtained from the numerical experiment are integrated spatially over a
> unit porous structure to determine the unknown model constants. The macroscopic
> turbulence model, thus established, is tested for the case of macroscopically unidirec-
> tional turbulent flow. The streamwise variations of the turbulence kinetic energy and
> its dissipation rate predicted by the present macroscopic model are compared against
> those obtained from a large scale direct computation over an entire field of saturated
> porous medium, to substantiate the validity of the present macroscopic model.

We now make some specific comments on the Nakayama and Kuwahara (1999) model. Having integrated the Reynolds averaged equations over an REV, Nakayama and Kuwa-hara (1999) obtained their momentum equation, equation (11). This is unexceptional. However, they then proceed to replace the last two terms of equation (11) by viscous-drag (Darcy) and form-drag (Forchheimer) terms to obtain equation (14). This is a standard procedure for laminar flows but it appears that this replacement is still highly question-able in the context of turbulence modeling. In the paragraph containing equation (14), Nakayama and Kuwahara (1999) wrote

In the numerical study of turbulent flow through a periodic array, Kuwahara *et al.* (1998) concluded that the Forchheimer-extended Darcy's law holds even in the turbulent flow regime in porous media.

That too is an acceptable statement but it does not justify the transition from equation (11) to equation (14). For one thing, equation (14) is substantially different from the standard Forchheimer-extended Darcy equation. Furthermore, there is a gap in the argument in proceeding from an equation for the turbulence regime in a bulk form, in which the total pressure-drop is related to the bulk fluid speed via an expression quadratic in the velocity, to an equation involving a differential expression. In summary, it seems that Nakayama and Kuwahara (1999) have made an assumption of a relationship between microscopic turbulence and macroscopic drag that cannot be justified except in the gross sense that for high Reynolds number the form-drag (Forchheimer) term will be dominant. Even in this case, this argument is not expected to be valid for high porosity porous media.

Further, there is a fundamental difficulty with any model in which time averaging (Reynolds averaging) is followed by space averaging. This procedure precludes the incorporation of the interaction between fluctuating quantities and the solid matrix of the porous medium, other than the minor effect of fluctuations in pressure and shear stresses along the interfacial solid-fluid area. This aspect was clearly stated by Antohe and Lage (1997) as the A–L method suffers from a similar handicap: volume averaging followed by time averaging precludes the incorporation of the turbulence effect by the space fluctuating quantities.

Finally, because of their assumption of periodic domain (periodic on the pore scale) when performing their numerical calculations, Masuoka and Takatsu (1996), Takatsu and Masuoka (1998), Nakayama and Kuwahara (1999), and Pedras and de Lemos (2001a) were unable to treat eddies on a scale larger than their period length. This means that global eddies were ruled out *a priori*.

8.4 MODELING: AVERAGING OPERATORS

We now present the mathematical foundation for deriving volume-averaged and time-averaged transport equations. Subsequently, the P–dL double-decomposition method, see Pedras and de Lemos (1999), used in deriving another turbulence model, is discussed in detail.

8.4.1 Volume averaging

The macroscopic governing equation for flow through a porous medium can be obtained by volume averaging the corresponding microscopic equations over a representative elementary volume, ΔV, see Figure 8.1. For a general fluid property, the intrinsic and volumetric averages are related through the porosity ϕ as, Bear (1972),

$$\langle \varphi \rangle^i = \frac{1}{\Delta V} \int_{\Delta V_f} \varphi \, dV, \quad \langle \varphi \rangle^v = \phi \langle \varphi \rangle^i, \quad \phi = \frac{\Delta V_f}{\Delta V}, \tag{8.1}$$

where ΔV_f is the volume of the fluid contained in ΔV. The property φ can then be defined as the sum of $\langle \varphi \rangle^i$ and a term related to its spatial variation within the REV, ${}^i\varphi$, as

$$\varphi = \langle \varphi \rangle^i + {}^i\varphi. \tag{8.2}$$

From equations (8.1) and (8.2) one derives $\langle {}^i\varphi \rangle^i = 0$. Figure 8.1 illustrates the idea underlined by equation (8.2) for the value of a property of vectorial nature (e.g., velocity)

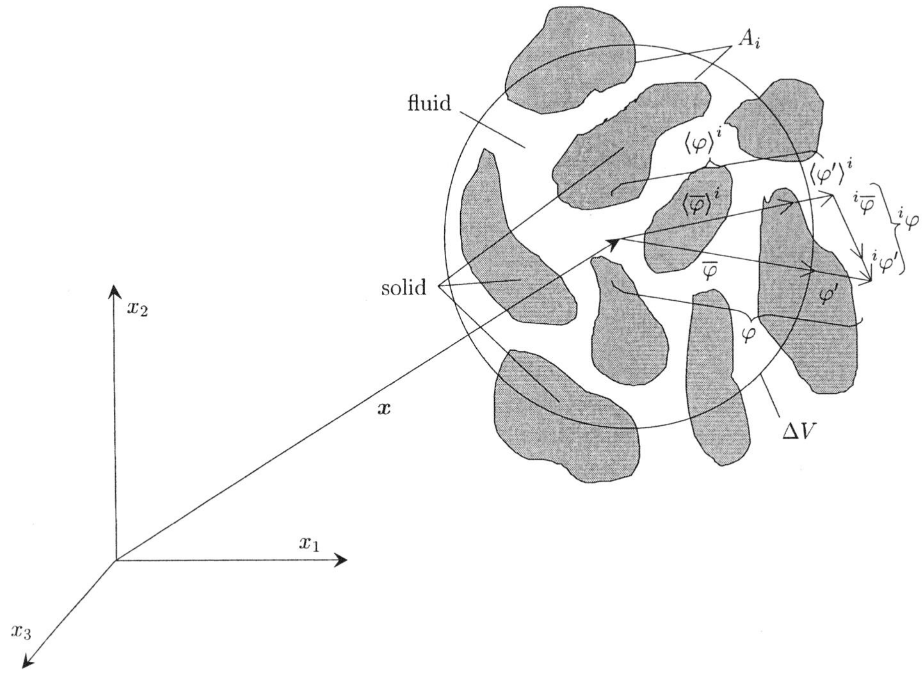

Figure 8.1 *Representative elementary volume (REV), intrinsic average, space and time fluctuations, see Pedras and de Lemos (2000a)*

at x. The spatial deviation is the difference between the real value (microscopic) and its intrinsic (fluid based average) value.

For deriving the governing flow equations, it is necessary to know the relationship between the volumetric average of derivatives and the derivatives of the volumetric average. These relationships, presented in Slattery (1967), Whitaker (1969), and Gray and Lee (1977), are known as the *theorem of local volumetric average*, namely

$$\langle \nabla \varphi \rangle^v = \nabla \left(\phi \langle \varphi \rangle^i \right) + \frac{1}{\Delta V} \int_{A_i} n \varphi \, dS, \tag{8.3}$$

$$\langle \nabla \cdot \boldsymbol{\varphi} \rangle^v = \nabla \cdot \left(\phi \langle \boldsymbol{\varphi} \rangle^i \right) + \frac{1}{\Delta V} \int_{A_i} \boldsymbol{n} \cdot \boldsymbol{\varphi} \, dS, \tag{8.4}$$

$$\left\langle \frac{\partial \varphi}{\partial t} \right\rangle^v = \frac{\partial}{\partial t} \left(\phi \langle \varphi \rangle^i \right) - \frac{1}{\Delta V} \int_{A_i} \boldsymbol{n} \cdot (\boldsymbol{u}_i \varphi) \, dS, \tag{8.5}$$

where A_i and $\boldsymbol{u}_i$ are the interfacial area and velocity of phase f and n is the unity vector normal to A_i. The area A_i should not be confused with the surface area surrounding volume ΔV in Figure 8.1. For single-phase flow, phase f is the fluid itself and $\boldsymbol{u}_i = 0$ if the porous substrate is assumed to be fixed. In developing equations (8.3) – (8.5), the only restriction applied is the independence of ΔV in relation to time and space. If the medium is further assumed rigid then ΔV_f is dependent only on space.

8.4.2 Time averaging

The need for considering time fluctuations occurs when turbulence effects are of concern. The microscopic time-averaged equations are obtained from the instantaneous microscopic equations. For that, the time-average value of property φ, associated with the fluid, is given by

$$\overline{\varphi} = \frac{1}{\Delta t} \int_t^{t+\Delta t} \varphi \, dt, \tag{8.6}$$

where Δt is the integration time interval. The instantaneous property φ can be defined as the sum of the time average, $\overline{\varphi}$, plus the fluctuating component, φ', and so is given by

$$\varphi = \overline{\varphi} + \varphi'. \tag{8.7}$$

Hence, $\overline{\varphi'} = 0$.

8.4.3 Commutative properties

From the definition of volume average, equation (8.1), and time average, equation (8.6), one can conclude that the time average of the volume average of property φ is given by

$$\overline{\langle \varphi \rangle^v} = \frac{1}{\Delta t} \int_t^{t+\Delta t} \left[\frac{1}{\Delta V} \int_{\Delta V_f} \varphi \, dV \right] dt. \tag{8.8}$$

On the other hand, the volume average of the time average is given by

$$\langle \overline{\varphi} \rangle^v = \frac{1}{\Delta V} \int_{\Delta V_f} \left[\frac{1}{\Delta t} \int_t^{t+\Delta t} \varphi \, dt \right] dV. \tag{8.9}$$

As mentioned, for a rigid medium, the volume of fluid, ΔV_f, is dependent only on space. If the time interval chosen for temporal averaging, Δt, is the same for all REVs, then the volumetric average commutes with time average because both integration domains in equations (8.8) and (8.9) are independent of each other. In this case, the order of application of the average operators is immaterial, and equations (8.8) and (8.9) lead to

$$\overline{\langle \varphi \rangle^v} = \langle \overline{\varphi} \rangle^v \quad \text{or} \quad \overline{\langle \varphi \rangle^i} = \langle \overline{\varphi} \rangle^i . \tag{8.10}$$

8.4.4 Double decomposition—space and time fluctuations

Figure 8.1 shows that for any point located at a certain position x, surrounded by a volume ΔV, a volume-average can be defined. This value will be different depending on the selected volume ΔV. Also, for this very same entity (point x), a time-average can be defined, according to equation (8.6), being dependent only on the time interval Δt. Further, from equations (8.1) and (8.7), we have

$$\langle \varphi \rangle^i = \frac{1}{\Delta V_f} \int_{\Delta V_f} \varphi \, dV = \frac{1}{\Delta V_f} \int_{\Delta V_f} (\overline{\varphi} + \varphi') \, dV = \langle \overline{\varphi} \rangle^i + \langle \varphi' \rangle^i , \tag{8.11}$$

and combining equations (8.2) and (8.6), we obtain

$$\overline{\varphi} = \frac{1}{\Delta t} \int_t^{t+\Delta t} \varphi \, dt = \frac{1}{\Delta t} \int_t^{t+\Delta t} \left(\langle \varphi \rangle^i + {}^i\varphi \right) dt = \overline{\langle \varphi \rangle^i} + \overline{{}^i\varphi}. \tag{8.12}$$

Further, the space-averaged value $\langle \varphi \rangle^i$ can also be decomposed into a time-mean and fluctuating component as follows:

$$\langle \varphi \rangle^i = \overline{\langle \varphi \rangle^i} + \langle \varphi \rangle^{i'} . \tag{8.13}$$

Using now the fact that the averages commute, equation (8.10), a comparison of equations (8.11) and (8.13) validates the following relationship:

$$\langle \varphi' \rangle^i = \langle \varphi \rangle^{i'}. \tag{8.14}$$

Equation (8.14) means that *the volume average of the time varying component of a single quantity is equal to the time fluctuation of the volume average of the same quantity.* Similarly, if we consider the time average component having also a spatial distribution, then

$$\overline{\varphi} = \langle \overline{\varphi} \rangle^i + {}^i\overline{\varphi}. \tag{8.15}$$

Likewise, comparing equations (8.12) and (8.15), in the light of equation (8.10),

$$ {}^i\overline{\varphi} = \overline{{}^i\varphi}, \tag{8.16}$$

or say, *the spatial deviation of the time-average quantity is equal to the time average of the spatial deviation.*

Further, since both time and space decompositions are based on the same value for φ, one can promptly write

$$\varphi = \overline{\varphi} + \varphi' = \langle \varphi \rangle^i + {}^i\varphi. \tag{8.17}$$

Space averaging the second and third terms of equation (8.17), and time averaging the fourth and fifth terms, we obtain

$$\varphi = \underbrace{\langle \overline{\varphi} \rangle^i + \langle \varphi' \rangle^i}_{\langle \varphi \rangle^i} + \overbrace{{}^i\overline{\varphi} + {}^i(\varphi')}^{{}^i\varphi} = \underbrace{\overline{\langle \varphi \rangle^i} + \overline{{}^i\varphi}}_{\overline{\varphi}} + \overbrace{\langle \varphi \rangle^{i'} + \left({}^i\varphi \right)'}^{\varphi'}. \tag{8.18}$$

Equation (8.18) presents two possible equivalent ways to double-decompose φ. It is interesting to note the meaning of the fifth and ninth terms in equation (8.18). The term ${}^i(\varphi')$ is *the spatial component of the time varying term* whereas $\left({}^i\varphi \right)'$ is *the time fluctuation of the spatial component.* However, using equations (8.10), (8.14), and (8.16), equation (8.18) can be simplified to

$$ {}^i(\varphi') = \left({}^i\varphi \right)', \tag{8.19}$$

and, for simplicity of notation, one can write both superscripts at the same level in the format ${}^i\varphi'$. Taking now the time average of the fluctuating component, written as

$$\varphi' = \langle \varphi \rangle^{i'} + {}^i\varphi' = \langle \varphi' \rangle^i + {}^i\varphi', \tag{8.20}$$

yields $\overline{{}^i\varphi'} = 0$. Likewise, volume averaging the spatial component, written as

$$ {}^i\varphi = {}^i\overline{\varphi} + {}^i\varphi' = \overline{{}^i\varphi} + {}^i\varphi', \tag{8.21}$$

results in $\langle {}^i\varphi' \rangle^i = 0$.

With these relationships in mind, integration of local (microscopic) flow governing equations applied to the domain of Figure 8.1 can be more easily treated by using the double-decomposition approach.

Figure 8.2, see Rocamora, Jr and de Lemos (2000a), is helpful in understanding the double-decomposition concept. The figure shows a three-dimensional diagram for a general vector variable φ. For a scalar, all the quantities shown would be drawn on a single line. Also, notice that the points B, C, D and E fall in the same plane, with segments BC and BE parallel to ED and CD, respectively. The path ACF represents the standard space-decomposition given by equation (8.2) whereas equation (8.7) is pictured by the path AEF. Further, equation (8.11) is represented by the path ABC and equation (8.15) by path ABE. Triangles EDF and CDF are associated with equations (8.20) and (8.21), respectively. Equation (8.10) is represented by segment AB and the two equations (8.14) and (8.16) by the equivalence between the parallel segments BC and ED and between segments BE and CD, respectively. Finally, equation (8.18) follows the sequence $ABCDF$, or the path $ABEDF$, both of them decomposing the same general variable φ.

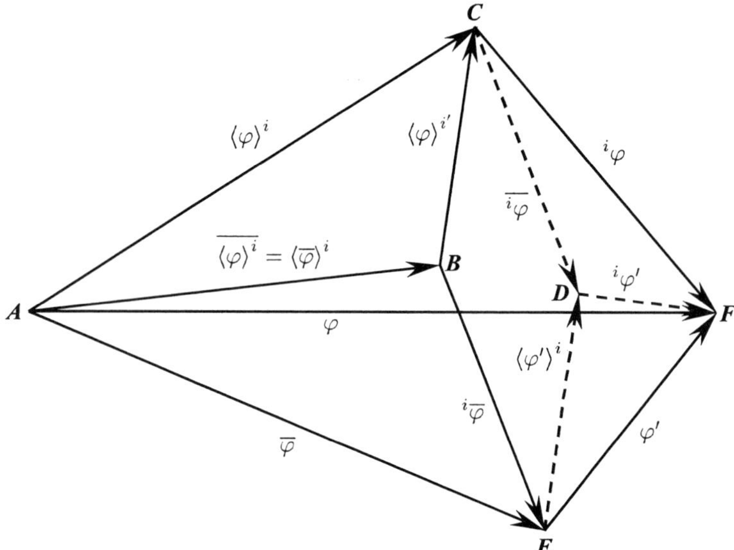

Figure 8.2 *General three-dimensional vector diagram for a quantity φ, see Rocamora, Jr and de Lemos (2000a)*

8.5 TRANSPORT EQUATIONS

In deriving the time- and space-averaged transport equations for modeling turbulence in porous media, we consider incompressible single-phase flow in a saturated, rigid porous medium. The microscopic continuity equation for a fluid flowing in a clear (of porous medium) domain is given by

$$\nabla \cdot \boldsymbol{u} = 0. \tag{8.22}$$

Recalling that the fluid speed at the solid–fluid interface is zero, equation (8.4) can be used to take the volume-average of equation (8.22). Then, taking the time-average, equation (8.6), of the resulting equation (recall that the time-integral commutes with the divergent operator) leads to

$$\nabla \cdot \left(\overline{\langle u \rangle^{i}} \right) = 0. \tag{8.23}$$

One can now take the time-average of equation (8.22), using equation (8.6), and then take the volume-average of the resulting equation using equation (8.4). The result is

$$\nabla \cdot \left(\langle \overline{u} \rangle^{i} \right) = 0, \tag{8.24}$$

which is exactly the same as equation (8.23), according to equation (8.10). Hence, the averaging order is immaterial for the continuity equation.

If we now expand the velocity field of equation (8.22), using the double-decomposition, equation (8.18), one obtains

$$\nabla \cdot \boldsymbol{u} = \nabla \cdot \left(\langle \overline{u} \rangle^{i} + \langle u' \rangle^{i} + {}^{i}\overline{u} + {}^{i}u' \right) = 0. \tag{8.25}$$

Comparing equation (8.25) and equation (8.24), the divergence of the sum of the last three terms within the parentheses in equation (8.25) must equal zero.

The microscopic momentum equation for a fluid with constant properties is given by the Navier–Stokes equation

$$\rho \left[\frac{\partial \boldsymbol{u}}{\partial t} + \nabla \cdot (\boldsymbol{u}\boldsymbol{u}) \right] = -\nabla p + \mu \nabla^{2}\boldsymbol{u} + \rho \boldsymbol{g}. \tag{8.26}$$

Taking the time-average using $\boldsymbol{u} = \overline{\boldsymbol{u}} + \boldsymbol{u}'$ gives

$$\rho \left[\frac{\partial \overline{\boldsymbol{u}}}{\partial t} + \nabla \cdot (\overline{\boldsymbol{u}}\,\overline{\boldsymbol{u}}) \right] = -\nabla \overline{p} + \mu \nabla^{2}\overline{\boldsymbol{u}} + \nabla \cdot \left(-\rho \overline{\boldsymbol{u}'\boldsymbol{u}'} \right) + \rho \boldsymbol{g}, \tag{8.27}$$

where the stresses $-\rho \overline{\boldsymbol{u}'\boldsymbol{u}'}$ are the well-known Reynolds stresses. On the other hand, the volumetric average of equation (8.26), using equations (8.3) – (8.5), results in

$$\rho \left[\frac{\partial}{\partial t} \left(\phi \langle u \rangle^{i} \right) + \nabla \cdot \left[\phi \langle uu \rangle^{i} \right] \right] = -\nabla \left(\phi \langle p \rangle^{i} \right) + \mu \nabla^{2} \left(\phi \langle u \rangle^{i} \right) + \phi \rho \boldsymbol{g} + \boldsymbol{R}, \tag{8.28}$$

where

$$R = \frac{\mu}{\Delta V} \int_{A_i} n\,(\nabla u)\,\mathrm{d}S - \frac{1}{\Delta V} \int_{A_i} np\,\mathrm{d}S \qquad (8.29)$$

represents the total-drag force per unit volume due to the presence of the porous matrix, being composed by both viscous-drag and form-drag. Further, using equation (8.2) to write $u = \langle u \rangle^i + {}^i u$ in the convective inertia term, one obtains

$$\rho \left[\frac{\partial}{\partial t} \left(\phi \langle u \rangle^i \right) + \nabla \cdot \left[\phi \langle u \rangle^i \langle u \rangle^i \right] \right]$$
$$= -\nabla \left(\phi \langle p \rangle^i \right) + \mu \nabla^2 \left(\phi \langle u \rangle^i \right) - \nabla \cdot \left[\phi \langle {}^i u {}^i u \rangle^i \right] + \phi \rho g + R. \qquad (8.30)$$

Hsu and Cheng (1990) pointed out that the third term on the right-hand side of equation (8.30), $\nabla \cdot \left(\phi \langle {}^i u {}^i u \rangle^i \right)$, represents the hydrodynamic dispersion due to spatial deviations. Note that equation (8.30), closed by replacing R with the viscous-drag (Darcy) term and the form-drag (Forchheimer) term, models typical porous media flow for $Re_p < 150$–200. However, when extending the analysis to turbulent flow the time varying quantities have to be considered.

The set of equations (8.27) and (8.30) are used when treating turbulent flow in clear fluid or low Re_p porous media flow, respectively. Each one of those equations was derived by applying only one averaging operator, either time or volume, respectively, into the microscopic equation. For modeling turbulent flow in porous media, it is necessary to apply both operators. Hence, the volume average of equation (8.27), giving for the time-mean flow in a porous medium, is given by

$$\rho \left[\frac{\partial}{\partial t} \left(\phi \langle \overline{u} \rangle^i \right) + \nabla \cdot \left(\phi \langle \overline{u}\,\overline{u} \rangle^i \right) \right]$$
$$= -\nabla \left(\phi \langle \overline{p} \rangle^i \right) + \mu \nabla^2 \left(\phi \langle \overline{u} \rangle^i \right) + \nabla \cdot \left(-\rho \phi \langle \overline{u'u'} \rangle^i \right) + \phi \rho g + \overline{R}, \qquad (8.31)$$

where

$$R = \frac{\mu}{\Delta V} \int_{A_i} n\,(\nabla \overline{u})\,\mathrm{d}S - \frac{1}{\Delta V} \int_{A_i} n\overline{p}\,\mathrm{d}S \qquad (8.32)$$

is the time-averaged total-drag force per unit volume, due to the solid matrix, composed of both, viscous-drag and form-drag terms.

Likewise, applying now the time average operation to equation (8.28), one obtains

$$\rho \left[\frac{\partial}{\partial t} \overline{\left(\phi \langle \overline{u} + u' \rangle^i \right)} + \nabla \cdot \overline{\left(\phi \langle (\overline{u} + u')(\overline{u} + u') \rangle^i \right)} \right]$$
$$= -\nabla \overline{\left(\phi \langle \overline{p} + p' \rangle^i \right)} + \mu \nabla^2 \overline{\left(\phi \langle \overline{u} + u' \rangle^i \right)} + \phi \rho g + \overline{R}. \qquad (8.33)$$

Dropping terms containing only one fluctuating quantity results in

$$\rho\left[\frac{\partial}{\partial t}\left(\phi\langle\overline{u}\rangle^i\right)+\nabla\cdot\left(\phi\langle\overline{u}\,\overline{u}\rangle^i\right)\right]$$
$$=-\nabla\left(\phi\langle\overline{p}\rangle^i\right)+\mu\nabla^2\left(\phi\langle\overline{u}\rangle^i\right)+\nabla\cdot\left(-\rho\phi\langle\overline{u'u'}\rangle^i\right)+\phi\rho g+\overline{R}, \quad (8.34)$$

where

$$\overline{R}=\frac{\mu}{\Delta V}\int_{A_i}n\left[\nabla(\overline{u}+u')\right]\mathrm{d}S-\frac{1}{\Delta V}\int_{A_i}n(\overline{p}+p')\,\mathrm{d}S$$
$$=\frac{\mu}{\Delta V}\int_{A_i}n\left(\nabla\overline{u}\right)\mathrm{d}S-\frac{1}{\Delta V}\int_{A_i}n\overline{p}\,\mathrm{d}S. \qquad (8.35)$$

Comparing equations (8.31) and (8.34) one can see that also for the momentum equation the order of the application of both averaging operators is immaterial as long as the total-drag term is not closed after the volume averaging operation. Pedras and de Lemos (2000a) propose that the term $\overline{R}$, appearing in both equations (8.31) and (8.34) (compare equations (8.32) and (8.35)) be modeled by the Hazen–Dupuit–Darcy (HDD) model extension (also known as the Darcy–Forchheimer model). This approach is consistent with the fact that the HDD model seems to correlate experimental data well even when the flow within the pores is turbulent. In other words, the HDD model results from the time and space averaging of the turbulent flow.

Consider now the convective inertia term and the Reynolds stress component of equation (8.31). These two terms can be decomposed using equation (8.15) as follows:

$$\nabla\cdot\left[\phi\left(\langle\overline{u}\,\overline{u}\rangle^i+\langle\overline{u'u'}\rangle^i\right)\right]=\nabla\cdot\left\{\phi\left[\left\langle\left(\langle\overline{u}\rangle^i+{}^i\overline{u}\right)\left(\langle\overline{u}\rangle^i+{}^i\overline{u}\right)\right\rangle^i+\langle\overline{u'u'}\rangle^i\right]\right\}$$
$$=\nabla\cdot\left\{\phi\left[\langle\overline{u}\rangle^i\langle\overline{u}\rangle^i+\langle{}^i\overline{u}\,{}^i\overline{u}\rangle^i+\langle\overline{u'u'}\rangle^i\right]\right\}. $$
$$(8.36)$$

Now, invoking equation (8.20) to write $u'=\langle u'\rangle^i+{}^iu'$, the term to the right of the equal sign in equation (8.36) becomes

$$\nabla\cdot\left\{\phi\left[\langle\overline{u}\rangle^i\langle\overline{u}\rangle^i+\langle{}^i\overline{u}\,{}^i\overline{u}\rangle^i+\langle\overline{u'u'}\rangle^i\right]\right\}$$
$$=\nabla\cdot\left\{\phi\left[\langle\overline{u}\rangle^i\langle\overline{u}\rangle^i+\langle{}^i\overline{u}\,{}^i\overline{u}\rangle^i+\left\langle\overline{\left(\langle u'\rangle^i+{}^iu'\right)\left(\langle u'\rangle^i+{}^iu'\right)}\right\rangle^i\right]\right\}$$
$$=\nabla\cdot\left\{\phi\left[\langle\overline{u}\rangle^i\langle\overline{u}\rangle^i+\langle{}^i\overline{u}\,{}^i\overline{u}\rangle^i+\overline{\langle u'\rangle^i\langle u'\rangle^i}\right.\right.$$
$$\left.\left.+\overline{\langle\langle u'\rangle^i\,{}^iu'\rangle}^i+\overline{\langle{}^iu'\langle u'\rangle^i\rangle}^i+\langle\overline{{}^iu'\,{}^iu'}\rangle^i\right]\right\}. $$
$$(8.37)$$

The fourth and fifth terms on the right-hand side of equation (8.37) are zero because they contain the volume average of one space varying quantity. Equation (8.36) then reduces to

$$
\begin{aligned}
\nabla \cdot \overline{\left(\phi \langle uu \rangle^i \right)} &= \nabla \cdot \left[\phi \left(\langle \overline{u}\,\overline{u} \rangle^i + \overline{\langle u'u' \rangle^i} \right) \right] \\
&= \nabla \cdot \left\{ \phi \left[\langle \overline{u} \rangle^i \langle \overline{u} \rangle^i + \overline{\langle u' \rangle^i \langle u' \rangle^i} + \langle {}^i\overline{u}\,{}^i\overline{u} \rangle^i + \overline{\langle {}^iu'\,{}^iu' \rangle}^i \right] \right\}.
\end{aligned}
\tag{8.38}
$$

Using the equivalence equations (8.10), (8.14), and (8.16), equation (8.38) can be further simplified into

$$
\nabla \cdot \overline{\left(\phi \langle uu \rangle^i \right)} = \nabla \cdot \left\{ \phi \left[\overline{\langle u \rangle^i}\,\overline{\langle u \rangle^i} + \overline{\langle u \rangle^{i'}}\,\overline{\langle u \rangle^{i'}} + \langle {}^i\overline{u}\,{}^i\overline{u} \rangle^i + \langle \overline{{}^iu'\,{}^iu'} \rangle^i \right] \right\}.
\tag{8.39}
$$

Another route to follow is to start with the application of the space decomposition in the convective inertia term, as usually done in classical mathematical treatment of porous media flow analysis, and then follow with the time average. The result is as follows:

$$
\begin{aligned}
\nabla \cdot \overline{\left(\phi \langle uu \rangle^i \right)} &= \nabla \cdot \overline{\left(\phi \left\langle \left(\langle u \rangle^i + {}^iu \right) \left(\langle u \rangle^i + {}^iu \right) \right\rangle^i \right)} \\
&= \nabla \cdot \overline{\left[\phi \left(\langle u \rangle^i \langle u \rangle^i + \langle {}^iu\,{}^iu \rangle^i \right) \right]}.
\end{aligned}
\tag{8.40}
$$

The time average of the right-hand side of equation (8.40), using equation (8.11) to express $\langle u \rangle^i = \langle \overline{u} \rangle^i + \langle u' \rangle^i$, is given by

$$
\begin{aligned}
\nabla \cdot \overline{\left[\phi \left(\langle u \rangle^i \langle u \rangle^i + \langle {}^iu\,{}^iu \rangle^i \right) \right]} &\\
= \nabla \cdot \left\{ \phi \left[\overline{\left(\langle \overline{u} \rangle^i + \langle u' \rangle^i \right) \left(\langle \overline{u} \rangle^i + \langle u' \rangle^i \right) + \langle {}^iu\,{}^iu \rangle^i} \right] \right\} &\\
= \nabla \cdot \left\{ \phi \left[\langle \overline{u} \rangle^i \langle \overline{u} \rangle^i + \overline{\langle u' \rangle^i \langle u' \rangle^i} + \overline{\langle {}^iu\,{}^iu \rangle^i} \right] \right\}.
\end{aligned}
\tag{8.41}
$$

With the help of equation (8.21), one can write ${}^iu = {}^i\overline{u} + {}^iu'$, and equation (8.41) becomes

$$
\begin{aligned}
\nabla \cdot \left\{ \phi \left[\langle \overline{u} \rangle^i \langle \overline{u} \rangle^i + \overline{\langle u' \rangle^i \langle u' \rangle^i} + \overline{\langle {}^iu\,{}^iu \rangle^i} \right] \right\} &\\
= \nabla \cdot \left\{ \phi \left[\langle \overline{u} \rangle^i \langle \overline{u} \rangle^i + \overline{\langle u' \rangle^i \langle u' \rangle^i} + \overline{\langle \left({}^i\overline{u} + {}^iu' \right) \left({}^i\overline{u} + {}^iu' \right) \rangle^i} \right] \right\} &\\
= \nabla \cdot \left\{ \phi \left[\langle \overline{u} \rangle^i \langle \overline{u} \rangle^i + \overline{\langle u' \rangle^i \langle u' \rangle^i} + \langle \overline{{}^i\overline{u}\,{}^i\overline{u} + {}^i\overline{u}\,{}^iu' + {}^iu'\,{}^i\overline{u} + {}^iu'\,{}^iu'} \rangle^i \right] \right\}.
\end{aligned}
\tag{8.42}
$$

The fourth and fifth terms on the right-hand side of equation (8.42) are zero for containing the time average of one time fluctuating component. In addition, recalling the equivalence

$\langle u' \rangle^i = \langle u \rangle^{i'}$, equation (8.14), and using equation (8.10) to write $\overline{\langle u \rangle^i} = \langle \overline{u} \rangle^i$, and equation (8.16) for ${}^i\overline{u} = {}^i u$, equation (8.42) becomes

$$\nabla \cdot \overline{\left[\phi \left(\langle u \rangle^i \langle u \rangle^i + \langle {}^i u \, {}^i u \rangle^i \right) \right]}$$

$$= \nabla \cdot \left\{ \phi \left[\underbrace{\langle \overline{u} \rangle^i \langle \overline{u} \rangle^i}_{\text{I}} + \underbrace{\overline{\langle u' \rangle^i \langle u' \rangle^i}}_{\text{II}} + \underbrace{\langle {}^i\overline{u} \, {}^i\overline{u} \rangle^i}_{\text{III}} + \underbrace{\left\langle \overline{{}^i u' \, {}^i u'} \right\rangle^i}_{\text{IV}} \right] \right\}, \quad (8.43)$$

which is identical to equation (8.38). Therefore, the averaging order is also immaterial when expanding the convective inertia term. The physical significance of each term in equation (8.43) is:

  I convective term of macroscopic mean velocity;

  II turbulent (Reynolds) stresses divided by density ρ due to the fluctuating component of the macroscopic velocity;

  III dispersion associated with spatial fluctuations of microscopic time mean velocity, note that this term is also present in laminar flow, or say, when $Re_p < 150$; and,

  IV turbulent dispersion in a porous medium due to both time and spatial fluctuations of the microscopic velocity.

Consider now the Reynolds stresses appearing in equation (8.27). For clear flow, the use of the eddy-diffusivity concept for expressing the *stress–rate of strain* relationship for the Reynolds stress gives

$$-\rho \overline{u'u'} = \mu_t 2\overline{D} - \frac{2}{3}\rho k I, \quad (8.44)$$

where $\overline{D} = \left[\nabla \overline{u} + (\nabla \overline{u})^\top \right]/2$ is the mean deformation tensor, $k = \overline{u' \cdot u'}/2$ is the turbulent kinetic energy per unit mass and I is the unity tensor. Applying equation (8.44) in equation (8.27) results in

$$\rho \left[\frac{\partial \overline{u}}{\partial t} + \nabla \cdot (\overline{u}\,\overline{u}) \right] = -\nabla \left(\overline{p} + \frac{2}{3}\rho k \right) + \mu \nabla^2 \overline{u} + \nabla \cdot (\mu_t 2\overline{D}) + \rho g. \quad (8.45)$$

To obtain an equivalent expression for the macroscopic Reynolds stress tensor, the volume-averaging operator is applied in both equations (8.27) and (8.45). Making use of equations (8.3) – (8.5), the several terms in equations (8.27) and (8.45) become

$$\left\langle \frac{\partial \overline{u}}{\partial t} \right\rangle^v = \frac{\partial}{\partial t} \left(\phi \langle \overline{u} \rangle^i \right), \quad (8.46)$$

$$\langle \nabla \cdot (\overline{u}\,\overline{u}) \rangle^v = \nabla \cdot \left(\phi \langle \overline{u}\,\overline{u} \rangle^i \right) + \frac{1}{\Delta V} \int_{A_i} n \cdot (\overline{u}\,\overline{u}) \, \mathrm{d}S, \quad (8.47)$$

$$\langle \nabla \overline{p} \rangle^v = \nabla \left(\phi \langle \overline{p} \rangle^i \right) + \frac{1}{\Delta V} \int_{A_i} n\overline{p} \, \mathrm{d}S, \quad (8.48)$$

8.5.1 Equations for the fluctuating velocity

The N–K model applied the time–space sequence and obtained an equation for $\langle k \rangle^i$. Using the double-decomposition concept, equations for $\langle u' \rangle^i$ and k_m were presented by Pedras and de Lemos (2001a), where the authors compared similar terms in both proposals for k. The starting point for an equation for the turbulent flow kinetic energy is an equation for the microscopic velocity fluctuation u'. Such a relationship can be written after subtracting the equation for the mean velocity $\overline{u}$ from the instantaneous momentum equation, see Hinze (1959), and Warsi (1998), as follows:

$$\rho \left\{ \frac{\partial u'}{\partial t} + \nabla \cdot \left[\overline{u} u' + u' \overline{u} + u' u' - \overline{u' u'} \right] \right\} = -\nabla p' + \mu \nabla^2 u'. \tag{8.62}$$

Now, the volume average of equation (8.62) is given by

$$\rho \frac{\partial}{\partial t} \left(\phi \langle u' \rangle^i \right) + \rho \nabla \cdot \left\{ \phi \left[\langle \overline{u} u' \rangle^i + \langle u' \overline{u} \rangle^i + \langle u' u' \rangle^i - \langle \overline{u' u'} \rangle^i \right] \right\}$$
$$= -\nabla \left(\phi \langle p' \rangle^i \right) + \mu \nabla^2 \left(\phi \langle u' \rangle^i \right) + R', \tag{8.63}$$

where

$$R' = \frac{\mu}{\Delta V} \int_{A_i} n \left(\nabla u' \right) \mathrm{d}S - \frac{1}{\Delta V} \int_{A_i} n p' \, \mathrm{d}S \tag{8.64}$$

is the fluctuating part of the total-drag due to the porous structure.

Expanding further the divergence operators in equation (8.63) by means of equations (8.15) and (8.20), the following equation for $\langle u' \rangle^i$ is found:

$$\rho \frac{\partial}{\partial t} \left(\phi \langle u' \rangle^i \right) + \rho \nabla \cdot \left\{ \phi \left[\langle \overline{u} \rangle^i \langle u' \rangle^i + \langle u' \rangle^i \langle \overline{u} \rangle^i + \langle u' \rangle^i \langle u' \rangle^i \right. \right.$$
$$\left. \left. + \langle {}^i\overline{u} \, {}^i u' \rangle^i + \langle {}^i u' \, {}^i \overline{u} \rangle^i + \langle {}^i u' \, {}^i u' \rangle^i - \overline{\langle u' \rangle^i \langle u' \rangle^i} - \langle \overline{{}^i u' \, {}^i u'} \rangle^i \right] \right\}$$
$$= -\nabla \left(\phi \langle p' \rangle^i \right) + \mu \nabla^2 \left(\phi \langle u' \rangle^i \right) + R'. \tag{8.65}$$

Another route to follow in order to obtain the same equation is to start out with the macroscopic general momentum equation (8.30) and use of the double-decomposition concept given by equation (8.18). The result is given by

$$\rho \left\{ \frac{\partial}{\partial t} \left[\phi \left(\langle \overline{u} \rangle^i + \langle u' \rangle^i \right) \right] \right.$$
$$\left. + \nabla \cdot \left[\phi \left\langle \left[\langle \overline{u} \rangle^i + \langle u' \rangle^i + {}^i\overline{u} + {}^i u' \right] \left[\langle \overline{u} \rangle^i + \langle u' \rangle^i + {}^i\overline{u} + {}^i u' \right] \right\rangle^i \right] \right\}$$
$$= -\nabla \left[\phi \left(\langle \overline{p} \rangle^i + \langle p' \rangle^i \right) \right] + \mu \nabla^2 \left[\phi \left(\langle \overline{u} \rangle^i + \langle u' \rangle^i \right) \right] + \phi \rho g + R, \tag{8.66}$$

which can be simplified to

$$\rho \left\{ \frac{\partial}{\partial t} \left[\phi \left(\langle \overline{u} \rangle^i + \langle u' \rangle^i \right) \right] \right.$$
$$+ \nabla \cdot \left[\phi \left(\langle \overline{u} \rangle^i \langle \overline{u} \rangle^i + \langle \overline{u} \rangle^i \langle u' \rangle^i + \langle u' \rangle^i \langle \overline{u} \rangle^i + \langle u' \rangle^i \langle u' \rangle^i \right. \right.$$
$$\left. \left. \left. + \langle {}^i\overline{u}\, {}^i\overline{u} \rangle^i + \langle {}^i\overline{u}\, {}^iu' \rangle^i + \langle {}^iu'\, {}^i\overline{u} \rangle^i + \langle {}^iu'\, {}^iu' \rangle^i \right) \right] \right\}$$
$$= -\nabla \left[\phi \left(\langle \overline{p} \rangle^i + \langle p' \rangle^i \right) \right] + \mu \nabla^2 \left[\phi \left(\langle \overline{u} \rangle^i + \langle u' \rangle^i \right) \right] + \phi \rho g + R.$$

(8.67)

Taking now the time average of equation (8.67), one obtains

$$\rho \left\{ \frac{\partial}{\partial t} \left(\phi \langle \overline{u} \rangle^i \right) + \nabla \cdot \left\{ \phi \left[\langle \overline{u} \rangle^i \langle \overline{u} \rangle^i + \overline{\langle u' \rangle^i \langle u' \rangle^i} + \langle {}^i\overline{u}\, {}^i\overline{u} \rangle^i + \left\langle \overline{{}^iu'\, {}^iu'} \right\rangle^i \right] \right\} \right\}$$
$$= -\nabla \left(\phi \langle \overline{p} \rangle^i \right) + \mu \nabla^2 \left(\phi \langle \overline{u} \rangle^i \right) + \phi \rho g + \overline{R}, \quad (8.68)$$

where

$$\overline{R} = \frac{\mu}{\Delta V} \int_{A_i} n \left(\nabla \overline{u} \right) \mathrm{d}S - \frac{1}{\Delta V} \int_{A_i} n \overline{p}\, \mathrm{d}S \tag{8.69}$$

represents the time-averaged value of the instantaneous total-drag given by equation (8.29).

An equation for the fluctuating macroscopic velocity is then obtained by subtracting equation (8.68) from (8.67) and one obtains

$$\rho \frac{\partial}{\partial t} \left(\phi \langle u' \rangle^i \right) + \rho \nabla \cdot \left\{ \phi \left[\langle \overline{u} \rangle^i \langle u' \rangle^i + \langle u' \rangle^i \langle \overline{u} \rangle^i + \langle u' \rangle^i \langle u' \rangle^i \right. \right.$$
$$\left. \left. + \langle {}^i\overline{u}\, {}^iu' \rangle^i + \langle {}^iu'\, {}^i\overline{u} \rangle^i + \langle {}^iu'\, {}^iu' \rangle^i - \overline{\langle u' \rangle^i \langle u' \rangle^i} - \left\langle \overline{{}^iu'\, {}^iu'} \right\rangle^i \right] \right\}$$
$$= -\nabla \left(\phi \langle p' \rangle^i \right) + \mu \nabla^2 \left(\phi \langle u' \rangle^i \right) + R'.$$

(8.70)

Here R' is also given by equation (8.64). Comparing equation (8.70) with equation (8.65) we observe that these two equations are identical.

8.5.2 Equation for turbulence kinetic energy

As mentioned previously, the determination of the flow macroscopic turbulent kinetic energy follows two different paths in the literature. In the A–L method (space–time averaging sequence) the turbulence kinetic energy is defined as $k_m = \langle u' \rangle^i \cdot \langle u' \rangle^i / 2$. On the other hand, the N–K method (time–space averaging sequence) defines the turbulence kinetic energy as $\langle k \rangle^i = \langle \overline{u' \cdot u'} \rangle^i /2$. The objective of this section is to derive the transport equation for both, k_m and $\langle k \rangle^i$, and compare the results.

From the instantaneous microscopic continuity equation, equation (8.24), one has

$$\nabla \cdot \left(\phi \langle u' \rangle^i \right) = 0. \tag{8.71}$$

Taking the scalar product of equation (8.63) and $\langle u' \rangle^i$, using equation (8.71) and time averaging the result, an equation for k_m is developed, having the following terms:

$$\overline{\rho \langle u' \rangle^i \cdot \frac{\partial}{\partial t} \left(\phi \langle u' \rangle^i \right)} = \rho \frac{\partial \left(\phi k_m \right)}{\partial t}, \tag{8.72}$$

$$\overline{\rho \langle u' \rangle^i \cdot \left\{ \nabla \cdot \left(\phi \langle u'u' \rangle^i \right) \right\}}$$
$$= \rho \langle u' \rangle^i \cdot \overline{\left\{ \nabla \cdot \left[\phi \langle \overline{u} \rangle^i \langle u' \rangle^i + \phi \langle {}^i\overline{u} \, {}^iu' \rangle^i \right] \right\}} \tag{8.73}$$
$$= \rho \nabla \cdot \left[\phi \langle \overline{u} \rangle^i k_m \right] + \overline{\rho \langle u' \rangle^i \cdot \left\{ \nabla \cdot \left[\phi \langle {}^i\overline{u} \, {}^iu' \rangle^i \right] \right\}},$$

$$\overline{\rho \langle u' \rangle^i \cdot \left\{ \nabla \cdot \left(\phi \langle u'u' \rangle^i \right) \right\}}$$
$$= \rho \langle u' \rangle^i \cdot \overline{\left\{ \nabla \cdot \left[\phi \langle u' \rangle^i \langle \overline{u} \rangle^i + \phi \langle {}^iu' \, {}^i\overline{u} \rangle^i \right] \right\}} \tag{8.74}$$
$$= \rho \phi \overline{\langle u' \rangle^i \langle u' \rangle^i} : \nabla \langle \overline{u} \rangle^i + \rho \langle u' \rangle^i \cdot \overline{\left\{ \nabla \cdot \left[\phi \langle {}^iu' \, {}^i\overline{u} \rangle^i \right] \right\}},$$

$$\overline{\rho \langle u' \rangle^i \cdot \left\{ \nabla \cdot \left(\phi \langle u'u' \rangle^i \right) \right\}}$$
$$= \rho \langle u' \rangle^i \cdot \overline{\left\{ \nabla \cdot \left[\phi \langle u' \rangle^i \langle u' \rangle^i + \phi \langle {}^iu' \, {}^iu' \rangle^i \right] \right\}} \tag{8.75}$$
$$= \rho \nabla \cdot \left[\phi \langle u' \rangle^i \frac{\langle u' \rangle^i \cdot \langle u' \rangle^i}{2} \right] + \overline{\rho \langle u' \rangle^i \cdot \left\{ \nabla \cdot \left[\phi \langle {}^iu' \, {}^iu' \rangle^i \right] \right\}},$$

$$\overline{\rho \langle u' \rangle^i \cdot \left\{ \nabla \cdot \left(-\phi \langle \overline{u'u'} \rangle^i \right) \right\}} = 0, \tag{8.76}$$

$$\overline{-\langle u' \rangle^i \cdot \nabla \left(\phi \langle p' \rangle^i \right)} = -\nabla \cdot \overline{\left[\phi \langle u' \rangle^i \langle p' \rangle^i \right]}, \tag{8.77}$$

$$\overline{\mu \langle u' \rangle^i \cdot \nabla^2 \left(\phi \langle u' \rangle^i \right)} = \mu \nabla^2 \left(\phi k_m \right) - \rho \phi \varepsilon_m, \tag{8.78}$$

$$\overline{\langle u' \rangle^i \cdot R'} \equiv 0, \tag{8.79}$$

where $\varepsilon_m = \nu \overline{\nabla \langle u' \rangle^i : \left(\nabla \langle u' \rangle^i \right)^\top}$. In handling equation (8.77) the porosity ϕ was assumed to be constant only for simplifying the manipulation to be shown next. However, this procedure does not represent a limitation in deriving a general transport equation for k_m since equation (8.77) requires further modeling.

Another important point is the treatment given to the scalar product shown in equation (8.79). Here, a different view from the work in Lee and Howell (1987), Wang and Takle (1995), Antohe and Lage (1997), and Getachew *et al.* (2000), is considered. The fluctuating drag form $\boldsymbol{R}'$ acts through the solid–fluid interfacial area and, as such, on fluid particles at rest. The fluctuating mechanical energy represented by the operation in equation (8.79) is not associated with any fluid particle movement and, therefore, is here considered to be zero.

A final equation for k_m gives

$$
\rho \frac{\partial (\phi k_m)}{\partial t} + \rho \nabla \cdot \left[\phi \langle \overline{\boldsymbol{u}} \rangle^i k_m \right]
$$
$$
= -\rho \nabla \cdot \overline{\left\{ \phi \langle \boldsymbol{u}' \rangle^i \left[\frac{\langle p' \rangle^i}{\rho} + \frac{\langle \boldsymbol{u}' \rangle^i \cdot \langle \boldsymbol{u}' \rangle^i}{2} \right] \right\}} \tag{8.80}
$$
$$
+ \mu \nabla^2 (\phi k_m) - \rho \phi \overline{\langle \boldsymbol{u}' \rangle^i \langle \boldsymbol{u}' \rangle^i} : \nabla \langle \overline{\boldsymbol{u}} \rangle^i - \rho \phi \varepsilon_m - D_m,
$$

where

$$
D_m = \rho \overline{\langle \boldsymbol{u}' \rangle^i \cdot \left\{ \nabla \cdot \left[\phi \left(\langle {}^i\overline{\boldsymbol{u}}\,{}^i\boldsymbol{u}' \rangle^i + \langle {}^i\boldsymbol{u}'\,{}^i\overline{\boldsymbol{u}} \rangle^i + \langle {}^i\boldsymbol{u}'\,{}^i\boldsymbol{u}' \rangle^i \right) \right] \right\}} \tag{8.81}
$$

represents the dispersion of k_m given by the last terms on the right-hand side of equations (8.73), (8.74) and (8.75). It is of interest to point out that this term can be either negative or positive.

The first term on the right-hand side of equation (8.80) represents the turbulent diffusion of k_m and is normally modeled via a diffusion-like expression resulting for the transport equation for k_m, see Antohe and Lage (1997) and Getachew *et al.* (2000), as follows:

$$
\rho \frac{\partial (\phi k_m)}{\partial t} + \rho \nabla \cdot \left[\phi \langle \overline{\boldsymbol{u}} \rangle^i k_m \right]
$$
$$
= \nabla \cdot \left[\mu + \frac{\mu_{t_m}}{\sigma_{k_m}} \nabla (\phi k_m) \right] + P_m - \rho \phi \varepsilon_m - D_m, \tag{8.82}
$$

where

$$
P_m = -\rho \phi \overline{\langle \boldsymbol{u}' \rangle^i \langle \boldsymbol{u}' \rangle^i} : \nabla \langle \overline{\boldsymbol{u}} \rangle^i \tag{8.83}
$$

is the production rate of k_m due to the gradients of the macroscopic time-mean velocity $\langle \overline{\boldsymbol{u}} \rangle^i$.

The A–L model uses the above equation for k_m considering $\boldsymbol{R}'$ as the HDD (Darcy–Forchheimer) model with macroscopic time-fluctuation velocities $\langle \boldsymbol{u}' \rangle^i$. The mode also does not include all dispersion terms that were here grouped into D_m equation (8.81). It should be noted that the averaging order in this case does matter.

The other procedure for composing the turbulence kinetic energy is to take the scalar product of equation (8.62) by the microscopic fluctuating velocity $\boldsymbol{u}'$. Then apply both time and volume averaging operators for obtaining an equation for $\langle k \rangle^i = \langle \overline{\boldsymbol{u}' \cdot \boldsymbol{u}'} \rangle^i / 2$. It is worth noting that in this case the order of application of both operations is immaterial since no additional mathematical operation is conducted in the averaging processes. Therefore, this is the same as applying the volume operator to an equation for the microscopic k.

The volume average of a transport equation for k has been carried out in detail by de Lemos and Pedras (2000a) and Pedras and de Lemos (2001a). Only the final resulting equation is presented here, namely

$$
\rho \left[\frac{\partial}{\partial t} \left(\phi \langle k \rangle^i \right) + \nabla \cdot \left(\overline{\boldsymbol{u}}_D \langle k \rangle^i \right) \right]
$$
$$
= \nabla \cdot \left[\left(\mu + \frac{\mu_{t_\phi}}{\sigma_k} \right) \nabla \left(\phi \langle k \rangle^i \right) \right] + P_i + G_i - \rho \phi \langle \varepsilon \rangle^i , \tag{8.84}
$$

where

$$
P_i = -\rho \langle \overline{\boldsymbol{u}'\boldsymbol{u}'} \rangle^i : \nabla \overline{\boldsymbol{u}}_D , \tag{8.85}
$$

$$
G_i = c_k \rho \phi \frac{\langle k \rangle^i |\overline{\boldsymbol{u}}_D|}{\sqrt{K}} \tag{8.86}
$$

are the production rate of $\langle k \rangle^i$ due to mean gradients of the seepage velocity and the generation rate of intrinsic k due the presence of the porous matrix.

A comparison between terms in the transport equation for k_m and $\langle k \rangle^i$ can now be conducted. Expanding the correlation forming the production term P_i by means of equation (8.2), a connection between the two generation rates can also be written as

$$
\begin{aligned}
P_i &= -\rho \langle \overline{\boldsymbol{u}'\boldsymbol{u}'} \rangle^i : \nabla \overline{\boldsymbol{u}}_D \\
&= -\rho \left(\langle \boldsymbol{u}' \rangle^i \langle \boldsymbol{u}' \rangle^i : \nabla \overline{\boldsymbol{u}}_D + \overline{\langle {}^i\boldsymbol{u}' \, {}^i\boldsymbol{u}' \rangle^i} : \nabla \overline{\boldsymbol{u}}_D \right) \\
&= P_m - \rho \overline{\langle {}^i\boldsymbol{u}' \, {}^i\boldsymbol{u}' \rangle^i} : \nabla \overline{\boldsymbol{u}}_D .
\end{aligned} \tag{8.87}
$$

One can note that all the production rate of k_m due to the mean flow constitutes only part of the general production rate responsible for maintaining the overall level of $\langle k \rangle^i$.

The dissipation rates also carry a similar correspondence if one expands

$$
\begin{aligned}
\langle \varepsilon \rangle^i &= \nu \left\langle \overline{\nabla \boldsymbol{u}' : (\nabla \boldsymbol{u}')^\top} \right\rangle^i \\
&= \nu \langle \nabla \boldsymbol{u}' \rangle^i : \left[\langle \nabla \boldsymbol{u}' \rangle^i \right]^\top + \nu \left\langle \overline{{}^i (\nabla \boldsymbol{u}') : {}^i (\nabla \boldsymbol{u}')^\top} \right\rangle^i \\
&= \frac{\nu}{\phi^2} \nabla \left(\phi \langle \boldsymbol{u}' \rangle^i \right) : \left[\nabla \left(\phi \langle \boldsymbol{u}' \rangle^i \right) \right]^\top + \nu \left\langle \overline{{}^i (\nabla \boldsymbol{u}') : {}^i (\nabla \boldsymbol{u}')^\top} \right\rangle^i .
\end{aligned} \tag{8.88}
$$

Considering further constant porosity,

$$\langle \varepsilon \rangle^i = \varepsilon_m + \nu \left\langle \overline{{}^i (\nabla \boldsymbol{u}') : {}^i (\nabla \boldsymbol{u}')^\top} \right\rangle^i . \tag{8.89}$$

This indicates that an additional dissipation rate is necessary to fully account for the energy decay process inside the REV.

8.6 MACROSCOPIC MODEL ADJUSTMENT

Pedras and de Lemos (2000b, 2001b, 2001c) performed numerical simulations on periodic REV cells, with the solid matrix represented by cylindrical, longitudinal and transversal elliptic rods. These simulations were used to calibrate the Pedras and de Lemos turbulence model. The microscopic (pore level) flow equations were numerically solved inside an elementary REV. In all cases, a periodic boundary condition was imposed along the axial direction. The Reynolds number Re_H based on the cell height H was varied from about 0.35 (creeping flow) to 1.2×10^6 (fully turbulent regime). A version of the k–ε model for low Re flow was also incorporated in the code, following the damping functions presented by Abe *et al.* (1997). The non-orthogonal grid was based on a generalized coordinate system, leading to a total of 150 by 100 irregular control volumes for the high Re model and 300 by 200 for the low Re cases. Samples of the numerical grid are shown in Figure 8.3. For $Re_H = 1.2 \times 10^5$, both k–ε models were calculated for comparison.

The numerical method SIMPLE was employed for relaxing the mean and turbulence equations within the domain. The dimensions of the periodic cell for the cases considered were $H = 0.1$ m, $S = 2H$, $D = 0.03$ m ($\phi = 0.8$), 0.05 m ($\phi = 0.6$) and 0.06 m ($\phi = 0.4$). The solutions were grid independent and all normalized residuals were reduced to 10^{-5}. Also, relaxation parameters for all variables were kept equal to 0.8. A summary of all relevant parameters for the circular rod case is presented in Table 8.1. The constant c_k introduced in the equation for $\langle k \rangle^i$, via equation (8.86), was determined for closure of the macroscopic mathematical model proposed by Pedras and de Lemos (2001a). In that work, a methodology was devised in order to obtain such value. Accordingly, the need of computing the fine flow properties in order to obtain the volume-integrated quantities has motivated the development of adequate numerical tools. As mentioned, those calculations were needed for adjusting the model and considered either the high Re k–ε closure, Rocamora, Jr and de Lemos (1998), as well as the low Reynolds version, Pedras and de Lemos (2001b). Heat transfer analysis was also the subject of additional research, Rocamora, Jr and de Lemos (1999). One of the outcomes of this development was the ability to treat hybrid computational domains with a single numerical tool, Rocamora, Jr and de Lemos (2000b, 2000c).

For macroscopic fully developed uni-dimensional flow in isotropic and homogeneous media the limiting values for $\langle k \rangle^i$ and $\langle \varepsilon \rangle^i$ in the additional terms introduced in equation (8.84) and its accompanying equation for $\langle \varepsilon \rangle^i$ (not shown here) are given the values k_ϕ and ε_ϕ, respectively. In this limiting condition, the transport equations for $\langle k \rangle^i$ and $\langle \varepsilon \rangle^i$

Rocamora, Jr, F. D. and de Lemos, M. J. S. (1998). Numerical solution of turbulent flow in porous media using a spatially periodic array and the k–ε model. In *Proceedings of the ENCIT98—7th Brazilian Congress Thermal Sciences*, Rio de Janeiro, Brazil, Vol. 2, pp. 1265–1271.

Rocamora, Jr, F. D. and de Lemos, M. J. S. (1999). Simulation of turbulent heat transfer in porous media using a spatially periodic cell and the k–ε model. In *Proceedings of the COBEM99—15th Brazilian Congress of Mechanical Engineering*, São Paulo, Brazil. (on CD-ROM).

Rocamora, Jr, F. D. and de Lemos, M. J. S. (2000a). Analysis of convective heat transfer for turbulent flow in saturated porous media. *Int. Comm. Heat Mass Transfer* **27**, 825–834.

Rocamora, Jr, F. D. and de Lemos, M. J. S. (2000b). Prediction of velocity and temperature profiles for hybrid porous medium-clean fluid domains. In *Proceedings CONEM2000—National Mechanical Engineering Congress*, Natal, Brazil. (on CD-ROM).

Rocamora, Jr, F. D. and de Lemos, M. J. S. (2000c). Laminar recirculating flow and heat transfer in hybrid porous medium-clear fluid computational domains. In *Proceedings of 34th ASME—National Heat Transfer Conference*, Pittsburgh, Pennsylvania. **ASME-HTD-I463CD**, paper NHTC2000-12317 (on CD-ROM).

Rocamora, Jr, F. D. and de Lemos, M. J. S. (2000d). Heat transfer in suddenly expanded flow in a channel with porous inserts. In *Proceedings of the IMECE2000 ASME International Mechanical Engineering Congress and Exposition*, Orlando, Florida. **ASME-HTD-366-5**, pp. 191–195.

Slattery, J. C. (1967). Flow of viscoelastic fluids through porous media. *Amer. Inst. Chem. Eng. J.* **13**, 1066–1071.

Takatsu, Y. and Masuoka, T. (1998). Turbulent phenomena in flow through porous media. *J. Porous Media* **1**, 243–251.

Travkin, V. S. and Catton, I. (1994). Turbulent transfer of momentum, heat and mass in a two-level highly porous medium. In *Heat Transfer, 1994: Proceedings International Heat Transfer Conference*, Brighton, UK, Vol. 6, pp. 399–404. Institute Chemical Engineers, Rugby, UK.

Travkin, V. S. and Catton, I. (1995). A two-temperature model for turbulent flow and heat transfer in a porous layer. *ASME J. Fluids Eng.* **117**, 181–188.

Travkin, V. S. and Catton, I. (1998). Porous media transport descriptions—non-local, linear and nonlinear against effective thermal/fluid properties. *Adv. Colloid Interface Sci.* **77**, 389–443.

Travkin, V. S., Hu, K., and Catton, I. (1999). Turbulent kinetic energy and dissipation rate equation models for momentum transport in porous media. In *Proceedings of 3rd ASME/JSME Joint Fluids Engineering Conference*, San Francisco, California. **FEDSM99-7275**, pp. 1–7.

Wang, H. and Takle, E. S. (1995). Boundary-layer flow and turbulence near porous obstacles. *Boundary-Layer Meteorol.* **74**, 73–88.

Warsi, Z. U. A. (1998). *Fluid Dynamics—Theoretical and Computational Approaches* (2nd edn). CRC Press, Boca Raton.

Whitaker, S. (1969). Advances in theory of fluid motion in porous media. *Ind. Eng. Chem.* **61**, 14–28.

Whitaker, S. (1999). *The Method of Volume Averaging*. Kluwer, Dordrecht.

9 TURBULENCE CHARACTERISTICS IN POROUS MEDIA

T. MASUOKA[*] and Y. TAKATSU[†]

[*]Department of Mechanical Engineering Science, Kyushu University, 6-10-1, Hakozaki, Higashi-ku, Fukuoka 812-8581, Japan

email: `masuoka@mech.kyushu-u.ac.jp`

[†]Department of Mechanical Engineering, Hiroshima Kokusai Gakuin University, 6-20-1, Nakano, Aki-ku, Hiroshima 739-0321, Japan

email: `takatsu@m.hkg.ac.jp`

Abstract

We experimentally and theoretically investigate the turbulence characteristics in porous media. Experiments are made on the flow through a bank of tubes in a narrow gap to examine the microscopic turbulent field in porous media and we recognize that the flow through porous media becomes turbulent at high Reynolds number and that flow distortions due to the obstructions in the solid matrix produces additional mixing to that of the interstitial vortex. Then we introduce the concept of the eddy diffusivity which is characterized by the pseudo vortex and the void vortex, and we propose the macroscopic momentum and energy equations for the turbulent flow through porous media The present 0-equation model describes well the Forchheimer flow resistance and the thermal dispersion, and we clarify that the void vortex mainly contributes to the Forchheimer flow resistance and that the pseudo vortex mainly contributes to the thermal dispersion. Furthermore, we discuss the mechanism of the production and thermal dissipation of the turbulence in porous media and we estimate Kolmogorov's micro-length scale.

Keywords: porous media, forced convection, turbulent flow, dispersion, Forchheimer flow resistance, turbulence model, flow visualization

9.1 INTRODUCTION

For the flow through porous media, we know that the Forchheimer flow resistance and dispersion occurs at high Reynolds number. However, most of the previous theoretical studies have been based on laminar flow theory. For example, the Forchheimer flow media. However, the non-dimensional pressure drop for the present data is lower than that for packed beds and this is attributed to the enhancement of the turbulence by the three-dimensional complicated structures of the solid matrix in the flow through packed beds.

Figure 9.3 indicates the time history of the output signal from the velocity anemometer and we examine the flow condition from the raw output signal. The time response of the velocity remains flat at $Re = 5 \times 10$ ($Re_d = 10^2$) in Figure 9.3(a) and the flow is laminar. The fluctuations appear at about $Re = 5 \times 10^2$ ($Re_d = 10^3$), see Figure 9.3(b), and the fluctuations increase with an increase in the Reynolds number as shown in Figure 9.3(c). These series of time histories clearly exhibit the existence of turbulence at high Reynolds number. As the present experimental data at $Re \approx 2 \times 10^3 - 6 \times 10^3$ in Figure 9.2 approximately satisfy the Forchheimer flow regime, where the drag term is proportional to the square of velocity, we can consider that the Forchheimer flow resistance becomes gradually dominant in the flow characteristics with an increase in the turbulence.

Figure 9.4 shows the photographs of the flow visualizations by dye emissions which correspond to the flow conditions in Figure 9.3. The tracers of the flow are from left to right. The streaks are steady at $Re = 5 \times 10$ ($Re_d = 10^2$), and the spreading phenomenon of the tracer, which is a feature of the dispersion, cannot be recognized. In the photograph at $Re = 5 \times 10^2$ ($Re_d = 10^3$), we observe the intermittent separation of the boundary-layer which develops along surfaces of the solid matrix and a pair of small vortices behind the left tube. At $Re = 5 \times 10^3$ ($Re_d = 10^4$), the turbulent boundary layers develop along the solid surface. Furthermore, we can remarkably recognize the spreading phenomenon of the tracer with an increase in the Reynolds number, Re. Thus we can confirm the transition from laminar to turbulent flow in porous media from not only probe measurements but through visualizations. We will focus on the organized motion of turbulent vortices in porous media and we consider that the obstruction due to the solid matrix plays an important role in the transport process of turbulent flow through porous media. Visualizations show that the existence of a solid matrix imposes spatial restrictions on the magnitude of the interstitial vortices in the pore region. This means that the existence of a solid matrix forces larger eddies than its representative length d_p to be dissipative into the interstitial eddy of the pore region. Furthermore, it is recognized that the obstruction due to a solid matrix induces the flow distortion. The interferences between the turbulent boundary layers due to the flow distortion form the mixing zone which is in contact with a pair of small vortices behind the left tube and the fluid in the mixing zone is transported by the flow distortion with the mixing length being of order d_p. In other words, the flow distortion produces the additional mixing which is of order d_p to that of the interstitial vortex.

The distribution of the power spectral density for temperature time series at $Re = 5 \times 10^3$ ($Re_d = 10^4$) is shown in Figure 9.5. We observe that there is a fast decay with a slope of $-5/3$ in the high frequency range and this range corresponds to the inertial–convective subrange for turbulent flow, see Tennekes and Lumley (1989). The dominant frequency is observed to be at $f \approx 20\,\mathrm{Hz}$, and the Strouhal number St at $Re = 2.5 \times 10^3 - 5 \times 10^3$ indicates a higher value of $1 - 1.1$ compared with that for a single tube where $St = 0.19 - 0.21$. Judging from the values of $St\,(= f d_p / U_D) \approx 1$ and $U_D \sim \sqrt{k}$, the mixing length is

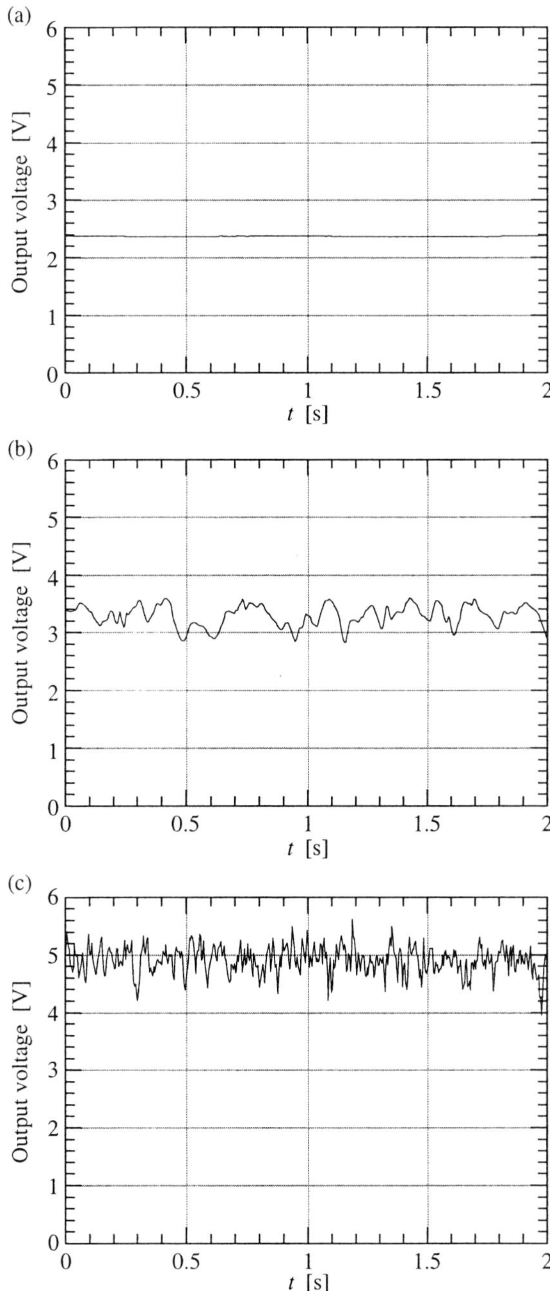

Figure 9.3 *Time history of the output signal from the velocity anemometer for (a)* $Re = 5 \times 10$, *(b)* $Re = 5 \times 10^2$, *and (c)* $Re = 5 \times 10^3$

(a)

(b)

(c)

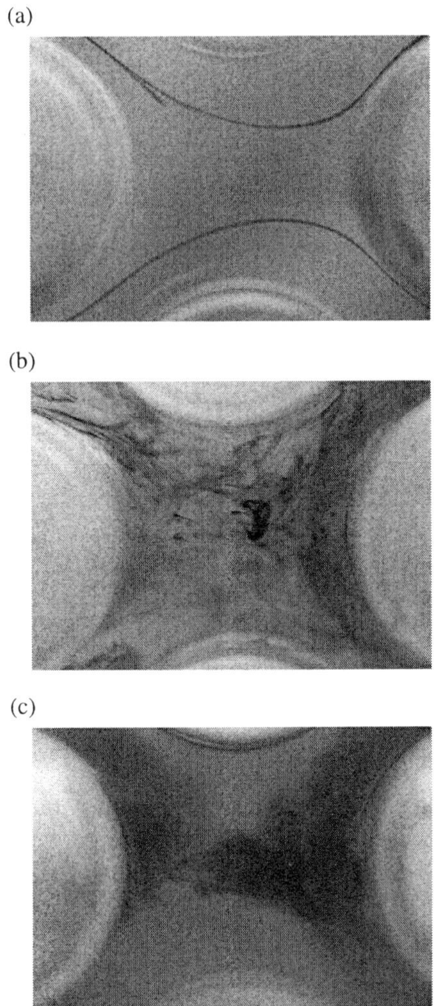

Figure 9.4 *Photographic illustrations of the dye emissions for (a) $Re = 5 \times 10$, (b) $Re = 5 \times 10^2$, and (c) $Re = 5 \times 10^3$*

of the order d_p, and the dominant frequency $f \approx 20\,\mathrm{Hz}$ is induced by the flow distortion. This fact suggests that the heat is transferred by the diffusion of the large eddy which corresponds to the mixing due to the flow distortion. Furthermore, the power spectral density converges to a small value at low frequencies which are less than $f \approx 20\,\mathrm{Hz}$, and we deduce that the larger eddy less than $f \approx 20\,\mathrm{Hz}$ is dissipated by the obstruction due to the solid matrix.

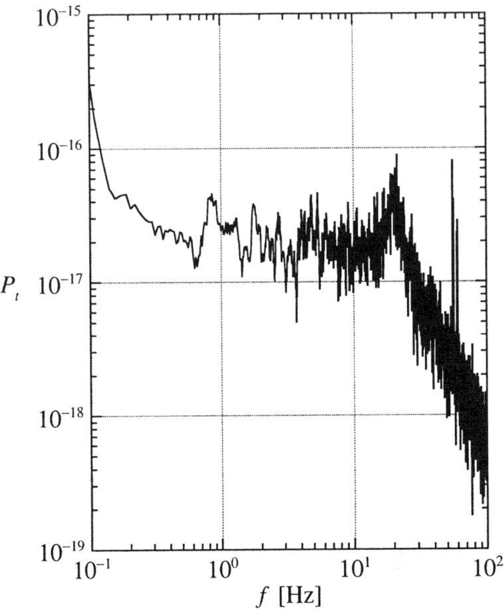

Figure 9.5 *Power spectral density for the temperature time series when* $Re = 5 \times 10^3$

9.4 MACROSCOPIC MOMENTUM EQUATION

Before we develop the macroscopic governing equations for the turbulent flow through porous media, we present a brief discussion on local volume averaging, see Slattery (1972, 1999). If B is any scalar, spatial vector, or second-order tensor, associated with the fluid phase, then the local volume average over V_f of a quantity B associated with the fluid phase is defined as follows:

$$\langle B \rangle^{(f)} \equiv \frac{1}{V_f} \int_{V_f} B \, \mathrm{d}V, \tag{9.4}$$

and the local volume average over V_s of a quantity B associated with the solid phase is defined as follows:

$$\langle B \rangle^{(s)} \equiv \frac{1}{V_s} \int_{V_s} B \, \mathrm{d}V, \tag{9.5}$$

where V_f and V_s are as indicated in Figure 9.6. Furthermore, the theorems for the volume average of a gradient and a divergence is expressed by the following equations:

$$\langle \nabla B \rangle^{(f)} = \nabla \langle B \rangle^{(f)} + \frac{1}{V_f} \int_{A_w} B\boldsymbol{n}\, dA, \tag{9.6}$$

$$\langle \operatorname{div} \boldsymbol{B} \rangle^{(f)} = \operatorname{div}\langle \boldsymbol{B} \rangle^{(f)} + \frac{1}{V_f} \int_{A_w} \boldsymbol{B} \cdot \boldsymbol{n}\, dA, \tag{9.7}$$

where A_w is the area of the interface between the fluid and solid phase in $V\ (= V_f + V_s)$.

In the turbulent flow through the porous media, the microscopic momentum equation can be given by the Reynolds equation coupled with the Boussinesq eddy viscosity formulation

$$\rho_f \left[\frac{\partial \boldsymbol{U}}{\partial t} + \operatorname{div}\left(\boldsymbol{U}\boldsymbol{U} \right) \right] = \operatorname{div}\left(\boldsymbol{S} + \boldsymbol{S}_t \right), \tag{9.8}$$

where the molecular and turbulent stress tensors, $\boldsymbol{S}$ and $\boldsymbol{S}_t$, are given by

$$\boldsymbol{S} = 2\mu \boldsymbol{D} - P\boldsymbol{I}, \tag{9.9}$$

$$\boldsymbol{S}_t = 2\mu_t \boldsymbol{D} - \frac{2}{3}k\boldsymbol{I}, \tag{9.10}$$

and

$$\boldsymbol{D} = \frac{1}{2}\left[\nabla \boldsymbol{U} + (\nabla \boldsymbol{U})^{\mathsf{T}} \right]. \tag{9.11}$$

If the eddy viscosity μ_t is constant within V_f, then the local volume average of equation (9.8) leads to the macroscopic Reynolds equation for the turbulent flow through porous

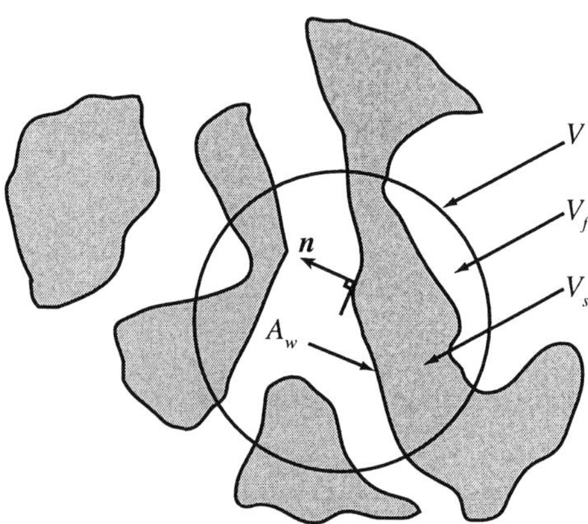

Figure 9.6 *Control volume of the local volume averaging of porous structure*

media, namely

$$\rho_f \left[\frac{\partial \langle \boldsymbol{U} \rangle^{(f)}}{\partial t} + \langle \mathrm{div}\,(\boldsymbol{U}\boldsymbol{U}) \rangle^{(f)} \right] = \langle \mathrm{div}\,(\boldsymbol{S} + \boldsymbol{S}_t) \rangle^{(f)}. \qquad (9.12)$$

The drag forces around the solid particles are derived from the right-hand side of equation (9.12), with the aid of equation (9.7), as follows:

$$\langle \mathrm{div}\,\boldsymbol{S} \rangle^{(f)} = \mathrm{div} \langle \boldsymbol{S} \rangle^{(f)} + \frac{1}{V_f} \int_{A_w} \boldsymbol{S} \cdot \boldsymbol{n}\, dA, \qquad (9.13)$$

$$\langle \mathrm{div}\,\boldsymbol{S}_t \rangle^{(f)} = \mathrm{div} \langle \boldsymbol{S}_t \rangle^{(f)} + \frac{1}{V_f} \int_{A_w} \boldsymbol{S}_t \cdot \boldsymbol{n}\, dA. \qquad (9.14)$$

The second term on the right-hand side of equation (9.13) is the drag force caused by the molecular stress tensor $\boldsymbol{S}$ and that of equation (9.14) is the drag force caused by the Reynolds (turbulent) stress tensor $\boldsymbol{S}_t$. We observe that the second term on the right-hand side of equation (9.13) is the original Darcy flow resistance, while Vafai and Tien (1981) formulate it by a linear combination of the Darcy flow resistance and the Forchheimer flow resistance, namely

$$\frac{1}{V_f} \int_{A_w} \boldsymbol{S} \cdot \boldsymbol{n}\, dA = -\phi \frac{\mu}{K} \langle \boldsymbol{U} \rangle^{(f)}. \qquad (9.15)$$

Attention should be given to the dominant regime of the Forchheimer flow resistance ($Re \gtrsim 10^3$) in order to clarify the nature of this flow resistance. The behavior of the turbulent flow is made explicit in the present experimental data at high Reynolds number and the contribution of the turbulence vortex must be reflected in the flow characteristics. If the laminar form drag $F_L U_D^2$ due to the spatial obstructions exists, except for the flow resistance $F_T U_D^2$ due to the turbulent vortex, the drastic variation should be recognized in the flow characteristics ($F_L U_D^2 \to (F_L + F_T)\, U_D^2$) when turbulence occurs. However, such a phenomenon has not ever been reported in previous studies, see Bird *et al.* (1976). Furthermore, the laminar form drag $F_L U_D^2$ increases with an increase in the wake region in the flow across a sphere or a circular cylinder, see Batchelor (1967), but the solid matrix imposes spatial restrictions on the development of the wake region in the flow through porous media. The present flow visualization recognizes a pair of very small vortices behind the left tube, see Figure 9.4, and such a phenomenon has been also reported in staggered tube banks, see Fujii *et al.* (1986). The laminar form drag $F_L U_D^2$ due to spatial obstructions may affect the deviations from the Darcy law. However, it is impossible to consider that this narrow wake region produces the large laminar form drag at high Reynolds numbers, and the Forchheimer coefficient F_L due to the laminar form drag is much less than the Forchheimer coefficient F_T due to the turbulent vortex at high Reynolds numbers. Thus the flow characteristics in porous media can be explained by the molecular drag force (the Darcy law which is proportional to the velocity) and the turbulent drag force (the Forchheimer law which is proportional to the square of the velocity) around the solid matrix.

Now we discuss the Reynolds stress tensor $\boldsymbol{S}_t$. As shown in Figure 9.7, it is expected that two types of vortices, namely, the void and pseudo vortices, play an important role in the transport mechanism of the turbulent flow through porous media. Taking notice of the flow around the solid particles in Figure 9.7, we can suppose that the interruption due to the solid particles induce the forced flow distortion. The flow distortion will transport the fluid far away and cause the associated exchange of momenta. So we refer to this momentum diffusion as the mixing of the pseudo vortex. The void vortex is the interstitial vortex which is formed in the pore between the solid particles. It can be estimated that the characteristic length scale of the pseudo vortex is the order of magnitude of the particle diameter d_p and that of the interstitial vortex is of the gap width $\sqrt{K}$. Thus, we consider the eddy viscosity μ_t in equation (9.10) as the algebraic sum of the eddy viscosities which are defined by the characteristic length scales of the pseudo and void vortices, namely

$$\mu_t = \mu_{t,P} + \mu_{t,V}, \tag{9.16}$$

where the first term on the right-hand side of equation (9.16) is the pseudo eddy viscosity $\mu_{t,P}$, and it is characterized by the pseudo vortex, and the second term is the void eddy viscosity $\mu_{t,V}$, and it is characterized by the void vortex. It is fair to say that the Reynolds stress tensor related to the void eddy viscosity, which is characterized by the interstitial vortices, contributes towards the drag force, because the pseudo vortex takes the role of a long distance momentum transport due to the forced flow distortion, while the void vortex directly determines the velocity profile of the turbulent shear flow along the solid particle due to the effect of its short-distance momentum exchange. Furthermore, equation (9.14) reduces to equation (9.17) on the grounds that the drag force caused by the molecular

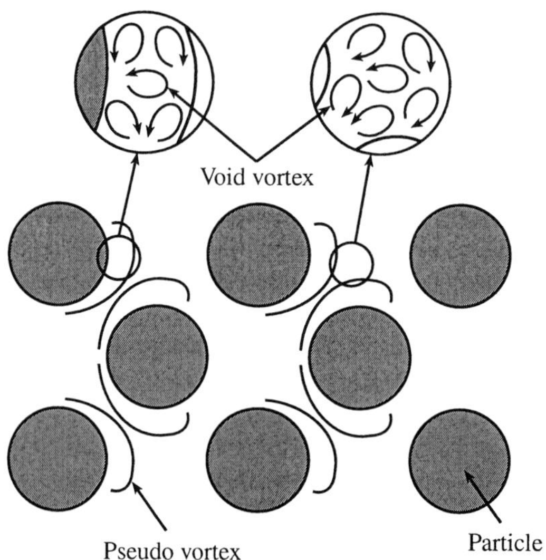

Figure 9.7 *Schematic model of the vortices in packed beds*

stress tensor S is expressed by equation (9.15) as follows:

$$\langle \mathrm{div}\, S_t\rangle^{(f)} = \mathrm{div}\langle S_t\rangle^{(f)} + \frac{1}{V_f}\int_{A_w} S_t \cdot n\, \mathrm{d}A = \mathrm{div}\langle S_t\rangle^{(f)} - \sigma\phi\frac{\mu_{t,V}}{K}\langle U\rangle^{(f)}, \quad (9.17)$$

where σ is the correction factor which is introduced in order to extend the concept of the hydrodynamic conductance defined by the Darcy law to the turbulent flow. We estimate the correction factor as $\sigma \approx 1$ by considering the similar contributions of the turbulent kinetic energy k and the pressure P to the stress tensors defined by equations (9.9) and (9.10). Here, we may note that the second term on the right-hand side of equation (9.17) represents the damping effect due to the void vortex associated with the local homogeneous and isotropic effects of turbulence.

Now we concentrate on the inertial term in equation (9.12). The microscopic velocity vector U can be decomposed into the sum of the mean velocity vector (spatial average) and the deviation velocity vector (spatial fluctuation), i.e.,

$$U \equiv \frac{1}{V_f}\int_{V_f} U\, \mathrm{d}V + U^V = \langle U\rangle^{(f)} + U^V. \qquad (9.18)$$

On substituting equation (9.18) into the inertial term of equation (9.12) we obtain

$$\mathrm{div}\langle UU\rangle^{(f)} = \mathrm{div}\left[\langle U\rangle^{(f)}\langle U\rangle^{(f)}\right] + \mathrm{div}\langle U^V U^V\rangle^{(f)}. \qquad (9.19)$$

The porous structures are commonly held periodic in the representative length scale related to the representative volume V, so that the spatial fluctuation may be considered as an almost periodic function of the representative length scale. As the divergence operator is valid for the representative length scale in principle, it is supposed that the second term on the right-hand side of equation (9.19) is negligible due to its periodic nature as compared to the first term. Therefore, the inertial term can be approximated by

$$\mathrm{div}\langle UU\rangle^{(f)} = \mathrm{div}\left[\langle U\rangle^{(f)}\langle U\rangle^{(f)}\right]. \qquad (9.20)$$

Using the above closure modeling for the drag force and the inertia, the macroscopic momentum equation for the turbulent flow through porous media becomes

$$\rho_f\left[\frac{\partial\langle U\rangle^{(f)}}{\partial t} + \mathrm{div}\langle UU\rangle^{(f)}\right] = \mathrm{div}\langle S + S_t\rangle^{(f)} - \phi\frac{\mu + \sigma\mu_{t,V}}{K}\langle U\rangle^{(f)}. \qquad (9.21)$$

9.5 MACROSCOPIC ENERGY EQUATION

The microscopic energy equations for the fluid and solid phases are given by

$$\rho_f c_f \left[\frac{\partial T}{\partial t} + \text{div}\,(\boldsymbol{U}T) \right] + \text{div}\,\boldsymbol{q}_f = 0, \tag{9.22}$$

$$\rho_s c_s \frac{\partial T}{\partial t} + \text{div}\,\boldsymbol{q}_s = 0, \tag{9.23}$$

where the heat flux vectors for the fluid and solid phases are given by

$$\boldsymbol{q}_f = -\left(\lambda_f + \lambda_t\right)\nabla T, \tag{9.24}$$

$$\boldsymbol{q}_s = -\lambda_s \nabla T. \tag{9.25}$$

With the aid of equations (9.4) and (9.5), the local volume averages of equations (9.22) and (9.23) yield the macroscopic energy equations for the fluid and solid phases, namely

$$\rho_f c_f \left[\frac{\partial \langle T \rangle^{(f)}}{\partial t} + \text{div}\langle \boldsymbol{U}T \rangle^{(f)} \right] + \text{div}\langle \boldsymbol{q}_f \rangle^{(f)} + \frac{1}{V_f} \int_{A_w} \boldsymbol{q}_f \cdot \boldsymbol{n}\,\mathrm{d}A = 0, \tag{9.26}$$

$$\rho_s c_s \frac{\partial \langle T \rangle^{(s)}}{\partial t} + \text{div}\langle \boldsymbol{q}_s \rangle^{(s)} + \frac{1}{V_s} \int_{A_w} \boldsymbol{q}_s \cdot \boldsymbol{n}\,\mathrm{d}A = 0. \tag{9.27}$$

As the heat flux vector is continuous at the interface between the fluid and solid phases, the interface condition is given by

$$\int_{A_w} \boldsymbol{q}_f \cdot \boldsymbol{n}\,\mathrm{d}A = \int_{A_w} \boldsymbol{q}_s \cdot \boldsymbol{n}\,\mathrm{d}A. \tag{9.28}$$

On substituting equations (9.26) and (9.27) into the above interface condition we have

$$\phi \left\{ \rho_f c_f \left[\frac{\partial \langle T \rangle^{(f)}}{\partial t} + \text{div}\langle \boldsymbol{U}T \rangle^{(f)} \right] + \text{div}\langle \boldsymbol{q}_f \rangle^{(f)} \right\}$$
$$+ \left(1 - \phi\right) \left[\rho_s c_s \frac{\partial \langle T \rangle^{(s)}}{\partial t} + \text{div}\langle \boldsymbol{q}_s \rangle^{(s)} \right] = 0. \tag{9.29}$$

Now we focus on the eddy thermal conductivity. Suppose that an analogy exists between the eddy viscosity and the eddy thermal conductivity, then the eddy thermal conductivity λ_t can be represented as the algebraic sum of the eddy thermal conductivities defined by the characteristic length scales of the pseudo and void vortices, namely

$$\lambda_t = \lambda_{t,P} + \lambda_{t,V}, \tag{9.30}$$

where the first term on the right-hand side of the above equation is the pseudo eddy thermal conductivity $\lambda_{t,P}$, which is characterized by the pseudo vortex, and the second term is the void eddy thermal conductivity $\lambda_{t,V}$, which is characterized by the void vortex.

The pseudo vortex contributes to the long distance heat transport due to the forced flow distortion, so that the thermal tortuosity cannot be produced from the heat flux vector related to the pseudo eddy thermal conductivity. Accordingly, the heat flux vectors can be reduced, with the aid of equations (9.4) and (9.5), as follows:

$$\langle q_f \rangle^{(f)} = - (\lambda_f + \lambda_t) \nabla \langle T \rangle^{(f)} - \frac{\lambda_f + \lambda_{t,V}}{V_f} \int_{A_w} T n \, dA, \tag{9.31}$$

$$\langle q_s \rangle^{(s)} = -\lambda_s \nabla \langle T \rangle^{(s)} - \frac{\lambda_s}{V_s} \int_{A_w} T n \, dA. \tag{9.32}$$

We define the local volume average of the temperature over the fluid and the solid phases as follows:

$$\langle T \rangle^{(m)} \equiv \frac{1}{V} \int_V B \, dV = \phi \langle T \rangle^{(f)} + (1 - \phi) \langle T \rangle^{(s)}, \tag{9.33}$$

and we make use of the local thermal equilibrium assumption, see Slattery (1972, 1999),

$$\langle T \rangle^{(m)} = \langle T \rangle^{(f)} = \langle T \rangle^{(s)}. \tag{9.34}$$

Now we concentrate on the advection term in equation (9.29). The microscopic velocity vector U and the temperature T can be decomposed into the mean component (spatial average) and the fluctuating component (spatial fluctuation), as follows:

$$U \equiv \frac{1}{V_f} \int_{V_f} U \, dV + U^V = \langle U \rangle^{(f)} + U^V, \tag{9.35}$$

$$T \equiv \frac{1}{V_f} \int_{V_f} T \, dV + T^V = \langle T \rangle^{(f)} + T^V. \tag{9.36}$$

On substituting equations (9.35) and (9.36) into the advection term of equation (9.29) we obtain

$$\mathrm{div}\langle UT \rangle^{(f)} = \mathrm{div}\left(\langle U \rangle^{(f)} \langle T \rangle^{(f)} \right) + \mathrm{div}\langle U^V T^V \rangle^{(f)}. \tag{9.37}$$

The second term on the right-hand side of equation (9.37) is the correlation term between the spatial fluctuations of the velocity and the temperature, and the previous studies have considered the correlation as the thermal dispersion effect, see Koch and Brady (1985), Levec and Carbonell (1985), Georgiadis and Catton (1988), Koch *et al.* (1989), and Hsu and Cheng (1990). The porous structures are commonly held periodic in the representative length scale related to the representative volume V, and the divergence operator is valid for the representative length scale. For simplicity, if we consider the fully-developed flow, then the spatial fluctuation should be the periodic function of the representative length scale and this leads to the following equation:

$$\mathrm{div}\langle U^V T^V \rangle^{(f)} = 0. \tag{9.38}$$

We can observe the dispersion even in the fully-developed flow, where no contribution due to the spatial fluctuations are expected. Therefore, the correlation term between the spatial fluctuations of the velocity and temperature quantities can be neglected, and the enthalpy transport term in equation (9.29) can be approximated by

$$\mathrm{div}\langle \boldsymbol{U}T\rangle^{(f)} = \mathrm{div}\left(\langle \boldsymbol{U}\rangle^{(f)}\langle T\rangle^{(m)}\right).\tag{9.39}$$

Furthermore, on introducing the concept of the effective thermal conductivity, as proposed by Kunii and Smith (1960), we obtain

$$\lambda_e \nabla\langle T\rangle^{(m)} = \left[\phi\lambda_f + (1-\phi)\lambda_s\right]\nabla\langle T\rangle^{(m)} + \frac{\lambda_f - \lambda_s}{V}\int_{A_w} T\boldsymbol{n}\,\mathrm{d}A.\tag{9.40}$$

Using the above closure modeling for the turbulent heat flux and the enthalpy transport terms, the macroscopic energy equation for the turbulent flow through porous media becomes

$$\left[\phi\rho_f c_f + (1-\phi)\rho_s c_s\right]\frac{\partial\langle T\rangle^{(m)}}{\partial t} + \phi\rho_f c_f\,\mathrm{div}\left(\langle \boldsymbol{U}\rangle^{(f)}\langle T\rangle^{(m)}\right) = \mathrm{div}\left(\lambda_p\nabla\langle T\rangle^{(m)}\right),\tag{9.41}$$

where

$$\lambda_p = \lambda_e + \phi\lambda_t + f_t\lambda_{t,V}\tag{9.42}$$

and

$$f_t = \frac{\lambda_e - \left[\phi\lambda_f + (1-\phi)\lambda_s\right]}{\lambda_f - \lambda_s}.\tag{9.43}$$

9.6   THE 0-EQUATION MODEL

We propose the 0-equation model for the eddy viscosity and the eddy thermal conductivity. For the fully-developed one-dimensional turbulent flow, the macroscopic momentum equation becomes

$$-\frac{\mathrm{d}P}{\mathrm{d}x} = \sigma\frac{\mu_{t,V}}{K}U_D,\tag{9.44}$$

and the empirical correlation for the flow resistance of packed beds at high Reynolds number, see Bird *et al.* (1976), is given by

$$-\frac{\mathrm{d}P}{\mathrm{d}x} = F\frac{\rho_f U_D^2}{\sqrt{K}},\tag{9.45}$$

where

$$F = \frac{1.75}{\sqrt{150\,\phi^3}}\quad\text{(packed bed)}.\tag{9.46}$$

Equation (9.45) is the so-called Forchheimer flow resistance equation which is observed at high Reynolds numbers. On comparing equations (9.44) and (9.45), we can find the void eddy viscosity $\mu_{t,V}$ as follows:

$$\mu_{t,V} = \frac{F}{\sigma}\rho_f U_D \sqrt{K}. \tag{9.47}$$

On the other hand, we can estimate the void eddy viscosity $\mu_{t,V}$ from the Kolmogorov–Prandtl expression, see Rodi (1982), as follows:

$$\mu_{t,V} = C_\mu \rho_f \sqrt{k} L. \tag{9.48}$$

If the velocity scale $\sqrt{k}$ is of the order U_D and the length scale L is of the order $\sqrt{K}$, we can derive the same order of magnitude as the void eddy viscosity $\mu_{t,V}$ obtained from the empirical correlation

$$\mu_{t,V} \sim \rho_f U_D \sqrt{K}. \tag{9.49}$$

Thus, it follows that the void vortex contributes to the Forchheimer flow resistance.

Now we discuss the mutual relation between the void and the pseudo eddy viscosities (thermal conductivities). The eddy viscosity ratio γ is defined as

$$\gamma = \frac{\mu_{t,P}}{\mu_{t,V}}. \tag{9.50}$$

If the turbulent Prandtl number Pr_t is independent of the vortex length scale, then the void and pseudo eddy thermal conductivities are given by

$$\lambda_{t,P} = \frac{c_f}{Pr_t}\mu_{t,P}, \tag{9.51}$$

$$\lambda_{t,V} = \frac{c_f}{Pr_t}\mu_{t,V}. \tag{9.52}$$

On substituting equations (9.50) – (9.52) into equation (9.42), we can rewrite the thermal dispersion λ_p as follows:

$$\frac{\lambda_p}{\lambda_f} = \frac{\lambda_e}{\lambda_f} + \phi\frac{\lambda_t}{\lambda_f} + f_t\frac{\lambda_{t,V}}{\lambda_f} = \frac{\lambda_e}{\lambda_f} + \frac{F\left[\phi\left(\gamma+1\right)+f_t\right]}{\sigma Pr_t}Pe, \tag{9.53}$$

where the Péclet number Pe is defined as follows:

$$Pe = \frac{U_D\sqrt{K}}{\lambda_f/\left(\rho_f c_f\right)} = \frac{U_D\sqrt{K}}{\alpha}. \tag{9.54}$$

If the turbulent Prandtl number Pr_t is independent of the Péclet number Pe in equation (9.53), we can find the relationship that the thermal dispersion λ_p is proportional to the Péclet number Pe, see Yagi $et\ al.$ (1960), Georgiadis and Catton (1988), and Hsu and Cheng (1990). The permeability of the packed beds is given by the Blake and Kozeny expression, see Bird *et al.* (1976), as follows:

$$K = \frac{\phi^3 d_p^2}{150 \left(1 - \phi\right)^2},$$ (9.55)

and we can estimate the eddy viscosity ratio γ by

$$\gamma \sim \frac{d_p}{\sqrt{K}} \approx 30,$$ (9.56)

where the porosity ϕ is treated as approximately 0.4, see Yagi *et al.* (1960).

The contribution of the eddy viscosity ratio γ to the thermal dispersion is shown in Figure 9.8, with the empirical correlation of Yagi *et al.* (1960) given by

$$\frac{\lambda_p}{\lambda_f} = 7.5 + 0.8 \, Pe_d \quad \text{(glass spheres)},$$ (9.57)

$$\frac{\lambda_p}{\lambda_f} = 13 + 0.7 \, Pe_d \quad \text{(steel spheres)},$$ (9.58)

where the particle Péclet number Pe_d is given by

$$Pe_d = Pe \frac{d_p}{\sqrt{K}} = \frac{U_D d_p}{\alpha}.$$ (9.59)

It can be seen from Figure 9.8 that the increase of the eddy viscosity ratio enhances the thermal dispersion and such a tendency is large in the region of high particle Péclet number.

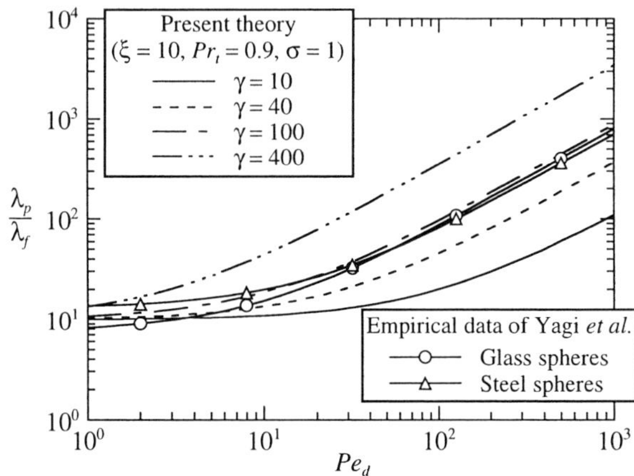

Figure 9.8 *Comparison of the present model with the empirical data of Yagi et al. (1960)*

We observe that the present result when $\gamma = 100$ is in good agreement with the empirical data of Yagi *et al.* (1960). Although this value of the eddy viscosity ratio, $\gamma = 100$, is somewhat greater than the value estimated from equation (9.56), both estimates can be considered to be of the same order of magnitude.

Table 9.1 indicates the relationship between the thermal conductivity ratio ξ ($\equiv \lambda_f/\lambda_s$) and the coefficient f_t defined by equation (9.43) at porosity $\phi = 0.4$. As the coefficient f_t is very small compared with the eddy viscosity ratio $\gamma = 100$ in equation (9.53), we can consider that the contribution of the void eddy thermal conductivity to the thermal dispersion is negligible. In other words, the mixing of the pseudo vortex mainly contributes to the thermal dispersion and equation (9.53) reduces to

$$\frac{\lambda_p}{\lambda_f} = \frac{\lambda_e}{\lambda_f} + \frac{F\phi\gamma}{\sigma Pr_t} Pe. \tag{9.60}$$

It has been suggested that the mixing of the void vortex mainly contributes to the Forchheimer flow resistance and the mixing of the pseudo vortex to the thermal dispersion. The turbulent drag force around a solid matrix, i.e., the Forchheimer flow resistance, is independent of the transport of the fluid due to the pseudo vortex as the inner region of the turbulent boundary-layer has a self-sustaining structure, see Maruyama and Tanaka (1987). However, the pseudo vortex can produce the energy exchange due to the temperature difference of the fluid. In the present visualization, the coherent structures due to the obstruction of a solid matrix in porous media are observed. One is the spatial restriction of the interstitial vortex and the other is the mixing induced by the flow distortion. Thus the interstitial vortex formed in the pore between the solid matrix is reflected in the void vortex in the model which has the effect of the short distance turbulent mixing due to the spatial restriction. On the other hand, the diffusion induced by the flow distortion is characterized by the long distance turbulent mixing of the pseudo vortex in the model. From the viewpoint of spectral dynamics, see Figure 9.5, the void vortex is characterized by the small vortices in the high frequency range and the pseudo vortex is characterized by the large vortex with the dominant frequency of the order of $St \approx 1$. This experimental evidence is well supported in our turbulence model which introduces the concepts of two types of vortices.

Table 9.1 *Relationship between thermal conductivity ratio ξ and the coefficient* f_t

ξ	0.01	0.1	1	10	100
f_t	1.02×10^{-3}	9.37×10^{-3}	2.00×10^{-1}	2.93×10^{-1}	5.15×10^{-1}

9.7   PRODUCTION AND DISSIPATION OF TURBULENCE

Now we look carefully into the k–ε model. We emphasize the experimental evidence that the fully-developed turbulent flow remains with no damping due to the turbulence in the direction of the main flow. This is in disagreement with the k–ε model of Antohe and Lage (1997) which has pointed out that k and ε must be zero in the fully-developed flow through porous media. Antohe and Lage (1997) and Getachew *et al.* (2000) derived the k–ε model from the time averaging of the momentum equation and this has been averaged by a representative elementary volume of the porous medium. This means that the resolution of the vortices for the k–ε model is larger than the representative elementary volume, because the k–ε model takes no account of the smaller eddies than the representative elementary volume in the spatial averaging process. In other words, the modeling of the production term, which is dependent only on the macroscopic average velocity gradient, brings about no turbulence in the fully-developed flow through porous media. However the order of time averaging and spatial averaging is independent of the final form of the turbulence model so long as the contribution of the microscopic eddies to the macroscopic turbulence field is justifiably modeled in the averaging (filtering) process, see Pedras and de Lemos (2000, 2001). Therefore, we need to focus on the behaviors of the microscopic eddies in a representative elementary volume of the porous medium to clarify the mechanism of the production and dissipation of the turbulence.

It is well known that the k equation of the k–ε model is a good approximation to many flow configurations, see Bradshaw *et al.* (1984), and this equation may stand even for the microscopic turbulent fields in porous media. We can write the microscopic k equation for the turbulent flow through porous media as follows:

$$\frac{\partial k}{\partial t} + \mathrm{div}\,(\boldsymbol{U}k) = \mathrm{div}\,\boldsymbol{d} + Pro - \varepsilon, \tag{9.61}$$

where the first term on the right-hand side of equation (9.61) represents the diffusion term, and Pro and ε are the production and the dissipation, respectively. The local volume average of equation (9.61) leads to the macroscopic k equation for the turbulent flow through porous media, namely

$$\frac{\partial \langle k \rangle^{(f)}}{\partial t} + \langle \mathrm{div}\,(\boldsymbol{U}k) \rangle^{(f)} = \langle \mathrm{div}\,\boldsymbol{d} \rangle^{(f)} + \langle Pro \rangle^{(f)} - \langle \varepsilon \rangle^{(f)}. \tag{9.62}$$

With the aid of a theorem for a volume average of a divergence, see Slattery (1972, 1999), the advection term in equation (9.62) leads to

$$\langle \mathrm{div}\,(\boldsymbol{U}k) \rangle^{(f)} = \mathrm{div}\langle \boldsymbol{U}k \rangle^{(f)} + \frac{1}{V_f}\int_{A_w} (\boldsymbol{U}k) \cdot \boldsymbol{n}\, \mathrm{d}A = \mathrm{div}\langle \boldsymbol{U}k \rangle^{(f)}, \tag{9.63}$$

where A_w is the interfacial area between the fluid and the solid phases. The porous structure is commonly periodic in the representative length scale associated with the representative elementary volume and the divergence operator is valid for the representative length scale.

This principle ensures the following expression:

$$\operatorname{div}\langle \boldsymbol{U} k\rangle^{(f)} = \operatorname{div}\left[\langle \boldsymbol{U}\rangle^{(f)}\langle k\rangle^{(f)}\right]. \tag{9.64}$$

Use of the divergence theorem with the diffusion term of equation (9.62) yields

$$\langle \operatorname{div} \boldsymbol{d}\rangle^{(f)} = \operatorname{div}\langle \boldsymbol{d}\rangle^{(f)} + \frac{1}{V_f}\int_{A_w} \boldsymbol{d}\cdot \boldsymbol{n}\, \mathrm{d}A, \tag{9.65}$$

where the vector $\boldsymbol{d}$ may be expressed by

$$\boldsymbol{d} = \frac{1}{\rho_f}\left(\mu + \frac{\mu_t}{\sigma_k}\right)\nabla k \tag{9.66}$$

and σ_k is the effective Prandtl–Schmidt number. The turbulent mixing in porous media can be characterized by the pseudo and void vortices, namely

$$\mu_t = \mu_{t,P} + \mu_{t,V}, \tag{9.67}$$

where the pseudo eddy viscosity and the void eddy viscosity exhibit the long distance turbulent mixing and the short distance turbulent mixing, respectively. We can estimate the second term on the right-hand side of equation (9.65) as follows:

$$\frac{1}{V_f}\int_{A_w} \boldsymbol{d}\cdot \boldsymbol{n}\, \mathrm{d}A \sim -\frac{\phi}{\rho_f K}\left(\mu + \frac{\mu_{t,V}}{\sigma_k}\right)\langle k\rangle^{(f)}, \tag{9.68}$$

and this shows that the obstruction of a solid matrix in porous media forces larger eddies than its representative length to be dissipative to the interstitial eddy. Introducing the correction factor σ_{k1}, equation (9.68) is reduced to

$$\frac{1}{V_f}\int_{A_w} \boldsymbol{d}\cdot \boldsymbol{n}\, \mathrm{d}A = -\sigma_{k1}\frac{\phi}{\rho_f K}\left(\mu + \frac{\mu_{t,V}}{\sigma_k}\right)\langle k\rangle^{(f)}. \tag{9.69}$$

We may estimate the correction factor as $\sigma_{k1} \sim 1$. Judging from $\langle k\rangle^{(f)} \geqslant 0$, equation (9.69) exhibits the dissipation intrinsic to the turbulent flow through porous media. This term physically means that the obstruction of a solid matrix forces larger eddies than its representative length to be dissipative to the interstitial eddy.

Now we focus on the production term in the macroscopic kinetic energy equation. The microscopic production term may be expressed as follows:

$$Pro = \frac{1}{\rho_f}\boldsymbol{S}_t\cdot \operatorname{grad} \boldsymbol{U} = \frac{1}{\rho_f}\left[\operatorname{div}\left(\boldsymbol{S}_t\boldsymbol{U}\right) - \operatorname{div}\boldsymbol{S}_t\cdot \boldsymbol{U}\right], \tag{9.70}$$

and the local volume average of equation (9.70) is given by

$$\langle Pro\rangle^{(f)} = \frac{1}{\rho_f}\left[\langle\operatorname{div}\left(\boldsymbol{S}_t\boldsymbol{U}\right)\rangle^{(f)} - \langle\operatorname{div}\boldsymbol{S}_t\cdot \boldsymbol{U}\rangle^{(f)}\right]. \tag{9.71}$$

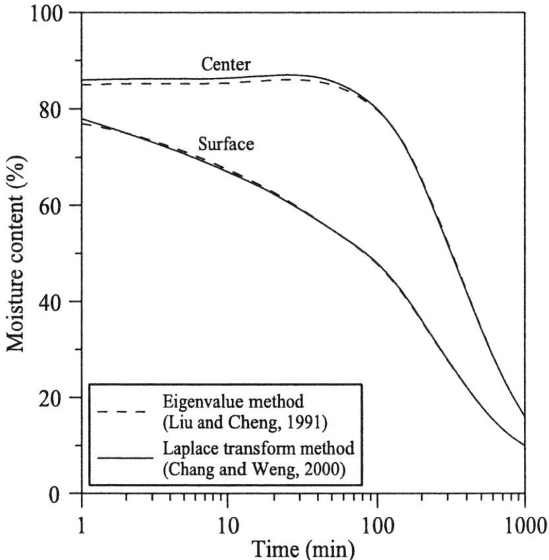

Figure 10.6 *Moisture content at the center and at the surface of the wood specimen*

$$\rho \frac{\partial m}{\partial t} = \frac{1}{r^2} \frac{\partial}{\partial r} \left(r^2 D_m \frac{\partial m}{\partial r} \right)$$
$$+ \frac{1}{r^2} \frac{\partial}{\partial r} \left(r^2 \delta D_m \frac{\partial T}{\partial r} \right) + \frac{1}{r^2} \frac{\partial}{\partial r} \left[r^2 \left(\frac{K_L}{\mu_L} + \frac{K_G}{\mu_G} \right) \frac{\partial P}{\partial r} \right], \tag{10.23b}$$

where r is the radius of the ceramic sphere, h_{LV} is the heat of evaporative phase change, K_G and K_L are the respective permeabilities of the gas and the liquid, μ_G and μ_L are the respective kinematic viscosities of the gas and the liquid, δ is the thermogradient coefficient, P is the pressure in the material, which satisfies the following thermodynamic relationship:

$$P = \Phi(m) P_S(T), \tag{10.24}$$

and Φ is the quotient of the actual pressure P and the saturation pressure P_S at the prevailing temperature. This equation was determined from using experimental data. In this simulation, the specific steam temperature is $175\,^\circ\text{C}$, the steam mass flow is $0.35\,\text{kg}\,\text{m}^{-2}\text{s}^{-1}$ and the initial pressure within the material is 1 bar. The transport coefficients used in the above equations were either measured experimentally or were derived theoretically from the pore size distribution of the material and they are given by, see Hager *et al.* (1997),

$$D_m = \frac{2 K_L \gamma(T)}{\mu_L r_z^2(m)} \left(\frac{\partial r_z(m)}{\partial m} \right)_T \quad \text{m}^2\text{s}^{-1}, \tag{10.25}$$

$$\delta = -\frac{1}{\gamma(T)}\frac{\partial \gamma(T)}{\partial T}\left\{\frac{\partial \ln[r_z(m)]}{\partial m}\right\}_T^{-1} \quad K^{-1}, \tag{10.26}$$

$$K_L = \frac{\rho}{8\tau\left[1+\rho V(r)_{\text{tot}}\right]}\int_{r_{\min}}^{r_z} r^2 \frac{\partial V}{\partial r}\, dr \quad m^2, \tag{10.27}$$

$$K_G = \frac{\rho}{8\tau\left[1+\rho V(r)_{\text{tot}}\right]}\int_{r_z}^{r_{\max}} r^2 \frac{\partial V}{\partial r}\, dr \quad m^2, \tag{10.28}$$

where V is the specific void volume, $V(r)_{\text{tot}}$ is the total void volume per unit mass of the dry solid, R is the radius of the sphere, which in the problem under investigation takes the value 5 mm, $\gamma(T)$ is the surface tension and is a function of the temperature, r_z is the equivalent radius which has a more detailed description, see Dullien (1979), and τ is the tortuosity which is the only adjustable parameter in this simulation by Hager *et al.* (1997) and a value of 20 was used for this fitting parameter.

The boundary conditions at the center of the sphere are written as equations (10.29a) and (10.29b) due to the spherical symmetry of the material. The surface boundary condition of the material, equation (10.29c), can be obtained by differentiating equation (10.24) with respect to time and using the relationship $P|_{r=R} = P_\infty$, and the second boundary condition at the surface, equation (10.29d), can be obtained from combining a heat and a mass balance over the surface of the sphere. The initial temperature and moisture content of the material, equations (10.30a) and (10.30b), are uniform and they are equal to the saturation temperature and the pressure of the surrounding steam. They are given by

$$\left.\frac{\partial T}{\partial r}\right|_{r=0} = 0, \tag{10.29a}$$

$$\left.\frac{\partial m}{\partial r}\right|_{r=0} = 0, \tag{10.29b}$$

$$\left.\frac{\partial T}{\partial t}\right|_{r=R} = -P_\infty\left(\frac{\partial P_S}{\partial T}\right)^{-1}\frac{1}{\Phi^2}\frac{\partial \Phi}{\partial m}\left.\frac{\partial m}{\partial t}\right|_{r=R}, \tag{10.29c}$$

$$\rho D_m h_{LV}\left.\frac{\partial m}{\partial r}\right|_{r=R} + \rho\delta D_m h_{LV}\left.\frac{\partial T}{\partial r}\right|_{r=R} + \frac{K_L}{\mu_L}h_{LV}\left.\frac{\partial P}{\partial r}\right|_{r=R}$$
$$-k\left.\frac{\partial T}{\partial r}\right|_{r=R} + h\left(T|_{r=R} - T_\infty\right) + \varepsilon^*\sigma^*\left(T^4|_{r=R} - T_\infty^4\right) = 0, \tag{10.29d}$$

$$T(r,0) = 100\,^\circ C, \tag{10.30a}$$

$$m(r,0) = 15\%, \tag{10.30b}$$

where ε^* is the emissivity and it is assumed to be unity and σ^* is the Stefan–Boltzmann constant.

Based on the above conditions, Hager *et al.* (1997) have obtained numerical simulations and experimental measurements which are compared in Figures 10.7 to 10.9. It can be observed from these figures that the drying process includes a period with a constant drying rate and a period with a decreasing drying rate. During the constant drying rate period, the material remains wet, the evaporation phenomena takes place only at the surface and

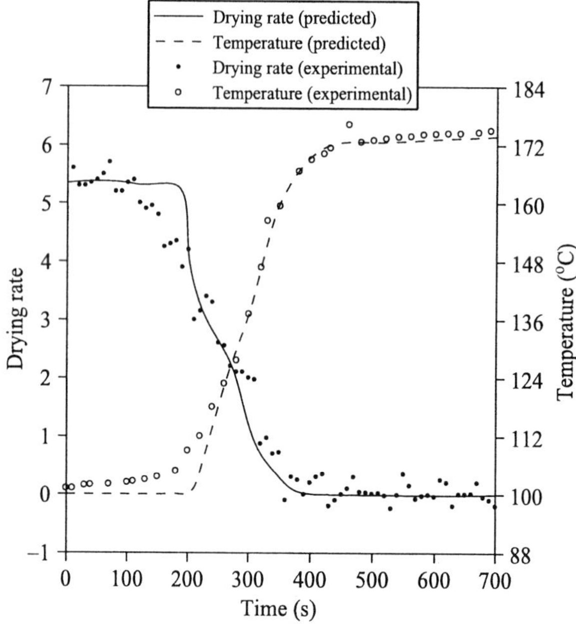

Figure 10.7 *Comparison of the numerical predictions with the experimental data*

the drying rate is controlled by the convective heat and mass transfer. Therefore, the material temperature remains at the wet bulb temperature, the moisture content drops linearly and the pressure within the material is equal to the surrounding pressure. In contrast, during the decreasing drying rate period, the evaporation front recedes from the surface, a sorption zone appears in the material next to the wet zone and evaporation takes place at the evaporation front as well as in the sorption region. The moisture content in the material drops, the temperature and the inner pressure increase, and the surface pressure remains equal to the surrounding pressure. As the moisture content of the material approaches equilibrium, the inner pressure also approaches the surrounding pressure and the temperature approaches the surrounding temperature.

10.4 CONCLUSIONS

This chapter has provided a review of coupled heat and moisture transfer in porous material. In the first part of this chapter the generalized governing equations for the coupled heat and moisture transfer are proposed and in the second part some practical engineering applications are analyzed by using these equations. The results show that the transport coefficients in the coupled system play an important role but it is difficult to accurately

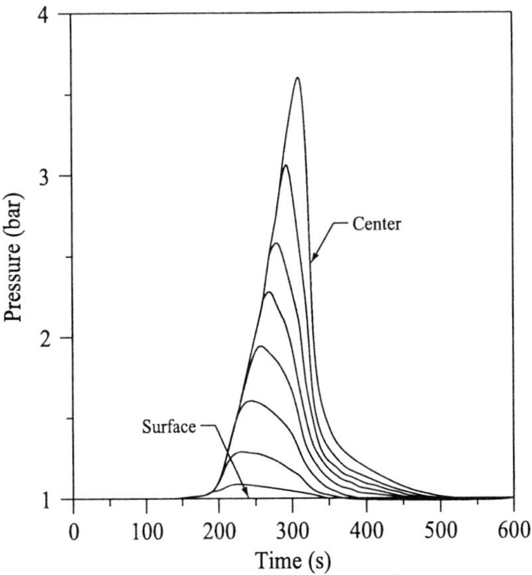

Figure 10.8 *Pressure at different locations in the ceramic sphere during drying*

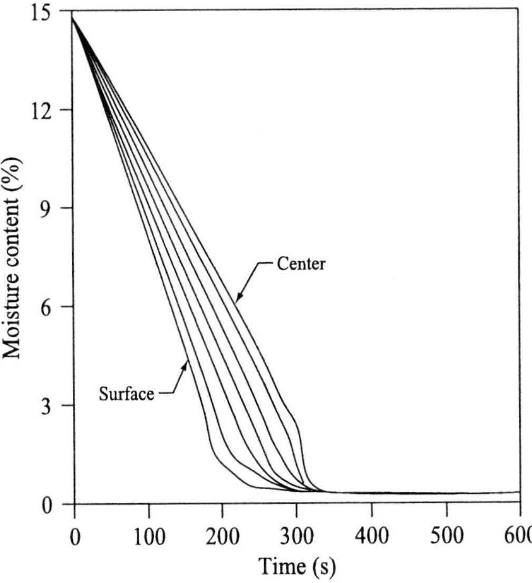

Figure 10.9 *Moisture content in the porous sphere during drying*

determine these coefficients. An inverse method of coupled heat and mass transfer for parameter estimation, or function estimation, with some temperatures measured on the surface or in the interior of the material, may be recommended for estimating these coefficients.

Furthermore, as the more efficient and advanced computer technology develops, the substantial insight into microscopic features at atomic or molecular levels is necessary to analyze the complicated behavior of heat and mass transfer. In the near future, the molecular dynamics simulation should be applied to the nanoporous materials involving the heat and mass transfer phenomena.

REFERENCES

ASAE Standards (1990). *Thermal Properties of Grain and Grain Products* (37th edn). ASAE, St Joseph, Michigan.

Budaiwi, I., El-Diasty, R., and Abdou, A. (1999). Modelling of moisture and thermal transient behaviour of multi-layer non-cavity walls. *Building Environ.* **34**, 537–551.

Chang, C. S., Converse, H. H., and Steele, J. (1994). Modeling of moisture content of grain during storage with aeration. *Trans. ASAE* **37**, 1891–1898.

Chang, W. J. and Weng, C. I. (2000). An analytical solution to coupled heat and moisture diffusion transfer in porous materials. *Int. J. Heat Mass Transfer* **43**, 3621–3632.

Chen, C. C. and Morey, R. V. (1989). Comparison of four EMC/ERH equations. *Trans. ASAE* **32**, 983–990.

Chen, P. and Pei, D. C. T. (1989). A mathematical model of drying process. *Int. J. Heat Mass Transfer* **32**, 297–310.

Chen, T. C., Weng, C. I., and Chang, W. J. (1992). Transient hygrothermal stresses induced in general plane problems by theory of coupled heat and moisture. *ASME J. Appl. Mech.* **59**, 10–16.

di Blasi, C. (1998). Multiphase moisture transfer in the high-temperature drying of wood particles. *Chem. Eng. Sci.* **53**, 353–366.

Dullien, F. A. L. (1979). *Porous Media: Fluid Transport and Pore Structure.* Academic Press, New York.

Eckert, E. R. G. and Faghri, M. (1980). A general analysis of moisture migration caused by temperature differences in an unsaturated porous medium. *Int. J. Heat Mass Transfer* **23**, 1613–1623.

Freer, M. W., Siebenmorgen, T. J., Couvillion, R. J., and Loewer, O. J. (1990). Modeling temperature and moisture content changes in bunker-stored rice. *Trans. ASAE* **33**, 211–220.

Hager, J., Hermansson, M., and Wimmerstedt, R. (1997). Modelling steam drying of a single porous ceramic sphere: Experiments and simulations. *Chem. Eng. Sci.* **52**, 1253–1264.

Häupl, P., Grunewald, J., Fechner, H., and Stopp, H. (1997). Coupled heat air and moisture transfer in building structures. *Int. J. Heat Mass Transfer* **40**, 1633–1642.

Huang, C. L. D., Siang, H. H., and Best, C. H. (1979). Heat and moisture transfer in concrete slabs. *Int. J. Heat Mass Transfer* **22**, 257–266.

Ilic, M. and Turner, I. W. (1989). Convective drying of a consolidated slab of wet porous materials. *Int. J. Heat Mass Transfer* **32**, 2351–2362.

Ingham, D. B. and Pop, I. (eds) (1998). *Transport Phenomena in Porous Media.* Pergamon, Oxford.

Irudayaraj, J. and Wu, Y. (1999). Numerical modeling of heat and mass transfer in starch systems. *Trans. ASAE* **42**, 449–455.

Khankari, K. K., Morey, R. V., and Patankar, S. V. (1994). Mathematical model for moisture diffusion in stored grain due to temperature gradients. *Trans. ASAE* **37**, 1591–1604.

Krischer, O. (1963). *Die Wissenschaftlichen Grundlagen der Trocknungstechnik.* Springer, Berlin.

Liu, J. Y. and Cheng, S. (1991). Solution of Luikov equations of heat and mass transfer problems in porous bodies. *Int. J. Heat Mass Transfer* **34**, 1747–1754.

Lobo, P. D., Mikhailov, M. D., and Özişik, M. N. (1987). On the complex eigenvalues of Luikov system of equations. *Drying Tech.* **5**, 273–286.

Luikov, A. V. (1966). *Heat and Mass Transfer in Capillary–Porous Bodies.* Pergamon, Oxford.

Luikov, A. V. (1975). Systems of differential equations of heat and mass transfer in capillary–porous bodies. *Int. J. Heat Mass Transfer* **18**, 1–14.

Nguyen, T. V. (1986). Modelling temperature and moisture changes resulting from natural convection in grain stores. In *Preserving Grain Quality by Aeration and In-Store Drying* (eds B. R. Champ and E. Highley). *ACIAR Procedures* **15**, 81–87.

Nield, D. A. and Bejan, A. (1999). *Convection in Porous Media* (2nd edn). Springer–Verlag, New York.

Obaldo, L. G., Harner, J. P., and Converse, H. H. (1991). Prediction of moisture changes in stored corn. *Trans. ASAE* **34**, 1850–1858.

Pandey, R. N., Srivastava, S. K., and Mikhailov, M. D. (1999). Solutions of Luikov equations of heat and mass transfer in capillary–porous bodies through matrix calculus: a new approach. *Int. J. Heat Mass Transfer* **42**, 2649–2660.

Perre, P., Moser, M., and Martin, M. (1993). Advances in transport phenomena during convective drying with superheated steam and moist air. *Int. J. Heat Mass Transfer* **36**, 2725–2746.

Philip, J. R. and de Vries, D. A. (1957). Moisture movement in porous materials under temperature gradients. *Trans. Amer. Geophys. Union* **38**, 222–232.

Plumb, O. A., Spolek, G. A., and Olmstead, B. A. (1985). Heat and mass transfer in wood during drying. *Int. J. Heat Mass Transfer* **28**, 1669–1678.

Stanish, M. A., Schajer, G. S., and Kayihan, F. (1986). A mathematical model of drying for hygroscopic porous media. *AIChE J.* **32**, 1301–1311.

Stewart, J. A. (1975). Moisture migration during storage of preserved, high moisture grains. *Trans. ASAE* **18**, 387–393.

Thomas, H. R., Morgan, K., and Lewis, R. W. (1980). A fully nonlinear analysis of heat and mass transfer problems in porous bodies. *Int. J. Numer. Meth. Eng.* **15**, 1381–1393.

Thorpe, G. R. (1980). Moisture diffusion through bulk grain. *J. Stored Prod. Res.* **17**, 39–42.

Vafai, K. (ed.) (2000). *Handbook of Porous Media.* Marcel Dekker, New York.

Whitaker, S. (1977). Simultaneous heat, mass, and momentum transfer in porous media: a theory of drying. *Adv. Heat Transfer* **13**, 119–203.

Wong, S. P. W. and Wang, S. K. (1990). Fundamentals of simultaneous heat and moisture transfer between the building envelope and the conditioned space air. *ASHRAE Trans.* **96**, 73–83.

11 ISOTHERMAL NUCLEATION AND BUBBLE GROWTH IN POROUS MEDIA AT LOW SUPERSATURATIONS

S. BORIES and M. PRAT

Institut de Mécanique des Fluides de Toulouse, UMR CNRS-INP/UPS N° 5502, Avenue du Professeur Camille Soula, F-31400 Toulouse, France

email: bories@imft.fr and prat@imft.fr

Abstract

Nucleation and growth of bubbles in porous media are important problems encountered in processes such as pressure depletion and boiling. After a recap of the basic principles of nucleation and bubble growth in the bulk, we discuss some of the results obtained during the last ten years within the framework of percolation and pore network models. In particular, results of experiments of liquid-to-gas phase change by pressure decline of supersaturated CO_2 solutions in 2D transparent etched networks are reported in order to understand the phenomena in porous media.

The observations confirm the heterogeneous nature of nucleation, i.e., the decisive role of the capillary roughness of the pore walls. Contrary to the bulk or Hele–Shaw cells, gas clusters have irregular and ramified shapes typical of invasion percolation patterns. As a result, the growth rate of a single gas cluster is different than the growth rate of an isolated single bubble in the bulk.

Numerical simulations of the growth pattern and of the growth rate of a single gas cluster are performed with a numerical automaton. Based on a pore network modelling technique and on a set of hypotheses derived from the observations, this automaton is first validated by comparing the numerical results with the experimental data. Then the automaton is used to explore the influences of the Jakob number, pressure decline rate, Bond number, and wettability.

In a last part, the closure of the macroscopic mass balance equations is discussed.

Keywords: nucleation, bubble growth, pore network, experiments, simulation, invasion percolation

11.1 INTRODUCTION

The liquid–gas phase change processes in porous media play a vital role in many technological applications. These include such areas as oil recovery from petroleum reservoirs,
geothermal systems, nuclear waste disposal, drying of materials and enhanced heat transfer systems, to name only a few. Two main different processes can be at the origin of a
liquid–gas phase change: the boiling nucleation and bubble growth which occurs in superheated liquids and the isothermal nucleation and bubble growth resulting from a pressure
reduction in supersaturated multicomponent liquid mixture. As emphasised by Firoozabadi and Kashchiev (1993), this latter process, i.e., the isothermal formation of a new
gas phase when the pressure is lowered below the saturation pressure, is a kinetic process
which plays a very important role in petroleum engineering. Called solution gas-drive,
this process is classified as one of the five main mechanisms of oil recovery from fractured
and unfractured reservoirs (the others four being gravity drainage, capillary imbibition,
viscous displacement and compressibility effects both from rock and fluid). Contrary to
external displacement processes, such as drainage, where the oil is produced by injecting
a gas inside the reservoir, in internal drives oil is pushed out of the pores by the growth of
gas bubbles coming from its own dissolved gas when pressure on the reservoir is reduced.
In direct relation with this particular process, many unresolved issues, such as the value
of the critical gas saturation, i.e., the maximum gas saturation existing before any flow of
gas may occur, and the influence of pressure decline rate, are of great importance.

Whatever the mechanisms involved in the phase change, the main problem in the modelling
of this non-equilibrium phenomenon in porous media at Darcy's scale is the determination
of the gas phase formation rate Γ_g which appears as a source term in the mass balance
equations

$$\frac{\partial}{\partial t}\left(\varepsilon \rho_i S_i\right) + \nabla \cdot \left(\rho_i \vec{U}_i\right) = \pm \Gamma_g, \tag{11.1}$$

where t is the time, ε the porosity, $\vec{U}_i$ the superficial velocity, ρ_i the density, and $i = l$ for
the liquid and g for the gas. The $+$ sign in the right-hand side of equation (11.1) is used
for the gas and the $-$ sign for the liquid. However, despite this key role played by Γ_g, few
studies have been devoted to the gas formation in porous media, and in most cases Γ_g is
simply derived from thermostatic conditions (dashed line in Figure 11.1). As can be seen
from Figure 11.1, which shows the kinetics of the formation of a gas phase by pressure
decline, such a description is clearly very questionable and therefore justifies the works
performed on this problem for some years.

However, as emphasised by Yortsos and Parlar (1989), the problem of bubble nucleation
and bubble growth in porous media has not only practical interests but also theoretical
ones. The interaction between nucleation, mass transfer and fluid flow and the effects of
pressure decline rate or superheat on the growth of the gas phase add indeed many novel
aspects not previously encountered in typical displacements, i.e., in drainage.

Without going into the details of nucleation, in this chapter we briefly describe the phe-
nomenology of this important process and we present the particular aspects of the phenomenon in porous media.

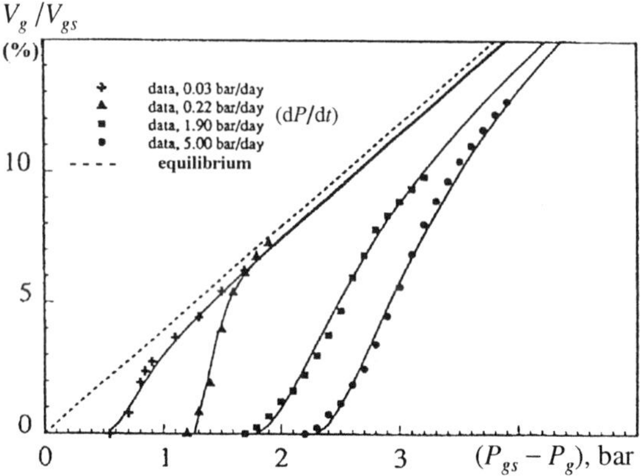

Figure 11.1 *Measured non-equilibrium gas saturation as a function of P for constant pressure decline rate data, see Moulu (1989)*

11.2 BASIC PRINCIPLES

Gas phase formation in supersaturated or superheated liquids is a first-order transition which requires two consecutive mechanisms:

(i) bubble nucleation, which appears when the liquid is brought to its saturation conditions, and

(ii) bubble growth, due to the mass or heat diffusion driven by pressure lowering or superheat and generally controlled by inertia, viscous and surface forces.

11.2.1 Nucleation

According to Zettlemoyer (1969) and Springer (1978), the liquid–gas nucleation is defined as the spontaneous formation of bubbles in the liquid. This phenomenon occurs when, for instance, at constant temperature and pressure lower than the saturation pressure of a pure liquid, see Figure 11.2(a), thermodynamic fluctuations of sufficient magnitude form growing molecular gas clusters which give rise to gas bubbles.

Let us consider for instance the vaporisation of a liquid in the isothermal, isometric system presented in Figure 11.2(b). If we assume that this system is isolated from its surroundings, so that its total energy remains constant in time, its density is uniform on a macroscopic scale. However, this is not true at the scale of individual molecules where the local density is changing constantly due to the molecules random motions and collisions. Due to collisions between vapour molecules, aggregates, embryos, nuclei and droplets form in the system. The size of an aggregate, embryo, etc., is characterised by the number

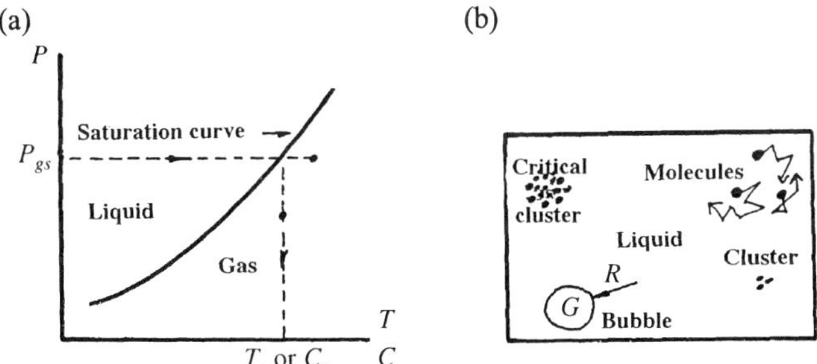

Figure 11.2 *(a) Liquid–gas equilibrium curve. (b) Schematic of homogeneous nucleation*

n of molecules contained in any of these molecular formations. An aggregate contains a minimum of two molecules and may contain as many as $n^* - 1$ molecules, the value of n^* being sufficiently small for

(i) the aggregate to disintegrate spontaneously into n single vapour molecules, and

(ii) the aggregate to be treated as gas like.

On the other hand, embryos contain a sufficiently high number of molecules ($n > n^*$) so that each embryo can be treated as a continuum medium which has the same properties of the bulk vapour. A particular important consequence of this definition is that the macroscopic flat film surface tension σ is assumed to have a meaning even for embryos consisting of only a few molecules.

According to classical nucleation theory, based on the previous hypothesis, a nucleus will only grow and become a bubble if it exceeds a certain critical size for a given degree of supersaturation. The critical size can be derived from the work of formation ΔG, i.e., the Gibbs free energy of a spherical cluster of radius R, see Frenkel (1946) and Springer (1978), namely

$$\Delta G = \frac{4}{3}\pi R^3 \sigma \left(\frac{3}{R} - \frac{2}{R^*} \right),$$
(11.2)

where R^* is the critical radius as explained hereafter. As can be seen from the right-hand side of equation (11.2), the work of formation depends on of the competition between two effects. The first one is the volume energy of the cluster and represents the decrease in chemical potential due to the forming of the gas cluster. The second one is the contribution of the surface free energy to ΔG and represents an increase in the chemical potential.

As shown in Figure 11.3, where the variation of ΔG with radius R is plotted for fixed values of supersaturation and temperature, ΔG first increases with R and reaches a maximum at some value of $R = R^*$ and then decreases continuously. According to the condition of equilibrium, at the maximum point $\Delta G^* (R^*)$ the nucleus is in equilibrium

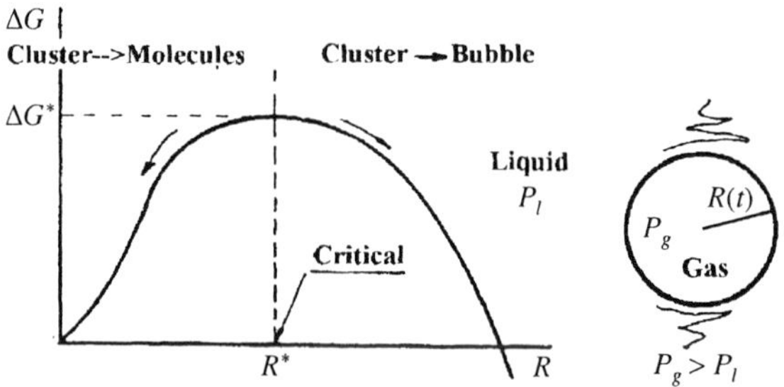

Figure 11.3 *Variation of the Gibbs free energy with bubble radius for a vapor bubble spontaneously formed in a superheated or supersaturated liquid*

with its surrounding liquid and at this equilibrium, which is a metastable one, correspond respectively to the critical radius R^* and the free energy barrier

$$\Delta G^* = \frac{4\pi R^{*2}\sigma}{3}. \tag{11.3}$$

When $R < R^*$ the nucleus tends to condense and disappears, while for $R > R^*$ it will tend to grow. Unless the free energy barrier ΔG^* is reached then the cluster cannot reach a critical size and cannot become a bubble.

We thus arrive at the important conclusion that for a cluster to become a bubble then the free energy barrier must first be reached. Once this barrier is reached then the addition of a single molecule puts the cluster over the top of the ΔG as a function of R curve. From this point, ΔG decreases very quickly and the nucleus (which is now called supercritical nucleus) grows extremely rapidly until the normal saturation pressure P_{gs} is established in the vapour and a bubble forms. Therefore the onset of phase change is taken to correspond to the condition $\Delta G = \Delta G^*$ at which, as we have seen, clusters of size R^* form.

When $R > R^*$, thermodynamic (mechanical and chemical) equilibrium defines the radius of the gas bubble. This radius is given by the Laplace equation

$$R = \frac{P_g - P_l}{2\sigma}, \tag{11.4}$$

where P_g and P_l, with $P_g > P_l$, are the pressure in the gas and the liquid, respectively. At equilibrium, the total gas pressure P_g is related to the temperature, concentration of the dissolved gas in the liquid and radius of curvature of the interface by the Kelvin equation. In many cases this equation can be approximated by

$$\frac{P_g}{P_{gs}} = \exp\left(-\frac{2\sigma}{\rho_l R \Re T}\right), \tag{11.5}$$

where P_g and P_{gs} are the gas pressure and the saturated gas pressure, i.e., the pressure at which the liquid and the corresponding or the dissolved gas phase can coexist at temperature T, see Defay and Prigogine (1951), and $\mathcal{R}$ is the constant for an ideal gas.

The difference $\Delta P = P_{gs} - P_l$ or $P_g - P_l$ that characterises the mechanical equilibrium for a given curvature of the gas–liquid interface is called supersaturation. As shown in Figure 11.4, this difference can help us to understand why a liquid pressure lower than the saturation pressure is needed to observe nucleation.

Three kinds of nucleation processes are generally distinguished, namely:

  (i) The homogeneous nucleation, see Frenkel (1946), Springer (1978) and Blander (1979), when nuclei form in the bulk of the liquid in the absence of foreign matter.

  (ii) The heterogeneous nucleation, see Jarvis (1975), Hwu *et al.* (1988), Lubetkin (1988) and Sheu *et al.* (1988), when the nuclei take place on a foreign matter, such as the walls of the container or at the interface between two immiscible fluid phases.

  (iii) The nucleation due to the capillary trapping of pre-existent stabilised microbubbles inside the roughness of the solid surface, see Bankoff (1958).

In homogeneous nucleation, the energy of cluster formation is given by equation (11.3). In heterogeneous nucleation this energy is reduced by the combined effects of surface roughness and wettability of the solid by the liquid phase. In that case the energy barrier is given by $(\Delta G^*)_{\text{heterogeneous}} = \phi (\Delta G^*)_{\text{homogeneous}}$, with $\phi < 1$. The factor ϕ takes into account the energy reduction and is a function of the shape and size of the heterogeneities of the solid surface, and of the liquid–solid wetting angle θ, see Moulu (1989) and Kashchiev and Firoozabadi (1993).

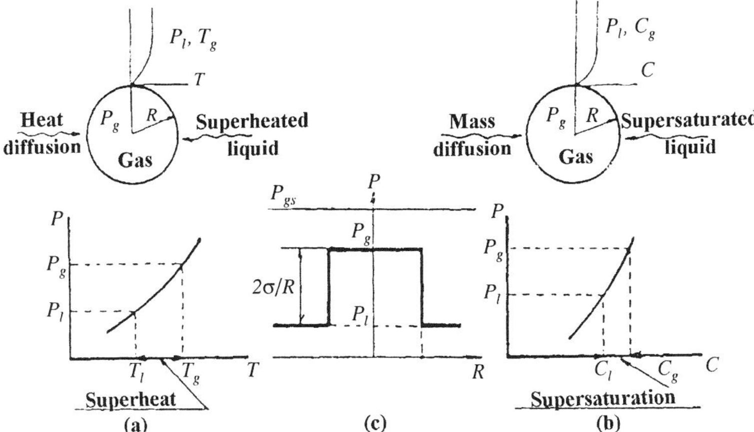

Figure 11.4 *(a) Superheated liquid. (b) Supersaturated liquid. (c) Influence of capillary and Kelvin effects on phase pressures*

In capillary trapping, the condition for nucleation is directly obtained by applying equation (11.4) with the radius of curvature evaluated as the conical pit mouth radius, see Drelich *et al.* (1996).

It is now well established that homogeneous nucleation requires a very high supersaturation ratio $\Delta P/P_{gs}$ and Wilt (1986) gave the example of 1100 to 1700 for CO_2 solutions near room conditions. Consequently, gas phase formation with low supersaturation, as very often observed, can only be explained by heterogeneous nucleation or capillary trapping.

11.2.2   Rate of bubble nucleation

The appearance of bubbles is a random process characterized by the rate of nucleation J, number of bubbles per unit time and unit volume of liquid. When the liquid is kept at a constant supersaturation ΔP, J is time-independent and may be expressed as follows:

$$J = J_0 = A^* \exp\left(-\frac{\Delta G^*}{kT}\right) = 2n\left(\frac{2\sigma}{\pi m^* b}\right)^{1/2} \exp\left(\frac{-16\pi\sigma^3}{3kT\left(\Delta P\right)^2}\right), \qquad (11.6)$$

where n is the number of molecules per unit volume, m^* the mass of a molecule, kT is the thermal energy, and b is a parameter close to $2/3$. Equation (11.6) applies to both single and multicomponent liquids, provided the pressure inside the nucleus bubble is practically equal to P_{gs}, see Reiss (1968), Lothe (1969) and Abraham (1974). The number N of bubbles nucleated in the liquid until time t is given by

$$N\left(t\right) = V_0 \int_0^t J\left(t'\right) \mathrm{d}t', \qquad (11.7)$$

and the corresponding total volume of gas is given by

$$V_g\left(t\right) = \int_0^t J\left(t'\right) v_b\left(t, t'\right) \mathrm{d}t', \qquad (11.8)$$

where $v_b\left(t, t'\right)$ is the volume at time t of a bubble nucleated at time $t' < t$. These formulae with $N = 1$ or $V_g\left(t\right)$ corresponding to an observable given value are generally used to determine J_0 experimentally.

When nucleation takes place at a variable supersaturation, J is, in general, a complicated function of time, see Kashchiev and Firoozabadi (1993). However, for slow enough changes of the supersaturation, J can be approximated by the quasi-stationary nucleation rate corresponding to the instantaneous values of the supersaturation. For bubble nucleation at time-dependent supersaturation $\Delta P\left(t\right)$ we therefore have $J\left(t\right) = J_0\left(\Delta P\left(t\right)\right)$. According to Kashchiev and Firoozabadi (1993), this formula is valid when the rate of supersaturation $\mathrm{d}\left(\Delta P\right)/\mathrm{d}t \ll 10^6\,\mathrm{Pa/s}$, a requirement which is usually met in practice.

In principle, equation (11.6), where ΔG^* is replaced by $\phi\Delta G^*$ and A^* by a specific prefactor A, taking into account the effects of the solid surface on the nucleation process, can also be used to determine the heterogeneous nucleation rate. However, contrary to the homogeneous case, until now very few experimental results are available to confirm the validity of this modelling. Nowadays the most popular approach to predict the vapour formation rate in heterogeneous nucleation is generally based on the capillary trapping nucleation, i.e., the existence of surface heterogeneities or potential nucleation sites (natural or machine-formed pits, scratches, gouges, grooves, etc.) that contain pre-existing or trapped gas, see Cole (1979). Provided that certain requirements on the conical and contact angles are met then these nucleation sites, randomly distributed at the surface, are activated (i.e., can grow to become bubbles) when the local supersaturation is exceeded as specified by equation (11.4), see Bankoff (1959), Marto *et al.* (1968), Ward and Forest (1976), Winterton (1977) and Tong *et al.* (1990).

In the foregoing situation, which corresponds to numerous practical situations, the issue of rate of nuclei formation is non-existent because the solid surface decreases the stability of supersaturated liquid. As a result, the energy needed to create clusters is many orders of magnitude less in the presence of crevices, cavities, etc. For instance, it is simple to show that in a site of conical geometry with an angle β, the interface between liquid and gas is flat when the contact angle is equal to $\pi/2 - \beta/2$. As there is no curvature of the interface, therefore there is no capillary pressure to collapse the bubble.

For the purpose of calculations, in capillary trapping, the condition for nucleation is obtained by directly applying equation (11.4) with the radius of curvature taken as the conical pit mouth radius $R = D$, see Figure 11.5, whereas the number of bubbles formed per unit time on unit surface is determined from microphotography and statistical analysis of crevices and cavities of the solid surface. In particular, Yang and Kim (1988) have shown that the active nucleation site density can be derived from the knowledge of the density probability functions of the cavities' mouth radius and conical angles. The resulting expression may be written as follows:

$$N = F \exp\left(-\frac{X}{\Delta P}\right), \tag{11.9}$$

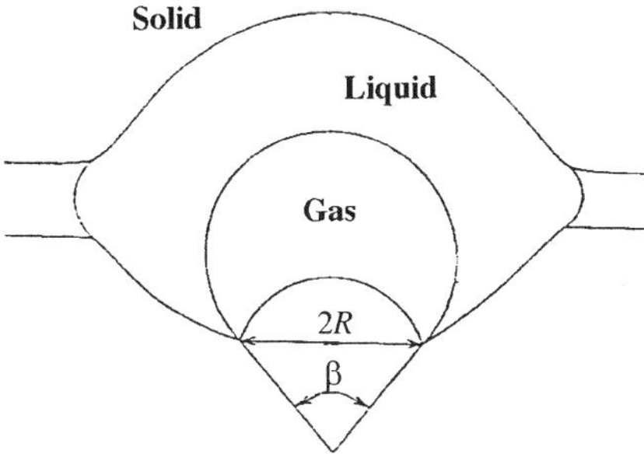

Figure 11.5 *Condition for nucleation in capillary trapping*

media is much less developed. In fact few studies have been conducted to quantify these processes in fluid saturated porous media and, in particular, to analyse the influence of geometry and topology of the microstructure on the gas phase formation. Hence many questions are still unresolved concerning this specific two-phase flow in which the liquid is pushed out by its own dissolved gas or vapor when the pressure is reduced. Moreover, among the studies devoted to this problem, most of them have been inconclusive until now, e.g., Hunt, Jr and Berry (1956), Moulu (1989) and Kashchiev and Firoozabadi (1993).

As summarised in Li and Yortsos (1995a), the origin of the difficulties encountered in analysing the previous investigations has been identified. They mainly result from

(i) the use of the macroscopic approach based on the identification of the porous medium with a continuum, and

(ii) the utilisation of models valid only in the bulk liquid phase to describe the nucleation and phase growth in porous media.

As discussed below, it is in fact crucial to take into account explicitly the presence of the solid microstructure that bounds the flow domain and the geometrical complexity of the pore space that constrains the growth of the gas phase.

In order to address porous media effects, certain aspects of nucleation and bubble growth have been recently studied experimentally, numerically and theoretically in simplified 2D porous structures, e.g., El-Yousfy (1992), Li and Yortsos (1995a, 1995b), Satik *et al.* (1995), Dominguez (1997) and Dominguez *et al.* (2000). These studies are based on visualisations of the liberation of CO_2 gas from various supersaturated carbonate liquids (via solute diffusion) in transparent micromodels, on pore network simulations and on scaling arguments. In particular, the scaling of a single bubble growth in a porous medium has been studied by Satik *et al.* (1995) under the condition of low supersaturation where diffusion predominates. Their study indicates that the following regimes develop in succession: a short duration early-time regime, where the growth is compact and classical scaling is applicable (see below), an invasion percolation regime, a transition to a viscous fingering regime and a diffusion-limited aggregation (DLA) regime where growth occurs at multiple sites at the same time and is controlled by viscous forces. In the present chapter, we are essentially concerned with the first two regimes. For the fractal regimes, i.e., invasion percolation and DLA, Satik *et al.* (1995) also derived scaling laws for the growth rate depending on the fractal dimension of the gas cluster, see equation (11.13) for the 2D version. Li and Yortsos (1995b) studied the growth of multiple bubbles evolving from different nucleation centres. These authors identified two growth regimes:

(i) a global percolation regime in which invasion percolation rules apply at the liquid–gas interface as a whole, and

(ii) a local percolation regime in which invasion percolation rules apply at the boundary of each cluster.

These findings were exploited by Du and Yortsos (1999) for studying the critical gas saturation in a porous medium.

In the present chapter, we are mainly interested in the growth of a single cluster and therefore we do not consider further the issue of the growth of multiple bubbles. The remainder of the chapter is devoted to the presentation of the main results obtained in the studies developed in our Institute on the subject. As mentioned before, these studies are based on a series of experiments performed in transparent pore network models saturated with different CO_2–liquid solutions aimed to understanding nucleation and gas cluster growth by pressure decline. Results derived from pressure measurements and from the analysis of images deduced from direct optical visualisations could be related to mechanisms of nucleation and bubble growth. A part of these results was used to build a network automaton intended to simulate the process numerically under various conditions. This automaton is briefly described and verified through direct numerical simulation of the experiments. Prior to this description, and to the discussion of simulations, we focus on the experimental study and the physical analysis of the underlying phenomena.

11.4   EXPERIMENTS

11.4.1   Experimental set-up

In order to get an insight into the micromechanics of the process and to test the theoretical predictions, a series of experiments were carried out. These experiments were performed on transparent pore network models, mostly resin micromodels and for some complementary results a glass micromodel, made using a standard procedure, see Bonnet and Lenormand (1977), initially saturated with solutions of CO_2 in various liquids in equilibrium at a pressure $P_0 = 3$ bar. The resin micromodel with a plan form of $15 \times 14\,\text{cm}^2$ consisted in a regular square lattice pattern of 42 000 rectangular ducts of $700\,\mu$m uniform thickness, and variable width. The lattice spacing is 1 mm and the widths of the ducts, varying in the range from $200\,\mu$m to $800\,\mu$m, obey a prescribed log–normal distribution law which corresponds to a typical pore size distribution function of a real porous medium. Details regarding the glass micromodel can be found in El-Yousfy (1992). A gas tank and a flow controller were used to impose the pressure decline rate. A video microcamera connected to an image processing apparatus was used to obtain quantitative data from the observation of the bubbles and clusters inside the micromodel. The schematic of the experimental set-up is shown in Figure 11.6.

The experimental procedure was as follows. CO_2 was first injected into the micromodel to displace the air. Pure liquid (water, n-octane, n-decane, water–glycol solutions, as listed in Table 11.1) was then injected to dissolve CO_2 and to pressurize the model typically at $P_{gs} = P_0 = 3$ bar. Finally, the pure liquid was displaced by the CO_2 solution. Once the liquid–CO_2 solution was in place inside the micromodel, the micromodel and the connected tanks were maintained at $P = 3.5$ bar for many hours in order that the solution homogenises. The experiments were then performed by reducing, either suddenly or with various constant rates ($m = -\mathrm{d}P/\mathrm{d}t$), the initial liquid pressure from $P_0 = 3$ bar to $P_0 - \Delta P$, by means of a microelectronic valve, denoted by m.e. in Figure 11.6. During the experiments, the pressure and volume of the solution flowing out of the micromodel

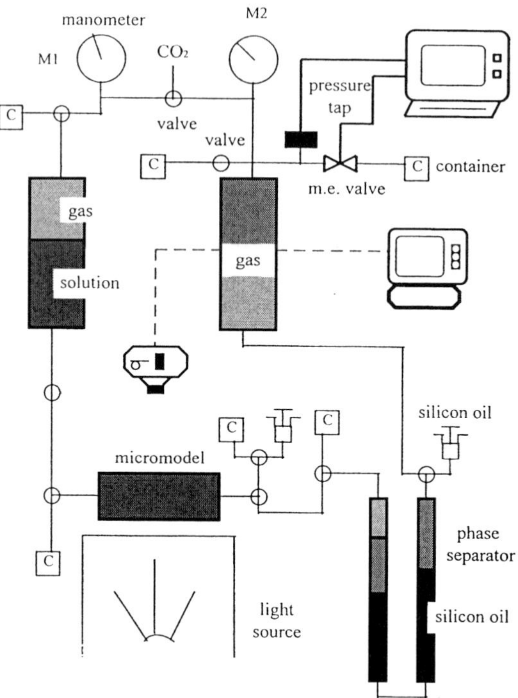

Figure 11.6 *Schematic diagram of the experimental set-up*

Table 11.1 *Fluid physical properties*

		$[He]$ $\mathrm{mole\,m^{-1}N^{-1}}$	$[\sigma]$ $\mathrm{N\,m^{-1}}$	$[D]$ $\mathrm{m^2 s^{-1}}$	$[\mu]$ Poiseuille	$[\rho]$ $\mathrm{kg\,m^{-3}}$
(A)	Octane	5×10^{-4}	21.80×10^{-3}	4.97×10^{-9}	0.510×10^{-1}	0.70×10^{3}
(B)	Decane	4×10^{-4}	23.43×10^{-3}	3.90×10^{-9}	0.850×10^{-1}	0.75×10^{3}
(C)	Xglycol $= 0.5$	3.21×10^{-4}	53.02×10^{-3}	0.60×10^{-9}	7.00×10^{-1}	1.10×10^{3}
(D)	Xglycol $= 0.2$	2.63×10^{-4}	58.06×10^{-3}	1.00×10^{-9}	3.00×10^{-1}	1.01×10^{3}

to the separator of phases through the two outlets situated in the middle of the two lateral sides, were measured as a function of time, and the processes of nucleation and growth were videotaped by means of a camera connected to a VCR and a PC. Experiments were stopped shortly after the gas started to come out from the micromodel.

The video recording was used to obtain the time for the first bubble to be observed, the number of bubbles produced, the final number and growth of various bubbles. Typical sequence of bubble growth at the pore level and inside the micromodel are shown in Figures

11.7 and 11.8. These images were also digitised and an image processing software, Visilog 4, was used to analyse the phase distribution.

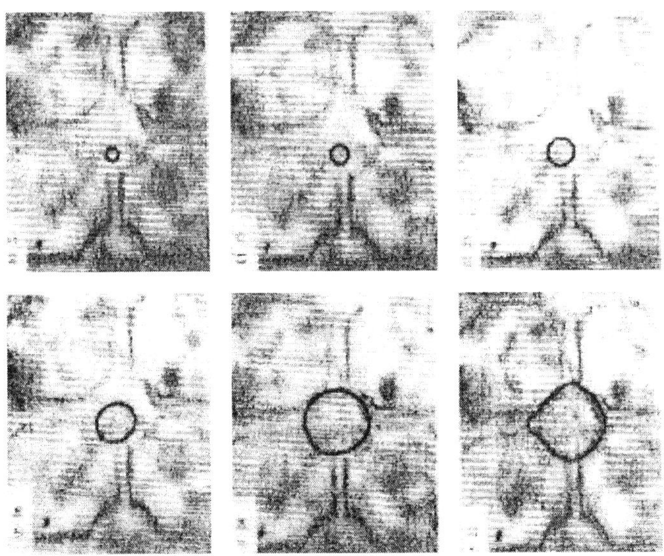

Figure 11.7 *Growth of a bubble within a pore of micromodel*

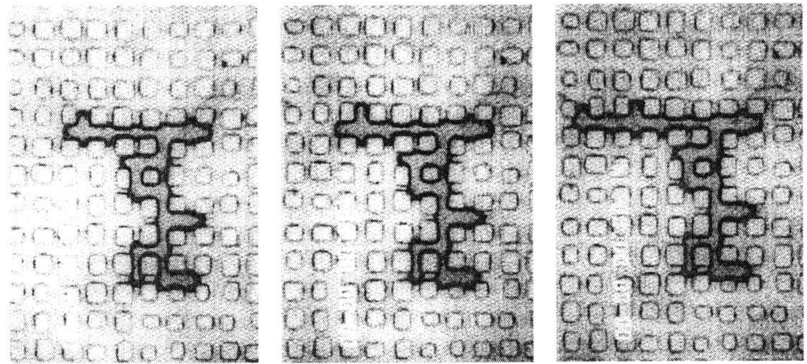

Figure 11.8 *Growth of a gas cluster within the micromodel*

equilibrium states. Contrary to the growth in the bulk or in an effective porous medium, such as a Hele–Shaw cell where the gas patterns are compact, for instance disk-shaped in the absence of gravity in a Hele–Shaw cell, visualisations show that in micromodel (which mimics real porous media) the shape of the formed gas clusters is clearly disordered and non-symmetric, see Figure 11.15. This complicated, but nevertheless reproducible, pattern is the result of the influence of the microstructure on the invasion process of the pore space. The menisci along the cluster perimeter are stationary or moving according to whether the capillary pressure is lower or greater than the capillary pressure threshold of the adjacent pore throat, i.e., $P_g - P_l = 2\sigma\cos\theta/r$, where r is the equivalent radius of the throat connecting the gas cluster to an adjacent liquid-occupied pore and $P_l = P_0 - \Delta P$. This regime corresponds to the second regime identified by Satik *et al.* (1995), i.e., the invasion percolation regime as discussed further below.

(c) The growth process is diffusion controlled. According to Szekely and Martins (1971), the growth of a vapor bubble in porous media is controlled by forces similar to those in bulk phase change, namely inertia, viscous, surface and pressure forces. In terms of dimensionless groups, this phenomenon can be characterised by the Jakob number, $Ja \propto \Delta P/P_0$, the pressure parameter, $G = a^2\Delta P/\rho_l D^2$, where a is the lattice spacing, the surface tension parameter, $\Phi = 2\sigma/a\Delta P$, the Schmidt number, $Sc = \eta/D$, and finally the parameter $B = Ja^2/\sqrt{G}$ which occurs in the momentum and mass transfer equations. This last number is a useful indicator of the mechanisms involved in the cluster growth. When $B \ll 1$ the process is diffusion-controlled, whereas for $B \gg 1$ the process is inertia-controlled. As it is well known, inertia control prevails only under severe conditions.

For the growth of a vapor cluster in porous media, a modified capillary number, $Ca = \rho_l \nu D/a\sigma\cos\theta$, which measures the relative importance of the viscous and

(a)   (b)

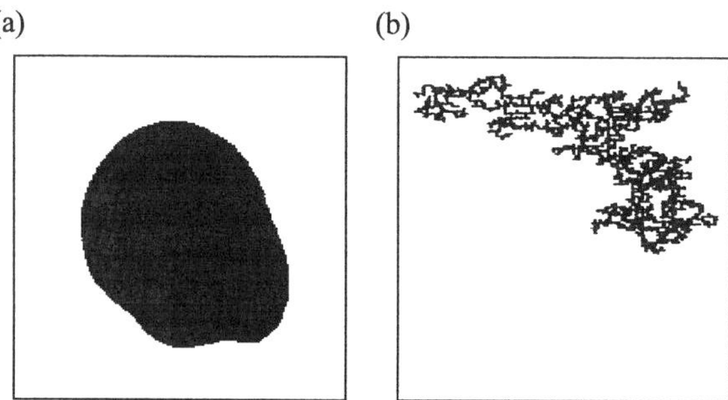

Figure 11.15   *(a) Typical shape of a bubble observed in a Hele–Shaw cell, and (b) in a micromodel (gas phase in black, liquid phase in white). With the Hele–Shaw cell, one expects a circular bubble. Small aperture variations may explain the non-perfectly circular shape depicted in (a)*

capillary forces, and the Bond number, $Bo = \rho_l a^2 g \sin \phi / \sigma \cos \theta$, which measures the relative importance of gravitation and capillarity, as well as the pore size function distribution of the microstructure must be taken into consideration. Concerning the pressure decline rate, which controls the evolution of supersaturation as a function of time, the corresponding dimensionless number is given by $M = ma^2 / P_0 D$.

For the experiments reported in the present chapter, i.e., with the different CO_2 solutions, the values of the dimensionless numbers are in the following ranges: $10 > Ja > 0.6$, $G > 10^{14}$, $\Phi \approx 10^{-2}$, $Ca \approx 10^{-6}$, $Bo \approx 5 \times 10^{-3}$, and $10^2 < Sc < 10^4$. Therefore the growth process is diffusion controlled. This result, which concerns the initial stages (as has been indicated above) as well as the advanced stages of the growth, was also confirmed experimentally in Hele–Shaw cell as reported in Dominguez *et al.* (2000).

(d) The confirmation of a non-trivial scaling for the growth rate of a gas cluster in the invasion percolation regime. For a diffusion-controlled regime, the growth rate of a single gas cluster can be readily estimated from a scale analysis of the diffusion equation. When the supersaturation Ja can be assumed constant in the far field, it is easy to show that the area (equivalent to the volume in micromodels) of a single gas cluster is given by $\Sigma \propto Ja^2 \tau \mathcal{F}$, where $\mathcal{F} = \Gamma^2 / 4\pi\Sigma$ is called the shape factor, and Γ and Σ are the perimeter of the cluster (proportional to the area of the gas–liquid interface) and the area of the cluster (proportional to the volume), respectively. While $\mathcal{F} = 1$ in the bulk or in an effective 2D porous medium, i.e., for a circular gas bubble growing in a Hele–Shaw cell the result is different in 2D real porous media. As can be seen from Figure 11.16, after the transition stage corresponding to the typical growth of the bubble in a single pore, the radius of the bubble varies as $\sqrt{t}$, see Figure 11.14, and this coefficient tends to increase as a function of time. Consequently, for the same values of Ja, the same boundaries conditions and $Bo = 0$, the volume of a single gas cluster tends to grow faster in a real porous medium than in a Hele–Shaw cell or an effective porous media.

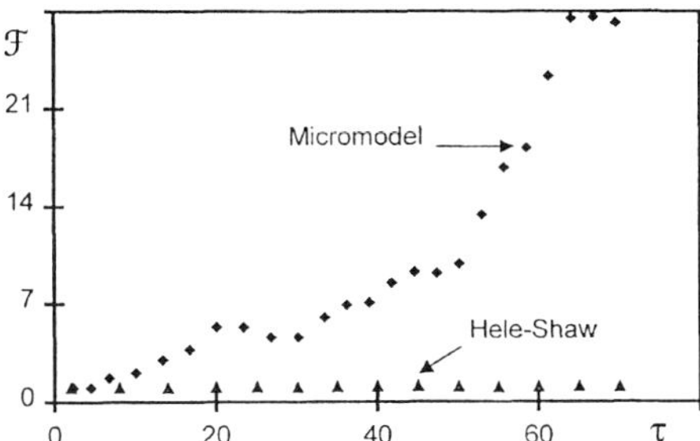

Figure 11.16 *Evolution of shape factor $\mathcal{F}$ in a micromodel and in a Hele–Shaw cell under identical thermodynamic conditions. These results have been obtained with fluid C and for rapid depressurizing*

Similar conclusions were derived by Satik *et al.* (1995). They concluded that the pattern exhibits a fractal structure and proposed an explicit scaling law to describe the gas cluster growth in the percolation domain. This law was derived under the assumption of quasistatic concentration fields and is expressed by

$$\left(\frac{R}{a}\right)^{\lambda}\left[1+\lambda\ln\left(\frac{Re}{R}\right)\right]=2\pi k\lambda\tau, \tag{11.13}$$

where here R is the cluster radius of gyration, see for example Stauffer and Aharony (1992), Re the outer boundary radius, λ is the fractal dimension of the gas cluster (equal to 1.82 in 2D) and $k=\mathcal{R}T/He\,M_s$, where M_s is the solute molecular weight. As shown in Figure 11.17, this law is in rather good agreement with our experimental results. This is in fact quite surprising since in our case the concentration fields are far from quasistatic (see below) and the size of the considered cluster is *a priori* too small for equation (11.13) to hold, see Satik *et al.* (1995). It is also worth mentioning that the relationship $\Sigma \propto Ja^2\tau\mathcal{F}$, with $\mathcal{F}=\Gamma^2/4\pi\Sigma$, is derived under the condition that the geometrical perimeter Γ can be assimilated to the effective perimeter for the mass transfer. In the case of fractal interfaces, it is well known that the effective (accessible) external perimeter is in fact smaller than the geometrical perimeter due to the screening of less advanced points of the interface by the most advanced ones, cf. Stauffer and Aharony (1992) and Sapoval (1994). The evolution of $\mathcal{F}$ for the micromodel experiment reported in Figure 11.16 is qualitatively consistent with a developing fractal interface. Although the concept of shape factor as defined above permits us to illustrate the striking difference between the growth of a bubble in a porous medium and in a Hele–Shaw cell, it is clearly of limited interest in the case of a fractal cluster owing to the screening effect. It may be observed that the screening is presumably less effective for transient diffusion regimes, as is the case in our experiments, than for quasistatic concentration fields. However, this specific aspect is still to be investigated.

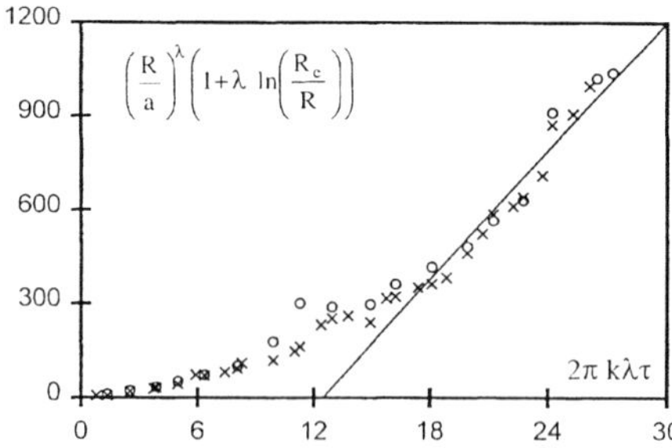

Figure 11.17 *Comparison of experimental gas cluster growth $(x,0)$ with the theoretical scaling law, equation (11.13) (straight line)*

Concerning the influence of the different parameters Ja, M, θ, and Bo on the growth of the gas cluster, the systematic experimental investigation carried out has led to the following main results:

(a) In agreement with the scale analysis, the volume of the gas formed is proportional to Ja^2, as shown in Figure 11.18, $NSB \propto Ja^2 \tau^\alpha$ (NSB = number of invaded bonds).

(b) The value of α depends on the pressure lowering law, as in the case of homogeneous fluids. For a rapid depressurization we have $1.3 < \alpha < 2.3$, with an average close to 1.5, see Figure 11.19, and for a linear decay $3.5 < \alpha < 4.7$, with an average close to $\alpha = 4$, see Dominguez (1997). Ranges of variation of α corresponding to these two

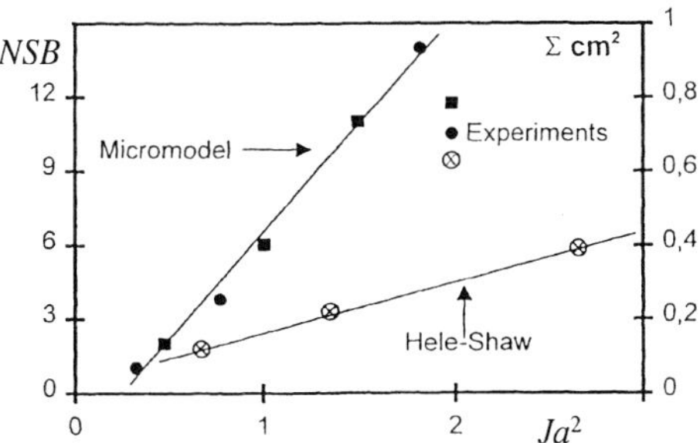

Figure 11.18 *Effects of micro-structure on bubble growth. Comparison of bubble growth in a Hele–Shaw cell with bubble growth in a micromodel as a function of Ja^2, NSB is the number of invaded pores*

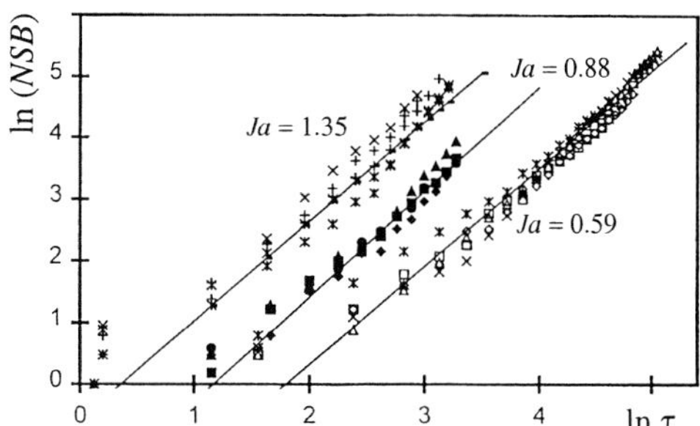

Figure 11.19 *Influence of Jacob number, Ja, on the growth law. The results of three experiments are reported for each Jacob number. $NSB \propto \tau^{1.62}$ for $Ja = 1.35$, $NSB \propto \tau^{1.61}$ for $Ja = 0.88$ and $NSB \propto \tau^{1.55}$ for $Ja = 0.59$. Each series of points with the same symbol corresponds to an experiment*

to continuum models), i.e., on a pore network model which is based on the modeling of mechanisms at pore scale.

In fact, pore network models have been used for studying a great variety of transport phenomena in porous media, including liquid–vapor phase change phenomena, see Prat (1993) and Laurindo and Prat (1996). Regarding the growth process studied in the present chapter, a pore network model was developed by Li and Yortsos (1995a, 1995b). The main differences between the network model of Li and Yortsos and the simulator presented in the present paper lie in the diffusive transport modelling (see below) and the fact that our model accounts additionally for the presence of liquid films along the walls of the invaded bonds. Also we would like to underline the fact that the effect of gravity is taken into account in our simulator and our version allows the simulation of correlated microstructures, see Dominguez *et al.* (2000). The porous medium is represented by a network of pores joined by bonds (throats), see Figure 11.22, and the pore and bond sizes are distributed randomly according to given distribution laws. The model is directly based on the observations of the cluster growth as explained below.

Observations from the transparent micromodel show that, after the initial gas bubble nucleates on the pore surface, it quickly detaches and migrates to the center of the pore. Then the bubble grows until it fills the entire pore body. At this stage, additional mass diffusion increases the gas pressure and the interface invades another pore. From these observations two main stages can be distinguished in the cluster growth:

(i) a slow pressurisation stage, during which the pressure in the cluster increases and the gas–liquid menisci move slightly to adjust their curvature to accommodate the pore geometry, and

(ii) a fast penetration stage immediately after a capillary barrier at a perimeter throat is exceeded.

Following this stage, that corresponds to an evolution at almost constant mass, the volume of the cluster increases and the gas pressure quickly reduces to the adjacent liquid pressure as the invaded pore is occupied. After completely occupying a pore, the meniscus can invade a neighbouring bond if the corresponding capillary barrier is exceeded. Moreover,

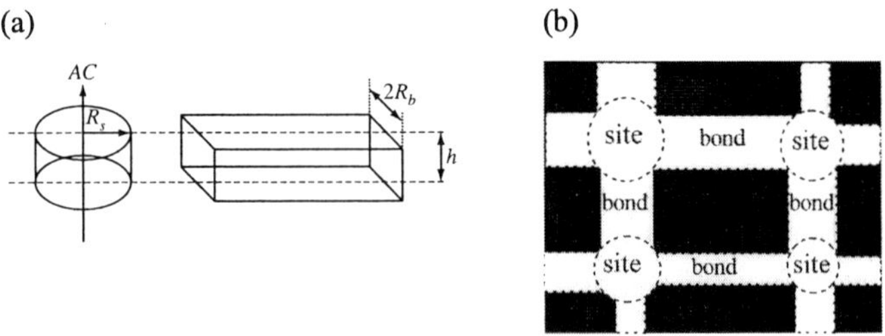

Figure 11.22   *Structure, sites and bonds of the numerical network*

for perfectly wetting liquids, it is also observed that a film of liquid remains in the corners of the pores and bonds invaded by the gas. Consistently with these observations we have developed a pore network automaton where both occupancy of pore and concentration fields are computed. The main features of the pore network automaton are listed as follows:

(a) The porous medium is modeled by a two-dimensional square lattice of pores and bonds. The pores provide the volumetric storage and the bonds control the capillary characteristics. Log–normal distributions are used to randomly assign bond and pore size on the numerical network. As in the micromodel, the lateral boundaries of the network are inaccessible to flow, with the exception of two producing sites at opposite ends. This network has the same characteristics as the micromodel, i.e., the same porosity $\varepsilon = 0.58$, the same length of bond (corresponding to the length of ducts $l_d = 0.1\,\mathrm{cm}$), the same thickness $h = 700\,\mu\mathrm{m}$, and the same pores and ducts sizes distributions. The geometry of pores and ducts is approximated by cylindrical sites and rectangular bonds, see Figure 11.22.

(b) As in the experiments, the network is initially occupied by a supersaturated liquid and the system has a uniformly distributed initial pressure P_0 and initial concentration of CO_2. The fluid physical properties are those of the fluids used in the experiments. The simulation starts by reducing the liquid pressure either by a sharp step ΔP or at a specified rate $m = -\mathrm{d}P/\mathrm{d}t$. In the simulations, we assume that we are after the nucleation of one bubble and that the pore that contains the nucleation site is fully occupied by the gas at pressure P_g, which is the same as the pressure of the liquid $P_l = P_0 - \Delta P$ (neglecting the capillarity of the pore body). At the same time, the bubble is surrounded by four liquid saturated bonds and the concentration C of the dissolved gas at the gas–liquid interface is given by $C = He\,P_l$, where P_l is the system pressure. As in Li and Yortsos (1995a, 1995b), the hypotheses that sustain these approximations are that thermodynamical equilibrium applies at the gas–liquid interface at all times (note that the Kelvin effect is neglected) and that the relation between the liquid pressure and the concentration is linear.

(c) The transient diffusive mass transfer of CO_2 in the liquid that drives the gas cluster growth, and the invasion of the network is computed by using explicit analytical mass transfer laws adapted from classic solutions of transient diffusion processes, cf. Crank (1957). One such law is sufficient for the less wetting liquids, since in this case the gas–liquid interface is restricted to the cross section of bonds. Two laws are considered in the case of the more wetting liquids because of the presence of liquid films along the bond walls. These laws, see Dominguez et al. (2000), allow one to compute the mass flux between the liquid and the gas in each interfacial bonds. A more traditional way for computing a concentration field governed by a diffusion equation in the network approach is to use a finite difference or finite volume discretization, see for example Prat (1993) or Li and Yortsos (1995a, 1995b). This requires the repeated solution of a linear system of equations, which is CPU time consuming. The method outlined in this chapter avoids this problem. However, this is clearly an approximation whose validity deserves to be explored in some detail (for instance through comparisons with the discretization approach). In this chapter we have simply proceeded by comparison with the experimental results.

(d) Invasion percolation rules are used to take into account the effect of capillarity. By expressing the mass flux balance at each gaseous node as a function of time, one obtains the evolution of the mass of the gas phase, and of the pressure in the gas cluster as a function of time, namely

$$\Delta n_T = \Sigma_B \Delta n \, (i,j)_B + \Sigma_F \Delta n \, (i,j)_F \, , \qquad (11.14)$$

$$\Delta P_g = \frac{\Delta n_T \mathcal{R} T}{V_g P_g \, (t + \Delta t)}, \quad P_g \, (t + \Delta t) = \Delta P_g + P_g \, (t) \, , \qquad (11.15)$$

where $\Sigma_B \Delta n \, (i,j)_B$, and $\Sigma_F \Delta n \, (i,j)_F$ are the CO_2 mass flux from the bonds B and the liquid films F belonging to the interface between the gas cluster and the surrounding liquid phase. $V_g = V_g \, (t)$ is the volume of the gas cluster at time t. If the pressurisation $P_g \, (t + \Delta t)$ is greater than $P_l \, (j) + P_C \, (i,j)$, where $P_C \, (i,j)$ is the capillary pressure of an arbitrary bond (i,j), the interface advances and occupies the site j (note that bonds belonging to isolated trapped liquid clusters cannot be invaded as in standard invasion percolation with trapping). The volume of the gas cluster at time $t + \Delta t$ is next calculated using the mass balance derived from the state equation: $\Delta n = 1/\mathcal{R} T \, \{(V_g P_g) \, (t + \Delta t) - (V_g P_g) \, (t)\}$. Based on the difference $\Delta t = \Delta n - \Delta n_T$, additional iterations are taken on the assumed gas volume until a convergence tolerance defined by $\Delta t < 10^{-13}$ is reached. In this mode of interface advance, the time required for the bond invasion is determined from the previous mass diffusion equations and the gas cluster grows according to the invasion percolation rules, i.e., the perimeter bonds are invaded one at a time in such a way that the largest perimeter bond (in the absence of gravity) is always invaded first.

(e) Gravity effects are taken into account using an appropriate bond invasion potential (as first proposed by Wilkinson, 1984 and subsequently used by several authors, see Laurindo and Prat, 1996 among others), which depends not only on the width of the throat (as in the no-gravity case) but also on the relative position of the throat in the gravity field.

Additional details concerning the topological rules governing the pore occupancy as well as the computational algorithm can be found in the thesis of Dominguez (1997).

11.5.2 Experiments versus numerical simulations

In this section, we present the results obtained by means of two-dimensional numerical simulations and the comparisons of these results with the experimental data.

Experimental and simulated results of pore volume–time curves $NSB \, (\tau)$, which represents the overall system behaviour, as well as of growth patterns, which provide local details, are compared in Figure 11.23 for an abrupt pressure decrease. As can be seen, a

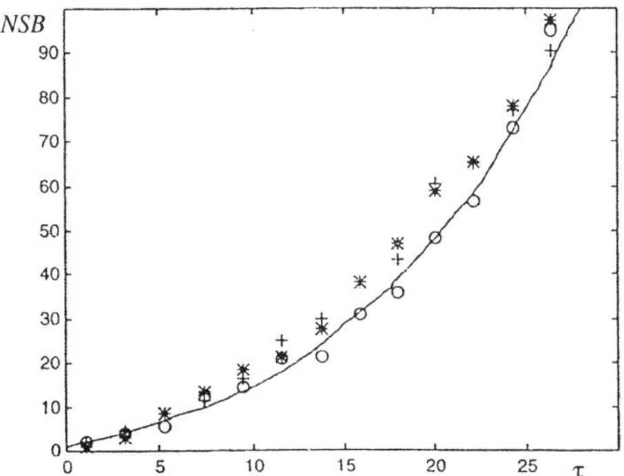

Figure 11.23 *Number of invaded bonds (NSB) as a function of time in a single cluster. Comparison between numerical results (solid line) and experimental data $(*, +, 0)$ ($Ja = 0.89$). $Ca = 4.81 \times 10^{-7}$, $\theta = 0.66$, fluid C. Rapid depressurizing*

rather good agreement is found between the experimental data and the simulations regarding the evolution of the volume of the gas phase as a function of time. A good agreement was also obtained for a constant pressure decline rate, see Dominguez *et al.* (2000).

Figure 11.24 shows the comparison between the experimental and numerical patterns for one of the experiments. The two patterns share many topological characteristics (irregular and ramified shapes, including disconnected liquid clusters) but are not identical. These differences in the microscale result from the differences between the spatial localization of pores and ducts on the experimental and numerical networks (only the pore and bond size distributions are identical in both networks). As emphasized in Li and Yortsos (1995a, 1995b), these differences in localization produce different occupancy sequences and therefore preclude the automaton from accurately matching the microscopic details of the bubble growth. Therefore, we conclude that the dominant mechanisms are correctly simulated in the range of parameters studied and that the automaton is able to match the most important aspects of phase growth. Consequently, it will be used to explore the sensitivity of the phenomenon to different parameters in the following.

11.5.3 Sensitivity study

This section concerns the investigation of the influence of the Jakob number, surface tension, wettability, Bond number, pressure decline rate and the microstructure on the growth of a single gas cluster.

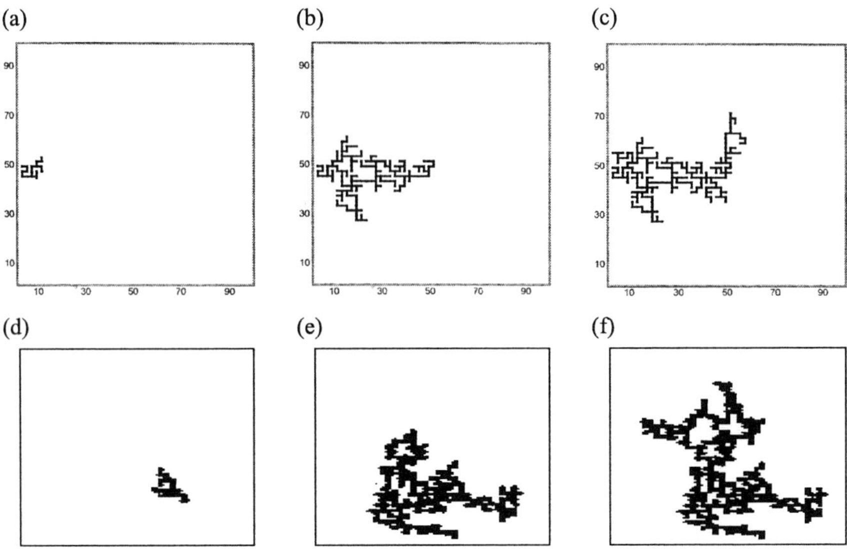

Figure 11.24 *Typical evolution of a gas cluster. Top: numerical results for (a)* $NSB = 16$, *(b)* $NSB = 146$, *(c)* $NSB = 201$. *Bottom: experimental results for (d)* $NSB = 16$, *(e)* $NSB = 146$, *and (f)* $NSB = 204$. $\Delta P_b/P_{\mathrm{sat}} = 0.41$, $M \approx 5 \times 10^{-3}$, *fluid* B, $Ca = 5.81 \times 10^{-7}$

Influence of Jakob number

The results concerning the influence of Jakob number on the gas saturation ($S \propto NSB/NT$, where NT is the total number of sites and bonds in the micromodel) are presented in Figure 11.25.

As can be seen, many aspects of these results are not only qualitatively but also quantitatively in agreement with the experimental data. In accordance with the experiments, the growth law takes the form

$$NSB \propto Ja^2 \tau^\alpha, \tag{11.16}$$

with α independent of Ja. The simulations also show the existence of a transition zone corresponding to the evolution of α from 1 to 2.13, which corresponds to the investigated range of variation of the gas cluster size NSB investigated. This last result is in agreement with the values obtained experimentally with the micromodel. It is worth noting that the transition zone time scales deduced from the simulations ($0 < \ln \tau < 3.5$) are also in agreement with the experimental data.

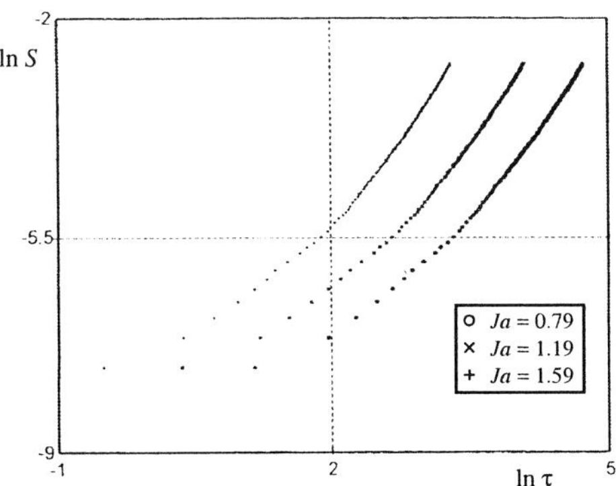

Figure 11.25 *Pore network simulation. Influence of the Jacob number on the growth law from numerical results ($\theta = 0.63$). $S \propto \tau^{2.13}$*

Influence of surface tension

As the capillarity directly controls all entrances of the interface bonds, the greater the surface tension, the higher is the pressure difference required to penetrate the bonds. As this difference is itself controlled by mass transfer, i.e., by the number of the CO_2 molecules contained in the gas phase, the effect of this parameter is then to modify the growth rate of the gas cluster, i.e., the values of the exponent α in equation (11.16) as illustrated in Figure 11.26.

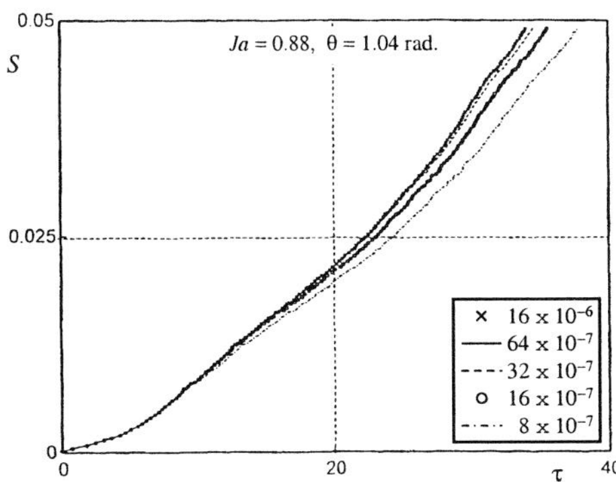

Figure 11.26 *Pore network simulation. Influence of capillary number. For a given value of τ, the value of S increases with the capillary number*

Influence of wettability

The influence of the wettability, which leads to the spreading of the liquid along the corners and surfaces of the pores occupied by the gas phase, is illustrated in Figure 11.27. As can be seen, this parameter does not modify the growth of the gas phase as a function of time; α remains constant and approximately equal to 2.1 irrespective of the value of θ. In fact the effect of the wettability is exclusively restricted to the monotonic increase of NSB. This result is in complete agreement with the experimental data. As a local (pore level) examination of mass transfer shows, this effect can be considered as a source effect owing to the liquid remaining in the extreme corners of the cross section of pores and micro-cavities of the walls.

Influence of pressure decline rate M

The results concerning the influence of the pressure decline rate M on the evolution of the volume of gas in the absence of gravity effect are presented in Figure 11.28. Again, in agreement with the experimental data, we note that the larger is the pressure decline rate, the faster is the bubble growth, which remains consistent with the results in the bulk.

Influence of Bond number

Figure 11.29 shows the influence of the Bond number on the growth phenomenon in the absence of bond/pore size correlations. As in the experiments, this parameter has a significant influence on the evolution of the volume of gas formed after the pressure lowering, the exponent α varying from 2 to 2.2 while the Bond number varies from 0 to 5.1×10^{-3}. The simulated patterns of gas clusters shown in Figure 11.30, and the

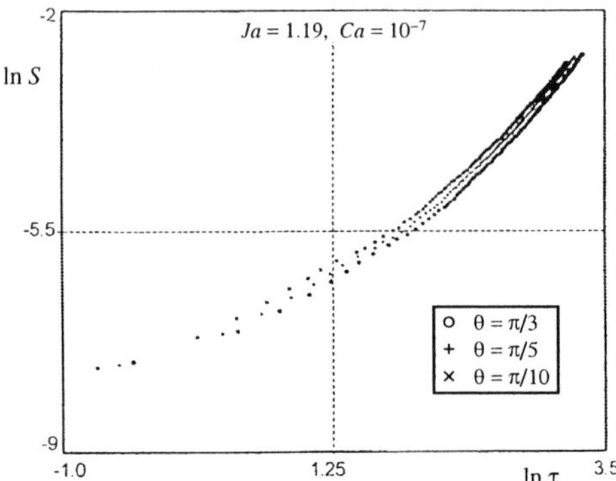

Figure 11.27 *Pore network simulation. Influence of the wetting properties at constant Ca and Ja. Best fits give $S \propto \tau^{2.18}$ for $\theta = \pi/3$ rad., $S \propto \tau^{2.11}$ for $\theta = \pi/5$ rad. and $S \propto \tau^{2}$ for $\theta = \pi/10$ rad. For a given value of τ, the value of S increases with θ*

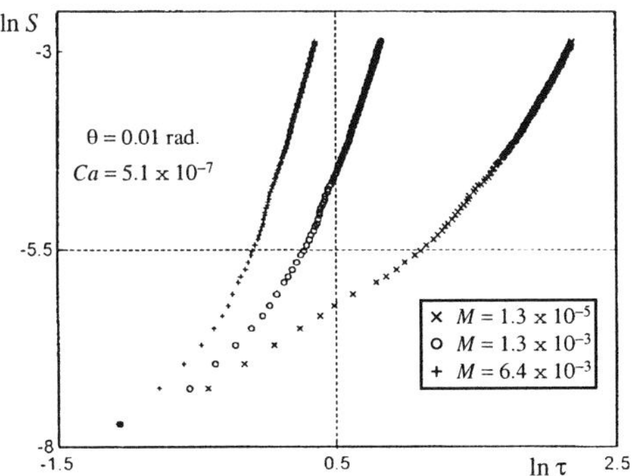

Figure 11.28 *Pore network simulation. Influence of pressure decline rate M on the growth law. Best fits give $S \propto \tau^{5.97}$, $S \propto \tau^{4.99}$ and $S \propto \tau^{2.71}$*

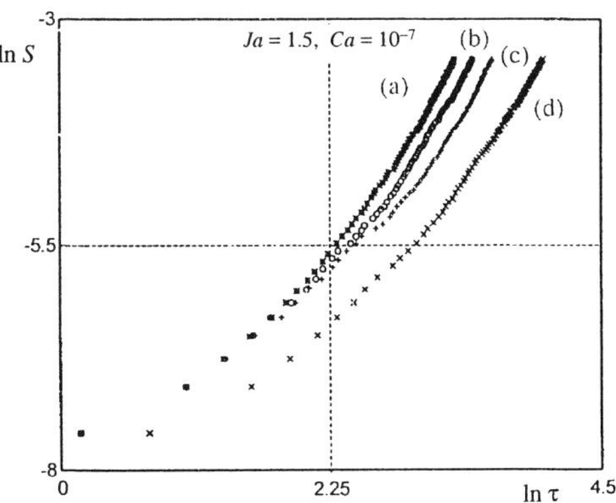

Figure 11.29 *Pore network simulation. Effect of gravity on the cluster growth law. Results are shown for (a) $Bo \times 10^3 = 5.15$, (b) $Bo \times 10^3 = 2.61$, (c) $Bo \times 10^3 = 1.31$, and (d) $Bo = 0$. Best fits give $S \propto \tau^{2.23}$ for (a), $S \propto \tau^{2.22}$ for (b), $S \propto \tau^{2.17}$ for (c), and $S \propto \tau^{2.06}$ for (d)*

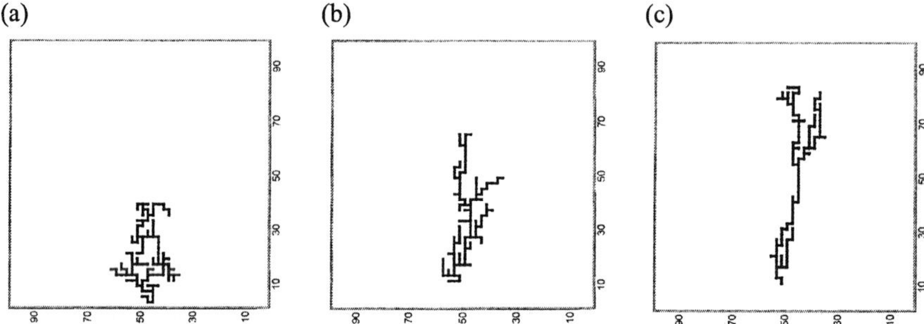

Figure 11.30 *Pore network simulation. Influence of Bond number on phase distribution. Results are shown for $NSB = 100$ with (a) $Bo = 0$, (b) $Bo \times 10^3 = 2.61$, and (c) $Bo \times 10^3 = 5.15$. (Gas phase in black, liquid phase in white)*

local analysis of mass transfer along the gas–liquid interface, assist us to understand this influence. The main factors are:

(i) the elongation of the cluster in the direction of the gravity when the Bond number increases, i.e., the increase of the effective accessible perimeter with the anisotropy of the cluster, and

(ii) the movement of the interface towards the region where the supersaturation is less affected by the mass transfer.

As confirmed numerically in Dominguez (1997), the latter effect indeed leads to greater local mass transfers through the liquid bonds adjacent to the region of the cluster that is the most advanced towards the upper edge of the network.

Influence of bond/pore size correlation

Preliminary results regarding the influence of bond/pore size correlations on the growth rate of gas clusters of small size: (from $NSB = 1$ to $NSB = 200$) were also obtained as a function of time using the procedure defined in Javier Cruz *et al.* (1989), see Dominguez *et al.* (2000). The results indicate that the growth rate is sensitive to the existence of correlations. This issue would deserve to be explored further, both experimentally and numerically, through a thorough statistical study.

11.6 CLOSURE OF MASS BALANCE EQUATIONS

As recalled in the introduction, the system of equations (11.1) is not closed and a gas generation law is necessary to express the source term Γ_i. For the initial stage of the process, corresponding to ours experiments (i.e., before the onset of bulk gas flow) the mass balance equation for the gas phase reduces to

$$\frac{\partial}{\partial t}\left(\varepsilon\rho_g S_g\right) = \Gamma_g \quad \text{with} \quad \Gamma_g = -\Gamma_l, \quad P_g - P_l = P_C\left(S_g\right). \tag{11.17}$$

Two ways can be used to determine Γ_i. Either we assume the instantaneous thermodynamic equilibrium or we take into account the two consecutive processes of nucleation and growth that control the evolution of the gas phase formation during depletion.

In the first case the prediction of Γ_i is immediate if we know the state laws for the different phases and the complementary equations governing the equilibrium conditions between these phases. For the liquid-to-gas transition studied here, if we assume that the CO_2 is a perfect gas and the liquid mixture an ideal solution, then Γ_i is easily derived from the state equation of the gas phase

$$V_g = \left(\varepsilon S_g\right) = \frac{n\mathcal{R}T}{P_g}, \tag{11.18}$$

$$P_g - P_l = P_C\left(S_g\right), \tag{11.19}$$

$$\Delta C = He\,\Delta P, \tag{11.20}$$

$$\frac{\partial}{\partial t}\left(\varepsilon\rho_g S_g\right) = \Gamma_g = M\frac{\mathrm{d}n}{\mathrm{d}t}, \tag{11.21}$$

where n and P_g are the mole content and the gas pressure derived from the equilibrium conditions (11.19) and (11.20).

The second way is derived from the approach used for the closure of the two-phase flow equations. It consists in finding the closed form solution of an elementary problem, defined by the local balance equations to express the local interfacial mass flux jump, then to resort to a topological law to express the averaged mass flux and the saturation on the Representative Elementary Volume (REV). In our study this information can be extracted from the observations and measurements previously presented.

Assuming that the liquid-to-gas transition in porous media is described by the two consecutive processes of nucleation and bubble growth then the local volume of the gas phase will be given by

$$\frac{\partial}{\partial t}\left(\varepsilon\rho_g S_g\right) = \Gamma_g = \left(\varepsilon S_l\right)\frac{M}{\mathcal{R}T}\frac{\mathrm{d}}{\mathrm{d}t}\left[P_g\int_0^t N\left(t'\right)V_g\left(t,t'\right)\mathrm{d}t'\right], \tag{11.22}$$

where V_g is the volume at time t of bubbles nucleated at time t'. It is given by equation (11.16) with α equal to the average experimental value, i.e., $\alpha = 1.5$, and $N\left(t'\right)$ is the active nucleation site density given by the Yang's law, i.e., equation (11.9).

Equations (11.21) and (11.22) have been tested through a comparison between the theoretical results derived from these equations and the experimental results related to the pressure–saturation correlation. As the experimental results (active nucleation sites, gas saturation) concern the micromodels as a whole the global mass balance describing the phenomenon at this scale must be previously established. This has been done by using the general transport theorem properties. In our case the application of this theorem consists of integrating the mass balance equations (11.1) with the associated boundary conditions over the volume of the micromodels. Assuming negligible concentration and velocity gradients, the result of this integration leads to the following differential equations which describe the mass balances for solute and total mass:

$$\frac{\mathrm{d}}{\mathrm{d}t}\left[C\left(V_p - V_g\right)\right] + \Pi_g = C\frac{\mathrm{d}Q}{\mathrm{d}t}, \tag{11.23}$$

$$\frac{\mathrm{d}Q}{\mathrm{d}t} = \frac{\mathrm{d}}{\mathrm{d}t}\left(V_p - V_g\right). \tag{11.24}$$

In these equations, $\Pi_g = \int_\Omega \Gamma_g \, \mathrm{d}\Omega$, where Γ_g is the source term defined by equation (11.21) or (11.22), depending on whether the thermodynamical equilibrium is considered or not, Ω is the volume of the micromodel, V_p is the total pore volume of the micromodel, V_g is the volume of the gas phase which develops because the pressure reduction, and Q is the volume of liquid displaced by the gas phase formation.

Equations (11.23), (11.24), and (11.18), (11.20), with (11.21) or (11.22) as closure equations for Γ_g, have been used in order to simulate numerically the pressure saturation experimental runs. In the simulation we assumed that depletion occurs in steps and that during each step the pressure decline, bubble formation, and growing of bubbles occur consecutively rather than simultaneously. Corresponding results and their comparison with the experimental data are shown in Figures 11.31 and 11.32.

Several conclusions can be drawn from these results:

- The satisfactory description of heterogeneous nucleation by Yang's law.

- The qualitative agreement between the experimental results and numerical simulations of the pressure–saturation relationship based on the use of Γ_g given by equation (11.22). Such is not the case when we assume that the thermodynamic equilibrium between the gas formed and the liquid is instantaneous and the source term described by equation (11.21). As expected, and experimentally observed, all curves are bounded above by the solution corresponding to quasi-static growth.

Regarding the discrepancy between the experimental data and the numerical simulations based on equation (11.22), different causes can be identified. In particular, the role of capillarity and coalescence, which were not taken into account, modifies the values of the CO_2 concentration in the liquid phase and the growing law of the gas phase volume as a function of time.

11.7 CONCLUSIONS

The works performed at the pore scale and the pore-network scale during the last ten years have led to significant improvements in the understanding of nucleation and bubble

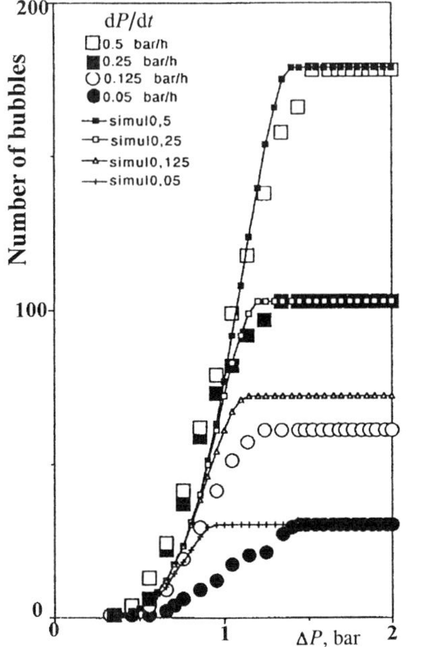

Figure 11.31 *Number of bubbles formed*

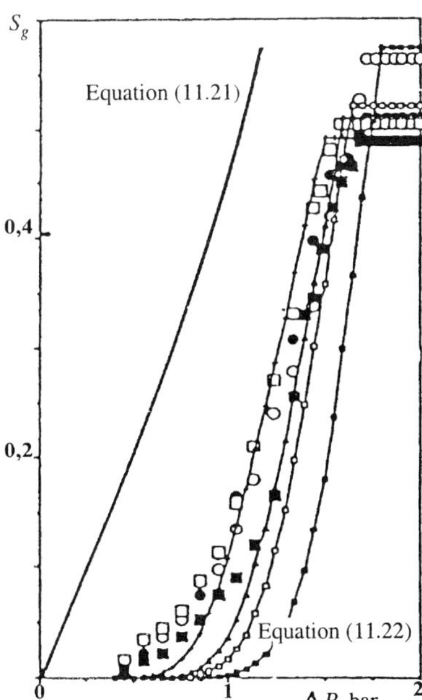

Figure 11.32 *Saturation as a function of supersaturation*

growth in porous media, at least for low supersaturations, where diffusion predominates. The most noticeable results include:

(i) The decisive role of the capillary trapping mechanism on the formation of bubbles, i.e., the pre-existence of microbubbles trapped in cavities or roughness on the solid surface.

(ii) The effect of the microstructure that leads to fractal patterns and a growth rate scaling which is different from the classical scaling, i.e., scaling for an effective medium and scaling in the bulk. In particular, the volume of a single gas cluster tends to grow faster in porous media than in the bulk.

The mechanisms of growth described in the present chapter are typical of invasion percolation (IP) in the absence of gravity forces and invasion percolation in a destabilizing gradient (IPDG) when gravity effects are present, see Xu *et al.* (1998). It is interesting to note that concepts of invasion percolation and invasion percolation in a gradient have been also extremely useful for studying other situations of liquid–gas phase change phenomena in porous media where capillary effects dominate, notably the slow drying of a porous medium, see Prat and Bouleux (1999) and references therein, and the vaporisation process occurring in the wick of a capillary evaporator, e.g., Figus *et al.* (1999). Regarding the particular process considered in the present chapter, gradient percolation ideas could be used to explore further the growth rate scaling in the presence of gravity forces, as well as to study the influence of viscous forces when the cluster becomes large.

The results for the growth reported in the present work are valid for spatially uncorrelated pore throats. Spatial correlations are expected to affect the development of the gas clusters and their influence would therefore deserve to be studied.

Although we have concentrated on isothermal nucleation driven by diffusion in a super-saturated liquid, the results presented and the conclusions reached are clearly of interest to boiling nucleation in a porous media at low superheat. However, there are specific aspects in boiling, such as the influence of the solid phase in the heat transfer, that would also deserve to be studied.

Finally, it is interesting to point out that the findings obtained with the pore network studies have been exploited to gain insight into an important practical macroscopic problem, such as the one of the critical gas saturation, e.g., Du and Yortsos (1999). Further results on this problem can be expected within the framework of invasion percolation and invasion percolation in a gradient concepts, e.g., influence of gravity or viscous forces. Naturally, percolation approaches are limited to situations where capillarity controls the invasion phenomenon locally, i.e., at small and moderate pressure decline rates. As emphasized by Du and Yortsos (1999), complex phenomena, such as snap-off of gas bubbles, bubble division in pore throats and ganglia motion and coalescence, will occur at larger pressure decline rates. These phenomena have not been considered in this chapter.

Acknowledgements

The authors wish to thank warmly Dr A. El-Yousfy, Dr A. Dominguez, Dr R. Lenormand and Dr C. Zarcone for their contributions to the studies reported in the present chapter.

REFERENCES

Abraham, F. F. (1974). *Homogeneous Nucleation Theory.* Adv. Theory San Antonio Chem., Suppl.1. Academic Press, New York.

Bankoff, S. G. (1958). Entrapment of gas in the spreading of a liquid over a rough surface. *AIChE J.* **4**, 24–26.

Bankoff, S. G. (1959). The prediction of surface temperature at incipient boiling. *Chem. Eng. Progress*, Symp. Series No. 29, **55**, 87–101.

Betata, S. A. (1998). *Les écoulements polyphasiques aux abords des puits dans les gisements d'hydrocarbures.* Ph.D. thesis. Université Paris 6, France.

Blander, M. (1979). Bubble nucleation in liquids. *Adv. Colloid Interface Sci.* **10**, 1–12.

Bonnet, J. and Lenormand, R. (1977). Réalisation de micromodèles pour l'étude des écoulements polyphasiques en milieu poreux. *Rev. Inst. Franç. Pétrol.* **42**, 447–480.

Cole, R. (1979). *Boiling Phenomena.* Advances in Heat Transfer, Vol. 1. Hemisphere, New York.

Cornwell, K. (1977). Naturally formed boiling site cavities. *Lett. Heat Transfer* **4**, 63–72.

Crank, J. (1957). *The Mathematics of Diffusion.* Clarendon Press, Oxford.

Crum, A. (1982). Nucleation and stabilisation of microbubbles in liquids. *Appl. Sci. Res.* **38**, 101–115.

Defay, R. and Prigogine, I. (1951). *Tension Superficielle et Adsorption.* Ed. Desoer, Liege.

Dominguez, A. (1997). *Formation d'une phase gazeuse par decompression d'une solution binaire en milieu poreux.* Ph.D. thesis. Institut National Polytechnique de Toulouse, France.

Dominguez, A., Bories, S., and Prat, M. (2000). Gas cluster growth by solute diffusion in porous media. experiments and automaton simulation on pore network. *Int. J. Multiphase Flow* **26**, 1951–1979.

Drelich, J., Miller, J. D., and Good, R. J. (1996). The effect of drop (bubble) size on advancing and receding contact angles for heterogeneous and rough solid surfaces. *J. Colloid Interface Sci.* **179**, 37–50.

Du, C. and Yortsos, Y. C. (1999). A numerical study of the critical gas saturation in a porous medium. *Transport in Porous Media* **35**, 205–225.

El-Yousfy, A. (1992). *Contribution à l'étude des mécanismes de formation d' une phase gaseuse par dé tente d'une solution binaire liquide gas en milieu poreux.* Ph.D. thesis. Institut National Polytechnique, Toulouse, France.

Epstein, M. (1994). A similarity solution for combined hydrodynamic and heat transfer controlled bubble growth in a porous media. *J. Heat Transfer* **116**, 516–521.

Figus, C., Le Bray, Y., Bories, S., and Prat, M. (1999). Heat and mass transfer with phase change in a porous structure partially heated. Continuum model and pore network simulations. *Int. J. Heat Mass Transfer* **42**, 2257–2569.

Firoozabadi, A. and Kashchiev, D. (1993). Pressure and volume evolution during gas phase formation in solution gas drive process. SPE 26826.

Florschuetz, L. W. and Chao, B. T. (1965). On the vapor bubble collapse. *J. Heat Transfer* **87**, 209–220.

Frenkel, J. (1946). *Kinetic Theory of Liquids.* Clarendon Press, Oxford.

Griffith, P. and Wallis, J. D. (1960). The role of surface conditions in nucleate boiling. *Int. J. Heat Mass Transfer* **30**, 49–60.

Hong, J. (1985). Application of fractional derivatives method to bubble growth: dissolution processes with or without first order chemical reactions. *AIChE J.* **31**, 1695–1706.

Hunt, Jr, E. B. and Berry, V. J. (1956). Evolution of gas from liquids flowing through porous media. *AIChE J.* **2**, 560–567.

Hwu, W. H., Sheu, J. S., and Maa, J. R. (1988). Adsorption and nucleation of water vapor on smooth surfaces. *J. Colloid Interface Sci.* **121**, 303–308.

Jarvis, T. J. (1975). Bubble nucleation mechanisms of liquid droplets superheated in ether liquids. *J. Colloid Interface Sci.* **50**, 359–372.

Javier Cruz, M., Mayagoitia, V., and Rojas, F. (1989). Mechanistics studies of capillary processes in porous media. Part 2. Construction of porous networks by Monte-Carlo methods. *J. Chem. Soc. Faraday Trans.* **85**, 2079–2086.

Jones, O. C. and Zober, N. (1978). Bubble growth in variable pressure fields. *J. Heat Transfer* **100**, 453–458.

Kashchiev, D. and Firoozabadi, A. (1993). Kinetics of the initial stage of isothermal gas phase formation. *J. Chem. Phys.* **98**, 4690–4699.

Laurindo, J. B. and Prat, M. (1996). Numerical and experimental network study of evaporation in capillary porous media. Phase distributions. *Chem. Eng. Sci.* **51**, 5171–5185.

Lee, H. S. and Merte, H. (1996). Spherical vapor bubble growth in uniformly superheated liquids. *Int. J. Heat Mass Transfer* **39**, 2447–2458.

Lenormand, R., Touboul, E., and Zarcone, C. (1988). Numerical models and experiments on immiscible displacements in porous media. *J. Fluid Mech.* **189**, 165–167.

Li, X. and Yortsos, Y. C. (1995a). Theory of multiple bubble growth in porous media by solute diffusion. *Chem. Eng. Sci.* **50**, 1247–1271.

Li, X. and Yortsos, Y. C. (1995b). Visualization and simulation of bubble growth in pore networks. *AIChE J.* **41**, 214–222.

Lothe, J. (1969). Statistical mechanics of nucleation. In *Nucleation* (ed. A. C. Zettlemoyer), Chapter 3, pp. 109–224. Marcel Dekker, New York.

Lubetkin, S. D. (1988). The kinetics of nucleation of adamant crystals from the vapour. *J. Chem. Phys.* **89**, 7502–7514.

Marto, P. J., Moulson, J. A., and Maynard, M. D. (1968). Nucleate pool boiling of nitrogen with different surface conditions. *J. Heat Transfer* **90**, 437–444.

Meakin, P., Feder, J., Frette, V., and Jossang, T. (1992). Invasion percolation in a destabilizing gradient. *Phys. Rev. A* **46**, 3357–3368.

Miyatake, O., Yamada, A., Tsutsui, Y., and Tanaka, I. (1994). Rate of bubble growth in a superheated binary solution of volatile components. *Int. Chem. Eng.* **34**, 377–385.

Moulu, J. C. (1989). Solution-gas drive: experiments and simulation. *J. Petroleum Sci. Eng.* **2**, 379–386.

Plesset, M. S. and Zwick, S. A. (1954). The growth of vapor bubble in superheated liquids. *J. Appl. Phys.* **25**, 493–500.

Prat, M. (1993). Percolation model of drying under isothermal conditions in porous media. *Int. J. Multiphase Flow* **19**, 691–704.

Prat, M. and Bouleux, F. (1999). Drying of capillary porous media with stabilized front in two dimensions. *Phys. Rev. E* **60 (5 PTB)**, 5647–5656.

Reiss, H. (1968). Translation–rotation paradox in the theory of nucleation. *J. Chem. Phys.* **48**, 5553–5558.

Sapoval, B. (1994). General formulation of Laplacian transfer across irregular surfaces. *Phys. Rev. Lett.* **73**, 3314–3316.

Satik, C., Li, X., and Yortsos, Y. C. (1995). Scaling of single-bubble growth in a porous medium. *Phys. Rev. E* **51**, 3286–3295.

Scriven, L. A. (1959). On the dynamics of phase growth. *Chem. Eng. Sci.* **10**, 1–12.

Sheu, J. S., Maa, J. R., and Katz, J. L. (1988). Adsorption and nucleation on smooth surfaces. *J. Statist. Phys.* **52**, 1143–1155.

Springer, G. S. (1978). Homogeneous nucleation. *Adv. Heat Transfer* **14**, 281–346.

Stauffer, D. and Aharony, A. (1992). *Introduction to Percolation Theory*. Taylor & Francis, London.

Szekely, J. and Fang, S. D. (1973). Non-equilibrium effects in the growth of spherical gas bubbles due to solute diffusion. *Chem. Eng. Sci.* **28**, 2127–2140.

Szekely, J. and Martins, G. P. (1971). Non equilibrium effects in the growth of spherical gas bubbles due to solute diffusion. *Chem. Eng. Sci.* **26**, 147–159.

Tong, W., Bar-Cohen, A., Simon, W., and You, S. (1990). Contact angle effects on boiling incipience of highly-wetting liquids. *Int. J. Heat Mass Transfer* **33**, 91–103.

Wang, C. H. and Dhir, V. K. (1993). On the gas entrapment and nucleation site density during pool boiling of saturated water. *J. Heat Transfer* **115**, 670–679.

Wang, Z. and Bankoff, S. G. (1991). Bubble growth on a solid wall in a rapidly-depressurized liquid pool. *Int. J. Multiphase Flow* **17**, 425–437.

Ward, C. A. and Forest, T. W. (1976). On the relation between platelet adhesion and the roughness of a synthetic biomaterial. *Annals Biomedical Eng.* **4**, 184–191.

Wilkinson, D. (1984). Percolation model of immiscible displacement in the presence of buoyancy forces. *Phys. Rev. A* **30**, 520–531.

Wilt, P. M. (1986). Nucleation rates and bubble stability in water carbon dioxide solutions. *J. Colloid Interface Sci.* **112**, 530–538.

Winterton, R. H. S. (1977). Nucleation of boiling and cavitation. *J. Phys. D: Appl. Phys.* **10**, 2041–2050.

Xu, B., Yortsos, Y. C., and Salin, D. (1998). Invasion percolation with viscous forces. *Phys. Rev. E* **57**, 739–751.

Yang, S. R. and Kim, R. H. (1988). A mathematical model of pool boiling nucleation site density in terms of surface characteristics. *Int. J. Heat Mass Transfer* **31**, 1127–1135.

Yortsos, Y. C. and Parlar, M. (1989). Phase change in binary systems in porous media. Application to solution gas drive. SPE Meeting, San Antonio, Texas. SPE 19697.

Zettlemoyer, A. C. (1969). *Nucleation*. Marcel Dekker, New York.

12 EFFECTS OF ROTATION ON CONVECTION IN A POROUS LAYER DURING ALLOY SOLIDIFICATION

D. N. RIAHI

Department of Theoretical and Applied Mechanics, 216 Talbot Laboratory, University of Illinois at Urbana–Champaign, 104 South Wright Street, Urbana, IL 61801, USA

email: d-riahi@uiuc.edu

Abstract

The topic of this chapter is that of the effects of an external constraint of rotation on convection, driven mainly by compositional buoyancy, in a porous layer adjacent to the solid–liquid interface during directional solidification of a binary alloy. In the solidification literature such a porous layer is referred to as a mushy layer. The analyses carried out by several studies in the past on the subject of this chapter and the subsequent results are reviewed first, the latest results unpublished elsewhere are presented briefly, and then in the second-half of the chapter the investigation of effects of rotation on convection in a horizontal porous layer of melt and dendrite solids is carried out subject to a simple model. The results based on this simple model indicate that the Coriolis force can have stabilizing effects on both the stationary and oscillatory mode of convection, while the oscillatory mode of convection experiences an additional destabilization due to the Coriolis force effect.

Keywords: convection, rotating convection, porous medium, alloy solidification, Coriolis effects, natural convection, porous layer, rotating fluid

12.1 INTRODUCTION

Convection effects during alloy solidification are known to be important, see for example Davis (1990). The convective flow affects the solid–liquid content within a porous layer, which exists adjacent to the solid–liquid interface, and influences the flow pattern and the critical conditions for the generation of flow instabilities in the solidification system. It is important to reduce the undesirable effects of convection as much as possible for the solidified system and also find ways to prevent the formation of localized chimneys within the porous layer for such systems, since it is known that chimney convection can lead to defects and imperfections in the final produced crystals, see for example Sample and Hellawell (1984).

Effect of an external constraint of rotation acting on convective flow in either an ordinary medium, see Chandrasekhar (1961), Hunter and Riahi (1975), Riahi (1977) and Busse (1978), or in a porous medium, see Riahi (1994b) and Vadasz (1998, 2000), is known to be generally non-trivial and significant for a sizeable range of values of the rotation rate.

The rotational effects on the convective flow instabilities during alloy solidification have been of interest to the crystal growers for a number of years. In industrial crystal growth processes it has been desirable to impose certain external constraints, such as rotation, in an optimized manner, upon the solidified system in order to reduce the effects of flow instabilities or oscillations which can lead to a microdefect density in the crystal and thus reduce the quality of the produced solidified material. Theoretical results on the effects of rotation about a vertical axis on the flow of melt during alloy solidification and in the normal gravity environment, see Riahi (1993, 1994a), indicated conditions under which rotation may stabilize the convective flow. Computational studies on the effects of rotation about a vertical axis of a horizontal layer on the flow of a melt during alloy solidification, see Neilson and Incropera (1993) indicated stabilization of vertical plumes and their lack of meandering due to such a rotational constraint. Sample and Hellawell (1982, 1984) performed an NH_4Cl alloy experiment in a cylindrical mold with a chilled bottom surface where solidification was induced. They applied a rotation and tilting technique to change the orientation of the force of gravity relative to the bottom surface of the cylinder. They observed that for slow and steady rotation of the mold about the vertical axis, which coincided with the axis of the cylinder, the chimney formation and development was about the same as in the case without rotation. However, for slow and steady rotation of a tilted mold about a vertical axis, the number of chimneys was reduced substantially and, under some conditions, they were completely eliminated. The fast rotation case is not considered beneficial to the crystal growers, as, for example, experimental results due to Kou *et al.* (1978) indicated that if the rate of rotation became too large then segregates were formed along a ring between the axis and the outer edge of the ingot in their experimental apparatus.

The experimental results referred to in the previous paragraph and the indication for the possible usefulness of inclined rotational constraint applied on the solidified system, where the axis of rotation is inclined with respect to the direction of the gravity force, led to further research studies. As described in some detail in the next section, some recent studies at the onset of convection in a layer of melt rotating about an axis, which was inclined with respect to the effective gravity vector, see Sayre and Riahi (1996, 1997) and Okhuysen and Riahi (2001), aimed at understanding centrifugal or Coriolis force effects and investigated, in particular, flow instabilities, due to either stationary or oscillatory mode of disturbances of the porous layer adjacent to the solidification front during alloy solidification. Results of the numerical computations of these studies indicated preference of particular modes of convection, and a rotational constraint was found to be effective only if the rotation axis was inclined with respect to the effective gravity vector. Here, by the effective gravity vector it is meant the resultant normal gravity vector and an average component vector of the centrifugal force in the direction perpendicular to the rotation axis. Another interesting result of these studies was that the spatial locations in the porous layer, which had a tendency to form chimneys, changes as the rate of rotation was changed. This result suggested a possible industrial operational procedure for the elimination of chimney's formation tendency to be the application of an external constraint of rotation with a variable rate of rotation acting on the solidified system.

A different type of study, based on a scaling analyses and asymptotic methods in the limit of sufficiently large solutal Rayleigh number, for the natural convection in the porous layer adjacent to the liquid–solid interface, were found to be an appropriate procedure in a high gravity environment, see Regel and Wilcox (1997), where the porous-Rayleigh number can become sufficiently large for sufficiently high rotation rate of a centrifuge system which can carry the solidified layer. Such studies, which are described in some detail in a later section of this chapter, were carried out by Riahi (1997, 1998, 1999) and provided qualitative results about the effects of centrifugal and Coriolis forces on the compositional convection in the porous layer during alloy solidification. A particular and notable result of these studies, which indicated that Coriolis' force effect can have different types of influence on the flow stability, depending on the rotation sense of the centrifuge, were found to agree with some recent experimental and computation studies on the subject by Ma *et al.* (1994) and Tao *et al.* (1994).

12.2   DOUBLE-LAYER MODEL

The investigation carried out by Sayre and Riahi (1996, 1997), and the corresponding results, are described briefly in this section and this is followed by some ongoing research investigations and preliminary results due to Okhuysen and Riahi (2001). A layer of a binary alloy melt of some composition C_0 and temperature T_∞ is considered which is solidified at a rate V_0, with the eutectic temperature T_e at the position $z = 0$ held fixed in a frame moving with the solidification speed in the z-direction, where the z-axis is assumed to be anti-parallel with the high gravity vector, see Figure 12.1.

The investigated double-layer model, which consists of a non-porous layer, referred to as the liquid layer, and a porous layer, referred to as the mushy layer, at normal gravity condition is based on the assumptions of the type considered by Worster (1992), and the extension under a high gravity condition is based on the assumptions of the type considered by Arnold *et al.* (1992) for the solidification system in a centrifuge. The porous layer adjacent to the solidifying surface is of thickness $h(x, y, t)$, where t is the time variable and the x- and y-axes are in a plane ($z = 0$) which is perpendicular to the z-axis. The solidifying system is placed in a centrifuge basket rotating at some angular velocity Ω about the centrifuge axis which makes an angle γ with respect to the z-axis. The centrifuge axis is anti-parallel to the earth gravity vector, see Figure 12.2.

Next, the equations for momentum, continuity, heat and solute are considered for both the non-porous layer ($z > h$) and the porous layer ($0 < z < h$) in the moving frame $Oxyz$ whose origin O is centered on the solid–mush interface ($z = 0$). The governing

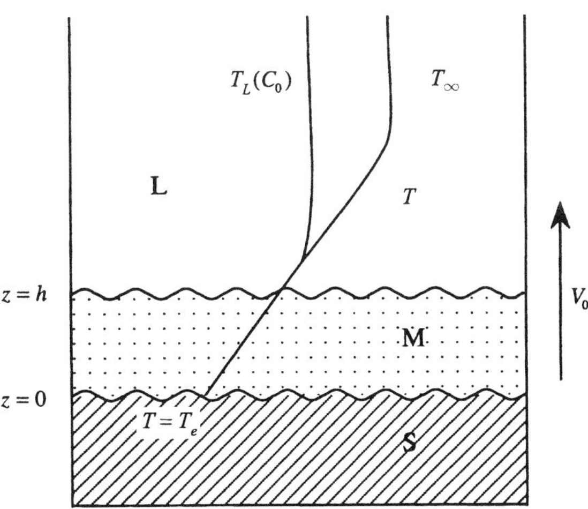

Figure 12.1 *A schematic diagram representing the directional solidification of an alloy at a speed V_0. A porous layer exists between a solid region, where $T < T_e$, and a liquid region. The profiles for the dimensional temperature and the local liquidus temperature T_L are also shown. L, M and S denote the liquid, mush (porous) and solid regions, respectively*

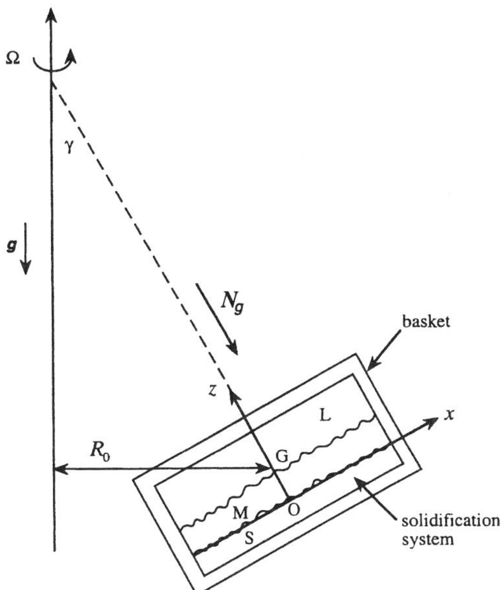

Figure 12.2 *Solidification system in a centrifuge, where G denotes the center of gravity of the centrifuge basket*

system of these equations for the solidifying system rotating with the centrifuge basket, see Arnold *et al.* (1992), and translating with the solidification front at speed V_0 is non-dimensionalized by using V_0, κ/V_0, κ/V_0^2, $\beta \Delta C \rho_0 g \kappa / V_0$, ΔC and ΔT as scales for the velocity, length, time, pressure, solute and temperature, respectively. Here κ is the thermal diffusivity, ρ_0 is a reference density, $\beta = \beta^* - \Gamma \alpha^*$, where α^* and β^* are the expansion coefficients for the heat and solute, respectively, and Γ is the slope of the liquidus curve, $\Delta C = C_0 - C_e$, C_e is the eutectic concentration of the alloy, $\Delta T = T_L(C_0) - T_e$ and T_L is the local liquidus temperature. Due to the variation of the density with respect to both the solute concentration and temperature, the centrifugal acceleration terms in the momentum equation cannot be converted into passive gradient terms and become important at significant rotation rates. The centrifugal acceleration term in the momentum equation is split into an average term, which is superimposed on the normal gravity term and a so-called gradient acceleration term, see Arnold *et al.* (1992). For the porous layer, Darcy's law is adopted in the governing equations.

The non-dimensional form of the equations for the momentum, continuity, temperature and solute concentration in the liquid layer ($z > h$) are given by

$$\frac{1}{Pr}\left(\frac{\partial}{\partial t} - \frac{\partial}{\partial z} + \boldsymbol{u} \cdot \nabla\right)\boldsymbol{u} = \nabla^2 \boldsymbol{u} - HR(\nabla P + C\boldsymbol{k}) + HT\boldsymbol{u} \times \boldsymbol{e} + HAC\boldsymbol{R},$$

$$\text{(12.1)}$$

$$\nabla \cdot \boldsymbol{u} = 0, \tag{12.2}$$

$$\left(\frac{\partial}{\partial t} - \frac{\partial}{\partial z} + \boldsymbol{u} \cdot \nabla\right)\theta = \nabla^2 \theta, \tag{12.3}$$

$$\left(\frac{\partial}{\partial t} - \frac{\partial}{\partial z} + \boldsymbol{u} \cdot \nabla\right)C = E\nabla^2 C, \tag{12.4}$$

while the corresponding non-dimensional equations in the porous layer ($0 < z < h$) are given by

$$-\frac{\boldsymbol{u}}{\Pi} = R(\nabla P + C\boldsymbol{k}) + \frac{T}{1 - \phi}\boldsymbol{e} \times \boldsymbol{u} + AC\boldsymbol{R}, \tag{12.5}$$

$$\nabla \cdot \boldsymbol{u} = 0, \tag{12.6}$$

$$\left(\frac{\partial}{\partial t} - \frac{\partial}{\partial z} + \boldsymbol{u} \cdot \nabla\right)\theta = \nabla^2 \theta + S_t\left(\frac{\partial}{\partial t} - \frac{\partial}{\partial z}\right)\phi, \tag{12.7}$$

$$\left(\frac{\partial}{\partial z} - \frac{\partial}{\partial t}\right)[(1 - \phi)(C_r - C)] + \boldsymbol{u} \cdot \nabla C = E\nabla \cdot [(1 - \phi)\nabla C]. \tag{12.8}$$

In the momentum equations (12.1) and (12.5), $\boldsymbol{e}$ is a unit vector along the rotation axis defined by

$$\boldsymbol{e} = \cos\gamma\,\boldsymbol{k} + \sin\gamma\,\boldsymbol{i}, \tag{12.9}$$

and $\boldsymbol{R}$ is a radial position vector from the rotation axis defined by

$$\boldsymbol{R} = (z\sin\gamma - x\cos\gamma)(\cos\gamma\,\boldsymbol{i} - \sin\gamma\,\boldsymbol{k}) - y\boldsymbol{j}. \tag{12.10}$$

Here i, j and k are unit vectors along the positive directions of the x-, y- and z-axes, respectively.

The non-dimensional form of the governing equations (12.1) – (12.8) are subjected to the appropriate boundary conditions of the type given in Worster (1992). For completing the system, these boundary conditions are given as follows:

$$\theta = -1, \quad u \cdot k = 0 \qquad\qquad \text{on} \quad z = 0, \qquad (12.11)$$

$$[\theta] = [n \cdot \nabla\theta] = [P] = 0 \qquad\qquad \text{on} \quad z = h, \qquad (12.12)$$

$$\theta - C = \tfrac{\partial}{\partial z}(\theta - C) = u - (n \cdot u)\,n = [n \cdot u] = 0 \quad \text{on} \quad z = h, \qquad (12.13)$$

$$\theta \to \theta_\infty, \quad C \to 0, \quad u \to 0 \qquad\qquad \text{as} \quad z \to \infty. \qquad (12.14)$$

Here the boundary conditions (12.11) and (12.12) are applied for the porous layer, while the boundary conditions (12.13) and (12.14) are applied for the non-porous layer, see Worster (1992). Simplifying assumptions of negligible temperature contribution in the buoyancy force, $\beta \approx \beta^*$ and $\kappa \gg D$ are already considered for the governing system, where D is the solute diffusivity. The non-dimensional system contains a number of variables and parameters which are defined as follows. The vector u is the volume flux of fluid per unit area, which is related to the actual fluid velocity u^* by $u = \psi u^*$, where ψ is the local liquid fraction, see Worster (1992), P is the pressure, C is the solute concentration, θ is the temperature, $Pr = \nu/\kappa$ is the Prandtl number, n is a unit vector normal to the interface between the porous and non-porous layers, ν is the kinematic viscosity, $R = \beta \Delta C N_g \kappa^2 / \left(V_0^3 \nu H \right)$ is the solutal Rayleigh number, $N_g = \left(g^2 + \Omega^4 R_0^2 \right)^{1/2}$ is the acceleration due to high gravity, $N_g = g$ corresponds to the normal gravity case while $N_g > g$ indicates the level of high gravity, g is the magnitude of the acceleration due to normal gravity, R_0 is the perpendicular distance from the center of gravity G of the centrifuge basket to the rotation axis, R_0 is a function of γ and $R_0 = 0$ where $\gamma = 0°$, $H = \kappa^2 / \left(V_0^2 \Pi_0 \right)$ is a non-dimensional parameter representing the ratio of an effective Rayleigh number in the liquid zone inside the chimneys, or above the porous layer to that in the porous layer outside chimneys, Π_0 is a constant reference value of the permeability $\Pi(\phi)$ of the porous medium, ϕ is the solid fraction of the porous zone $(\phi = 1 - \psi)$, $T = 2\Omega\kappa^2 / \left(V_0^2 \nu H \right)$ is the Coriolis parameter, which is the square root of a Taylor number, $A = \beta \Delta C \Omega^2 \kappa^3 / \left(V_0^4 \nu H \right)$ is the gradient acceleration parameter due to the centrifugal force, $\theta_\infty = T_\infty / \Delta T$, $E = D/\kappa$ is the inverse of the Lewis number, $S_t = L/\left(\tilde{C} \Delta T \right)$ is the Stefan number, $\tilde{C}$ is the specific heat per unit volume, L is the latent heat of solidification per unit volume, $C_r = (C_s - C_0)/\Delta C$ is a concentration ratio, and C_s is the composition of the solid phase forming the dendrites in the porous layer. Due to the liquidus relationship, which holds to a good approximation in the porous layer, $\theta = C$ in the porous zone outside the chimneys.

The flow solution examined by Sayre and Riahi (1996, 1997) was in the limit of small rotation rate and for zero Coriolis effect. The solution, as the sum of base flow solutions, which was at most a function of z, see Worster (1992), and a normal mode type solution for disturbances whose dependence in the plane perpendicular to the z-axis was of the form $\exp(B)$, where $B = \mathrm{i}(wt + \alpha_1 x + \alpha_2 y)$. Here i is the pure imaginary number $(\sqrt{-1})$,

ω is the frequency of the disturbances and $\boldsymbol{\alpha} = (\alpha_1, \alpha_2)$ is the wave number vector of the disturbances. After the pressure gradient terms in the momentum equations have been eliminated, by taking the curl of these equations, it was assumed that the solution of the governing linear system of equations (12.1) – (12.8) and the boundary conditions (12.11) – (12.14) was of the following form:

$$\begin{pmatrix} \theta \\ C \\ u \\ \phi \end{pmatrix} = \begin{pmatrix} \theta_0 (z) \\ C_0 (z) \\ 0 \\ \phi_0 (z) \end{pmatrix} + \begin{pmatrix} \theta_1 (z) \\ C_1 (z) \\ u_1 (z) \\ \phi_1 (z) \end{pmatrix} \exp (B) . \tag{12.15}$$

Here quantities with subscript '0' denote base flow quantities, and the disturbance quantities with subscript '1' are assumed to be small in comparison with the base flow quantities. The solution of the form (12.15) is then used in the resulting governing system, and each equation and boundary condition for the base flow quantity is subtracted from the corresponding equation and boundary condition for the total (base flow plus disturbance) quantity. Hence the linear system for the disturbance quantities is considered in the limit of sufficiently small amplitude of disturbances. Each equation and boundary condition is then multiplied by $\exp(-B)$ and then the average of the resulting system is obtained, with respect to the x and y variables in both layers. The solution of the resulting system for θ_1, C_1, u_1 and ϕ_1 were then determined numerically for both stationary, see Sayre and Riahi (1996), and oscillatory, see Sayre and Riahi (1997), disturbances.

Due to the complexity of the resulting disturbance system for θ_1, C_1, u_1 and ϕ_1, a numerical code, of the type applied by Worster (1992), was developed by Sayre and Riahi (1996), to solve the eigenvalue problem and determine the eigenfunctions and eigenvalues of the linear system in the neutral stability limit. Here the numerical approach is described briefly. A new independent variable $\tau = \theta_\infty - \theta_0$, for $0 \leqslant \tau \leqslant \tau_e$, is defined, where θ_0 is the base flow solution for θ, $\tau_e \equiv 1 + \theta_\infty$ and $\tau = 0$ and $\tau = \tau_e$ correspond to $z = \infty$ and $z = 0$, respectively. Using this new variable, the governing system becomes a system of ordinary differential equations for the disturbance variables as functions of the independent variable τ. The new form of the stability system is conveniently over a finite domain in τ but it has a regular singular point at $\tau = 0$. To take into account this feature of the system, any disturbance variable is assumed to be a product of τ^m and a function of τ in the liquid region, $0 < \tau < \tau_i$, where $\tau_i \equiv \theta_\infty / (1 - E)$ corresponds to the value of $z = h_0$ and m is a root of the indicial equation. Here h_0 is the constant value of h in the absence of disturbances. Using this numerical procedure in the liquid region, four linearly independent solutions for the dependent variables were found which satisfy the boundary conditions at $\tau = 0$ and with the corresponding values m_i $(i = 1, 2, 3, 4)$ of m. When $m = m_i$, the corresponding boundary values of the scaled dependent variables at $\tau = 0$ are found from the governing system for these variables. In addition, for each value of $m = m_i$, a Taylor series expansion of the governing equations for these variables about $\tau = 0$ was applied to determine the first three derivatives of these variables at $\tau = 0$. These results allowed the numerical evaluation of the governing equations for the scaled variables in the liquid region to be started from the asymptotic expressions for the scaled dependent variables near $\tau = 0$. Sayre and Riahi (1996, 1997) applied an efficient fourth-order Runge–Kutta scheme for this purpose. For each value of m_i, the governing equations were integrated from $\tau = 0$ to $\tau = \tau_i$. Next, the interface boundary conditions between the liquid and mushy regions were used to relate the dependent variables in the porous layer at $\tau = \tau_i$ to the dependent variables in the non-porous layer. These values were used to start the numerical integration of the equations for the dependent variables in the porous layer at $\tau = \tau_i$, which continued until $\tau = \tau_e$. However, it was found that the resulting solution does not, in general, satisfy all the boundary conditions. Thus the remaining boundary conditions were used to compute the so-called residuals r_{ij} $(i, j = 1, 2, 3, 4)$ corresponding to the index m_i, see Worster (1992). The determinant, det, of the matrix $[r_{ij}]$ is then computed and R is varied until det $= 0$. The corresponding solutions are eigenfunctions of the stability system which represent the marginally stable states of the system.

To determine the results for the stationary disturbances, Sayre and Riahi (1996) set $\gamma = 30°, S_t = C_r = \theta_\infty = 1, Pr = 10$ unless otherwise stated and $\Pi(\phi) = 1$. The eigenvalue relation det $= 0$ then provided a marginal stability curve $R(\alpha)$, $\alpha \equiv (\alpha_1^2 + \alpha_2^2)^{1/2}$, for each choice of the parameters E, H and A. The parameter E is the inverse of the Lewis number and is typically very small. The parameter H is a representative of the square of the ratio of the thermal length scale, on which the depth h of the porous layer depends, to the average spacing between the dendrites within the porous layer, see Worster (1992). This parameter is typically very large. The gradient acceleration parameter A was assigned the values 0, 0.1, 0.3 and 0.6 and the values chosen for S_t, C_r, θ_∞, Pr, E and H are similar to those values chosen by Worster (1992). The results for the neutral stability curve, R versus α, were found as a function of A and for values of $E = 0.025$ and $H = 10^5$. The system is unstable in the region above the neutral curve and stable below the curve. Just as in the case of zero rotation, see Worster (1992), the marginal curve for each value of A has two minima, corresponding to two distinct modes of convection. The first mode, corresponding to the smaller α value, was called the long wavelength mode since its wavelength was comparable to h and causes the flow throughout the porous layer, while the second mode, corresponding to the larger α value, is called the short wavelength mode since its wavelength is comparable to the depth of the compositional boundary-layer ahead of the mushy–liquid interface and leaves the fluid within the interstices of the porous medium essentially stagnant, see Worster (1992). These properties were confirmed by the observation of the streamlines to be discussed below. It was found that R increases as A increases for a given α and the wave numbers of the two modes increase with increasing A. Hence, rotation has the familiar effect of increasing the critical values of R and α, see Chandrasekhar (1961). Streamlines for the two convection modes, corresponding to the local minima of the neutral stability curves, were determined as functions of A. It was found from these results that, for the $A = 0$ case, fluid flows deeply in the two layers for the long wavelength mode, while the flow is restricted to a thin region about the interface between the porous and non-porous zones for the short wavelength mode. For the $A \neq 0$ cases, fluid flows in both layers and it is stronger in the porous layer for the long wavelength mode. As the rotation increases, the flow is stabilized more strongly in the non-porous layer. For the short wavelength mode, the flow is restricted to the porous layer only, but such flow is stronger close to the interface between the porous and non-porous zones for the larger rotation rate. Vertical and horizontal velocity data were also determined in both layers for both of the convection modes. It was found that the flow speed in the non-porous layer is generally larger than the corresponding one in the porous layer, for the zero rotation case, while the opposite is generally true for the nonzero rotation cases. Also for each case, the flow speed for the long wavelength mode is generally larger than the corresponding one for the short wavelength mode. The density data of the solid fraction in the porous layer for both convective modes and for different values of A were also determined. It was found that the short wavelength mode causes less perturbation to the solid fraction than the long wavelength mode. Thus, the long wavelength mode was mostly associated with the solid fraction perturbations. For this mode, there was a substantial decrease in the solid fraction in the interior of the porous layer in regions of up flow and this indicates a tendency to form chimneys. The information provided by these data as a function of A indicated that the spatial locations in the porous regions which have a tendency to form chimneys, i.e., regions corresponding to a negative perturbation to the solid fraction that represent the local melting of the dendrites, changed as the rotation rate changed. This interesting result suggested an important operational procedure for the possible elimination of chimneys formation tendency to be a variable rotational rate constraint applied on the solidifying system. The relative stability of the two convection modes were found to vary considerably with the values of H and E. A particular interpretation of H is as a measure of the relative mobility of the fluid in the melt region to that in the porous layer. Thus, increasing H causes the melt region to be more unstable relative to the porous layer. The results from the marginal stability curves, for various values of H with $E = 0.025$ held fixed, confirmed such interpretation of H. It appeared that the long wavelength mode was the most critical (unstable) one for sufficiently strong rotation, or for sufficiently small H, while the short wavelength mode was the most unstable one for weak rotation, provided H was sufficiently large. In regard to the variation of E, it should be stated that the main effect of such a variation is to change the thickness of the compositional boundary-layer ahead of the mushy–liquid interface relative to the depth of the porous layer, and the thickness of the compositional boundary-layer decreased with decreasing E. The results for the marginal stability curves, for various values of E and A with $H = 10^5$, indicated that the convection modes become more stabilized as E decreased. The wavelength of the short wavelength mode decreased with decreasing E and rotational effects were seen to be minimized for some intermediate values of E. The results described so far were for the cases of aqueous solutions where $Pr = 10$ is representative. However, metallic alloys, which are of commercial interest, have a representative value of 0.02 for Pr. Marginal stability curves for the two cases of $Pr = 10$ and $Pr = 0.02$ indicated that the system is more stable at lower Pr and that the wavelength of the long wavelength mode for the lower Pr is smaller than the corresponding one for the higher Pr. Since the inverse Prandtl number $(1/Pr)$ measures the strength of the advection or the diffusion of the vorticity generated by buoyancy, due to the growth rate velocity V_0 of the solidification and thus, the smaller the value of Pr, the larger is the advection of such vorticity towards the solid boundary, which acts as a sink of vorticity due to the no-slip condition. Thus the system tends to be more stable for smaller Pr.

To determine the results for the oscillatory disturbances, Sayre and Riahi (1995, 1997) set $\gamma = 30°$, $S_t = C_r = \theta_\infty = 1$ and $\Pi(\phi) \equiv 1$. The eigenvalue relation det $= 0$ then provided a marginal stability curve $R(\alpha)$ or $R(\omega)$ for each choice of the values of the other parameters. The results for the neutral stability curves, R versus α, in the absence of rotation indicated that the most critical modes are non-oscillatory for $Pr = 10$. However, for $Pr = 0.02$, the marginal stability curve was found to have a minimum which corresponded to an oscillatory mode of convection. This oscillatory mode was found to be preferred over the stationary modes and corresponded to $\omega = 1.6$ or $\omega = -1.6$, since the stability system was found numerically to be insensitive with respect to the sign of ω as long as the rotational effects were absent. The system for $Pr = 0.02$ was found to be more unstable than the one for $Pr = 10$ in a certain range of α of the convection modes. For a nonzero rotation rate it was found that there are two preferred oscillatory modes of convection, with distinct different wavelengths and periods. It was found that rotation was destabilizing for the oscillatory modes and that R values for the marginal curve were no longer symmetric with respect to the $\omega = 0$ axis. The preferred oscillatory mode had a small wavelength and α increases with A. The case with $Pr = 0.02$ corresponded to smaller R without rotation, while the case with $Pr = 10$ corresponded to smaller R with rotation. The rotational effect made the convection cells slightly inclined with respect to the z-axis. The preferred oscillatory modes tended to be more concentrated close to the solid–mushy interface. The fluid was found to flow in both layers for the preferred oscillatory mode with longer wavelength, while flow was restricted to the porous zone for the preferred oscillatory mode with a shorter wavelength. The flow speed was found to be more significant in the porous zone, unless the wavelength of the preferred oscillatory mode was sufficiently long. The rotational constraint, low Pr value, and short wavelength modes all led to some decrease in the amount of negative perturbations to the solid fraction, which meant less tendency for the chimney formation in the porous zone. The spatial location in the porous layer, which tended to form chimneys, changed as the rotation rate changed. The rotational constraint reduced the destabilizing effect of H, and the preferred oscillatory modes of convection became more stabilized as E decreased.

Okhuysen and Riahi (2001) have developed a more sophisticated numerical code to determine the eigenvalue and eigenfunction of the neutral stability system with higher accuracy and they included the effects of the Coriolis force. Their preliminary investigation for the stationary disturbances considered the case $A = 0, \gamma = 30°, Pr = 10, E = 0.025$, $H = 10^5, S_t = C_r = \theta_\infty = 1$ and $\Pi(\phi) \equiv 1$. They found, in particular, that the rotation resulted in a higher critical value R_c, a local increase in the solid fraction near the liquid–mushy interface was reduced from the normal gravity case, and regions of reduced solid fraction that were surrounded by regions of increased solid fraction in the porous zone were 'opened up' under rotation.

12.3 CHIMNEY MODEL

The investigations carried out by Riahi (1997, 1998, 1999, 2001a) and the corresponding analysis and the results are described briefly in this section. The same type of double-layer model as described in the previous section is considered. However, here we are concerned mainly with the convection in any cylindrical chimney, whose axis is assumed to be parallel to the z-axis, within the porous layer. Thus the governing equations are, considered in a cylindrical coordinate system whose axial direction is along the z-axis. Riahi (1997) considered weakly non-axisymmetric convection in a cylindrical chimney, whose axis coincided with the z-axis, see Figure 12.3, and in the porous layer in the asymptotic limit of the strong compositional buoyancy force, negligible thermal buoyancy and sufficiently large Pr and E^{-1}.

The governing equations for the flow of the melt in the chimneys are equations (12.1) – (12.4), while the governing equations for the flow in the porous layer outside the chimneys are equations (12.5) – (12.8). The boundary conditions are those given by (12.11) – (12.13). However, the expressions for e and $\boldsymbol{R}$, which are expressed in cylindrical coordinate system and, hence, are different from those given by (12.9) and (12.10), are given by

$$e = \cos\gamma\,\boldsymbol{k} + \sin\gamma\left(\cos\xi\,\boldsymbol{r} - \sin\xi\,\boldsymbol{\xi}\right), \qquad (12.16)$$

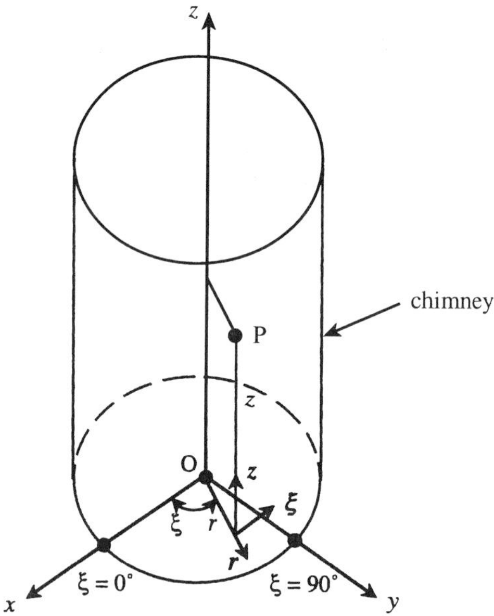

Figure 12.3 *A schematic sketch of a cylindrical chimney in the porous layer, the coordinate system, whose z-axis coincides with the axis of the chimney, the cylindrical coordinates (r, ξ, z) of a point P of the chimney, unit vectors (r, ξ, z) and the two locations of the chimney where the azimuthal angle ξ equals $0°$ and $90°$*

$$\boldsymbol{R} = \left(r \cos^2 \xi \cos^2 \gamma + r \sin^2 \xi - z \sin \gamma \cos \gamma \cos \xi\right) \boldsymbol{r}$$
$$+ \left(z \sin \gamma \cos \gamma \sin \xi + r \sin \xi \cos \xi - r \sin \xi \cos \xi \cos^2 \gamma\right) \boldsymbol{\xi} \tag{12.17}$$
$$+ \left(z \sin^2 \gamma - r \sin \gamma \cos \gamma \cos \xi\right) \boldsymbol{k},$$

where r is the radial variable, $\boldsymbol{r}$ is a unit vector in the radial r-direction, ξ is the azimuthal variable, and $\boldsymbol{\xi}$ is a unit vector in the azimuthal ξ-direction. Asymptotic and scaling analyses were applied to chimney convection to determine the results qualitatively about the effects of both the Coriolis and centrifugal forces on the convective motion within chimneys in the porous layer.

The analysis begins by assuming that the radius $a\,(\xi, z)$ of the chimney is small ($a \ll 1$) and that the orders of magnitude of r, ξ and z are a, 1 and 1, respectively. Under the assumption that the magnitude of $\boldsymbol{u}$ is of order one in the mushy zone, it is found from equation (12.5) that to the leading term in the pressure field in the porous zone is unaffected by the flow velocity and $\theta = C$ is independent of r and ξ in this zone. The steady-state form of equations (12.7) and (12.8) for the r- and ξ-independent variables $\theta_0\,(z)$, $\phi_0\,(z)$ and $w_0\,(z) \equiv \boldsymbol{u}_0 \cdot \boldsymbol{k}$ then imply

$$(w_0 - 1) \frac{\mathrm{d}\theta_0}{\mathrm{d}z} = \frac{\mathrm{d}^2\theta_0}{\mathrm{d}z^2} - S_t \frac{\mathrm{d}\theta_0}{\mathrm{d}z}, \tag{12.18}$$

$$-\frac{\mathrm{d}\phi_0}{\mathrm{d}z}\,(C_r - \theta_0) + \frac{\mathrm{d}\theta_0}{\mathrm{d}z}\,(w_0 - 1) = 0. \tag{12.19}$$

It is also assumed that $C_r \gg \theta$ in the porous zone, see Worster (1991).

Examining the non-axial terms in equation (12.2) or equation (12.6), it is found that a weak non-axisymmetric flow case is based on the conditions that

$$v \leqslant o\,(u), \qquad \frac{\partial F}{\partial \xi} \ll F, \tag{12.20}$$

where $u \equiv \boldsymbol{u} \cdot \boldsymbol{r}$, $v \equiv \boldsymbol{u} \cdot \boldsymbol{\xi}$ and F can be any dependent variable. Using the conditions (12.20) in equation (12.2) implies that, to the leading terms, a stream function $S\,(r, \xi, z)$ for the flow in the chimney can be introduced, so that

$$(u, w) = \left(-\frac{1}{r}\frac{\partial S}{\partial z}, \frac{1}{r}\frac{\partial S}{\partial r}\right). \tag{12.21}$$

Considering the flow in the chimney described by equations (12.1) – (12.4), and assuming that $C \sim 1$ and $w \gg 1$ hold in the chimney, see Worster (1991), equation (12.1) then implies that

$$w \sim HRa^2, \quad u \sim HRa^3, \quad S \sim HRa^4. \tag{12.22}$$

Further analysis, based on the scaling (12.22) and the condition

$$\frac{1}{a^2} \ll HR \ll \frac{1}{a^4}, \tag{12.23}$$

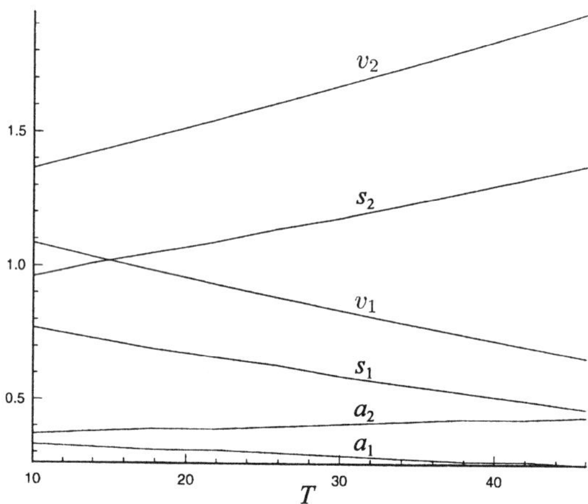

Figure 12.9 *Case of large axial velocity and weak Coriolis force for $R = 10^6$, $H = 10^4$ and $A_c = -0.025$. Orders of magnitude of scaled chimney's radius a_i ($100a$), scaled volume flux s_i ($100S_a$) and scaled non-azimuthal flow speed v_i ($0.001\,V_a$), for $\gamma = 150°$ ($i = 1$) and $\gamma = 210°$ ($i = 2$), as a function of T*

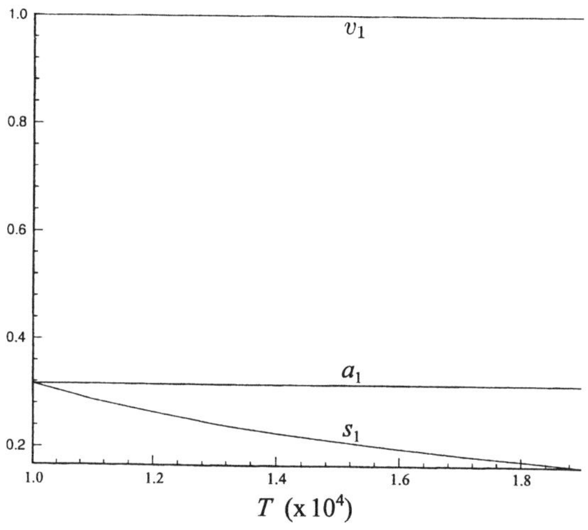

Figure 12.10 *Case of large axial velocity and strong Coriolis force for $R = 10^6$, $H = 10^4$ and $A_c = 0.001$. Orders of magnitude of scaled chimney's radius a_i ($100a$), scaled volume flux s_i ($1000S_a$) and scaled non-azimuthal flow speed v_i ($0.001\,V_a$), for $\gamma = 150°$ ($i = 1$), as a function of T*

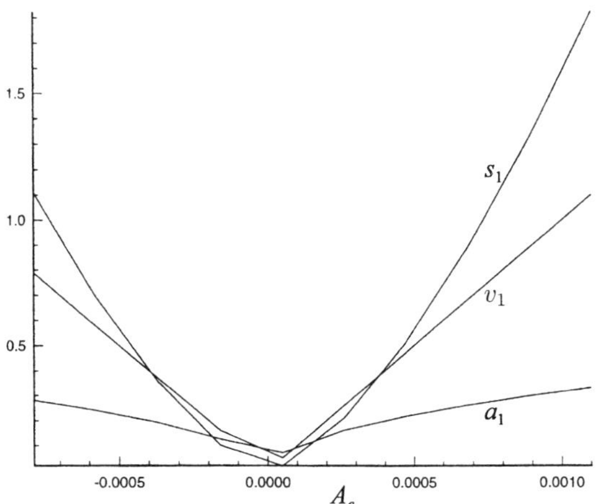

Figure 12.11 *Case of large axial velocity and strong Coriolis force for $R = 10^6$, $H = 10^4$ and $T = 20000$. Orders of magnitude of scaled chimney's radius a_i ($100a$), scaled volume flux s_i ($10000S_a$) and scaled non-azimuthal flow speed v_i ($0.001 V_a$), for $\gamma = 150°$ ($i = 1$), as a function of A_c*

γ affected the chimney convection. Within a certain range, which is controlled by the axial component of the Coriolis force in the (Pr, T, A)-parameter space for small Pr, the melt convection in the chimneys decreased with increasing T if $\gamma < 180°$, while chimney convection increased with T if $\gamma > 180°$, keeping all the other parameter values the same as in the $\gamma < 180°$ case. These results agreed with some earlier experimental studies for small Pr, see Ma *et al.* (1994), under centrifugation, where the Coriolis effect was found to have such different types of influence on the flow stability, depending on the rotation sense of the centrifuge. These results were also in agreement with some computational studies for $Pr = 0.02$, see Tao *et al.* (1994), under centrifugation, where the enhancement of the convection was found if centrifuge rotated counter-clockwise ($\sin \gamma < 0$), while convection was reduced if the centrifuge rotated clockwise ($\sin \gamma > 0$).

Riahi (2001a) extended his asymptotic and scaling analyses presented above to the case of weakly unsteady mode of non-axisymmetric chimney convection in a high gravity environment and determined new analytical results for the leading order magnitudes of the azimuthal averages for the radial velocity, axial velocity and the volume flux of the vertical flow in the chimney as functions of R, Pr, T, A and γ, and for given small values of a.

12.4 SINGLE-LAYER MODEL

The oscillatory instability, detected by Anderson and Worster (1996) in a porous layer during alloy solidification and in the absence of any external constraint, was based on a simple one-layer model developed earlier by Amberg and Homsy (1993) in which the dynamics of the porous layer were decoupled from the dynamics of the overlying liquid layer. Here, we extend the model treated by Anderson and Worster (1996) by imposing an external constraint of rotation on the solidification system and study the effects of the Coriolis force on the linear convective instability which is present once R exceeds its critical value R_c.

It should be noted that the single-layer model described in this section, which takes into account the rotational effects through the presence of the Coriolis force only, is relevant both in the geophysical applications, where the centrifugal force is insignificant, and in the engineering applications, where understanding the Coriolis effects alone can be of primary interest before the combined Coriolis and centrifugal force effects can be understood.

We consider a binary alloy melt that is cooled from below and it is solidified at a constant speed V_0. The solidifying system is assumed to be rotating at a constant speed Ω in the vertical direction which is anti-parallel to the normal gravity vector. Following Amberg and Homsy (1993) and Anderson and Worster (1996), we consider a porous layer of thickness h adjacent and above the solidification front to be physically isolated from the overlying liquid and underlying solid zones. Thus, it is assumed that the horizontal porous layer is bounded from above and below by rigid and isothermal boundaries. We consider the governing system in a moving frame translating at the speed V_0 with the solidification front and rotating with the speed Ω along the vertical axis.

The non-dimensional form of the governing equations are equations (12.5) – (12.8) with $A = 0$, and the boundary conditions are (12.11) and

$$\theta = \boldsymbol{u} \cdot \boldsymbol{k} = \phi = 0 \quad \text{on} \quad z = \delta, \tag{12.33}$$

where $\delta = hV_0/K$ is the dimensionless depth of the porous layer.

The non-dimensional form of the governing system (12.5) – (12.8), (12.11) and (12.33) contains the non-dimensional parameters R, T, S_t and C_r and, in addition, the present model contains the non-dimensional thickness δ of the porous layer as a parameter which is assumed to be small, see Amberg and Homsy (1993).

Following Anderson and Worster (1996) in reducing the model asymptotically, we follow their formulation, rescale R, the dependent and independent variables, based on δ ($\delta \ll 1$), in the following form:

$$\tilde{R}^2 = \delta R, \quad (x, y, z) = \delta\left(\tilde{x}, \tilde{y}, \tilde{z}\right), \quad t = \delta^2 \tilde{t}, \quad \boldsymbol{u} = \frac{\tilde{R}\tilde{\boldsymbol{u}}}{\delta}, \quad P = \tilde{R}\tilde{P}, \tag{12.34}$$

and assume that $\delta C_r = \tilde{C}$ and $\delta S_t = \tilde{S}$ are order one quantities as $\delta \to 0$. As discussed in Anderson and Worster (1996), the assumption of a thin porous layer ($\delta \ll 1$) is associated with large non-dimensional far-field temperature $\theta_\infty \gg 1$, which can occur when the initial concentration is close to C_e. The assumption of an order one quantity $\tilde{C}$ corresponds to the near-eutectic approximation, see Fowler (1985), which allows us to describe the porous layer of constant permeability to the leading order. The assumption of an order one quantity $\tilde{S}$ allowed Anderson and Worster (1996) to detect an oscillatory instability from their analytical porous-layer model.

The above rescalings were used in the governing system (12.5) – (12.8), (12.11) and (12.33). This system admits a motionless steady basic state of the form given as follows:

$$\theta_0 = (z - 1) + \frac{\delta\lambda}{2}\left(z - z^2\right) + O\left(\delta^2\right), \tag{12.35}$$

$$\phi_0 = -\frac{\delta}{\tilde{C}}(z - 1) + O\left(\delta^2\right), \tag{12.36}$$

$$\tilde{P}_0 = P^* + \tilde{R}\left(z - \frac{z^2}{2}\right) + \frac{\tilde{R}\delta\lambda}{2}z^2\left(\frac{1}{2} - \frac{z}{3}\right) + O\left(\delta^2\right), \tag{12.37}$$

where P^* is a constant, $\lambda \equiv \tilde{S}/\tilde{C} + 1$, and as can be seen from expressions (12.35) – (12.37), the basic state variables are designated by the subscript '0' quantities. The basic state solutions (12.35) – (12.37) contain the parameters δ, $\tilde{C}$, R and λ. Since the basic state solid fraction is found to be small, an expansion for $\Pi(0)/\Pi(\phi)$ in powers of ϕ is assumed, see Amberg and Homsy (1993), where ϕ is ϕ_0 plus the solid fraction of the infinitesimal disturbances superimposed on the motionless basic state. The system for such disturbances admits a normal mode type solution is of the form

$$\left(\boldsymbol{u}, \tilde{P} - \tilde{P}_0, \theta - \theta_0, \phi - \phi_0\right) =$$
$$\left[\boldsymbol{u}(z), \tilde{P}(z), \tilde{\theta}(z), \tilde{\phi}(z)\right]\exp\left(\sigma\tilde{t} + iw\tilde{t} + i\alpha_1\tilde{x} + i\alpha_2\tilde{y}\right), \tag{12.38}$$

where σ is the real growth rate of the disturbances. Applying such a normal mode form in the disturbance system yields a system of ordinary differential equations in the variable z for the z-dependence coefficients of the disturbance's dependent variables. This system contains the parameters $\tilde{R}$, T, δ, ω, α, λ and $\tilde{C}$. The presence of the small parameter δ in the system suggests an expansion of the dependent variables, ω and R in powers of δ. The systems, up to order δ, are then solved to determine the results for the neutral stability system and its critical conditions.

The results for the single-layer model are given as follows. The solution $\omega = 0$ always satisfies the eigenvalue relation, so that the stationary (non-oscillatory) disturbances can always be admitted by the single-layer model. The neutral stability system implies that the case $\omega_r \neq 0$ can always be possible for particular nonzero values of T, regardless of the value that a parameter combination $\lambda_0 \equiv (\lambda - 1)/\left(\tilde{C}\lambda^2\right)$ may take. Here ω_r is the real frequency and it turns out that λ_0 enters the eigenvalue relation as a parameter which combines both $\tilde{C}$ and λ. Hence, in this sense the presence of a rotational constraint enhances the existence of an oscillatory mode, since Anderson and Worster (1996) showed that $\omega_r \neq 0$ cannot exist for $\lambda_0 \lesssim 0.4$ in their non-rotating system. The critical value

Riahi, D. N. (1994a). Effect of Coriolis and centrifugal forces on the melt during directional solidification of a binary alloy. In *Materials Processing in High Gravity* (eds L. L. Regel and W. R. Wilcox), pp. 133–137. Plenum Press, New York.

Riahi, D. N. (1994b). The effect of Coriolis force on nonlinear convection in a porous medium. *J. Math. and Math. Sci.* **17**, 515–536.

Riahi, D. N. (1997). Effects of centrifugal and Coriolis forces on chimney convection during alloy solidification. *J. Crystal Growth* **179**, 287–296.

Riahi, D. N. (1998). High gravity convection in a mushy layer during alloy solidification. In *Nonlinear Instability, Chaos and Turbulence* (eds L. Debnath and D. N. Riahi), Vol. I, pp. 301–336. WIT Press.

Riahi, D. N. (1999). Effects of rotation on a non-axisymmetric chimney convection during alloy solidification. *J. Crystal Growth* **204**, 382–394.

Riahi, D. N. (2001a). Non-axisymmetric chimney convection in a mushy layer under a high gravity environment. In *Materials Processing in High Gravity* (eds L. L. Regel and W. R. Wilcox). Plenum Press, New York. In press.

Riahi, D. N. (2001b). Effects of Coriolis force on nonlinear convection in a mushy layer. In preparation.

Riahi, D. N. (2001c). Effects of centrifugal and Coriolis forces on a hydromagnetic chimney convection in a mushy layer. *J. Crystal Growth.* In press.

Sample, A. K. and Hellawell, A. (1982). The effect of mold procession on channel and macro-segregation in ammonium chloride–water analog castings. *Metall. Trans. B* **13**, 495–501.

Sample, A. K. and Hellawell, A. (1984). The mechanisms of formation and prevention of channel segregation during alloy solidification. *Metall. Trans. A* **15**, 2163–2173.

Sayre, T. L. and Riahi, D. N. (1995). Oscillatory instability of the liquid and mushy layers during alloy solidification under rotational constraint. TAM Report No. **808, UILU-ENG-95-6013**, 18 pages.

Sayre, T. L. and Riahi, D. N. (1996). Effect of rotation on flow instabilities during solidification of a binary alloy. *Int. J. Eng. Sci.* **34**, 1631–1645.

Sayre, T. L. and Riahi, D. N. (1997). Oscillatory instabilities of the liquid and mushy layers during solidification of alloys under rotational constraint. *Acta Mechanica* **121**, 143–152.

Tao, F., Zheng, Y., Ma, W. J., and Xue, M. L. (1994). Unsteady thermal convection of melt in a 2-D horizontal boat in a centrifugal field with consideration of the Coriolis effect. In *Materials Processing in High Gravity* (eds L. L. Regel and W. R. Wilcox), pp. 67–79. Plenum Press, New York.

Vadasz, P. (1998). Free convection in rotating porous media. In *Transport Phenomena in Porous Media* (eds D. B. Ingham and I. Pop), pp. 285–312. Pergamon, Oxford.

Vadasz, P. (2000). Flow and thermal convection in rotating porous media. In *Handbook of Porous Media* (ed. K. Vafai), pp. 395–439. Marcel Dekker, New York.

Worster, M. G. (1991). Natural convection in a mushy layer. *J. Fluid Mech.* **224**, 335–359.

Worster, M. G. (1992). Instabilities of the liquid and mushy regions during solidification of alloys. *J. Fluid Mech.* **237**, 649–669.

13 CHEMICALLY DRIVEN CONVECTION IN POROUS MEDIA

I. POP[*], J. H. MERKIN[†] and D. B. INGHAM[†]

[*]Faculty of Mathematics, University of Cluj, R-3400 Cluj, CP 253, Romania

email: `popi@math.ubbcluj.ro`

[†]Department of Applied Mathematics, University of Leeds, Leeds LS2 9JT, UK

email: `amtjhm@amsta.leeds.ac.uk` and `amt6dbi@amsta.leeds.ac.uk`

Abstract

This chapter is concerned with the analysis of free convection boundary-layer flows on cylindrical bodies embedded in fluid-saturated porous media where the flow results from the heat released by an exothermic catalytic reaction on the surface converting a reactive component within the convective fluid to an inert product. The reaction is modelled by first-order kinetics with an Arrhenius temperature dependence. A simple model for homogeneous–heterogeneous reactions in which the homogeneous (bulk) reaction is given by isothermal cubic autocatalytic kinetics and the heterogeneous (surface) reaction by first-order kinetics is also considered. As an application, the steady-state boundary-layer flows near the stagnation point on these catalytic surfaces are analysed in detail. Multiple solution branches and critical points arising from a hysteresis bifurcation are identified by analytical and numerical solutions. It is shown that the form that these solution branches take depends on whether or not the effects of reactant consumption are included. A short description of the heat transfer and reaction characteristics of a chemically reactive forced convection flow near the plane stagnation point of a catalytic porous bed with finite thickness is also presented in this chapter. This analysis predicts conditions for the onset of convective flow driven by catalytic surface reactions and the solutions demonstrate that the models considered give reliable results which can be used with great confidence in different practical situations in which convective flows in porous media are driven by chemically reactive systems.

Keywords: surface reaction, porous media, convective Darcy boundary-layer flow, first-order Arrhenius kinetics, homogeneous reaction, cubic autocatalysis

13.1 INTRODUCTION

Convective flows in fluid-saturated porous media are important in many technological applications such as in geothermal energy, cooling of nuclear reactors and underground disposal of nuclear wastes, petroleum reservoir operations, building insulation, irrigation systems, the cooling of electronic components, to name just a few applications of this area in contemporary technology. The large number of recently published papers and review articles on fluid flow through porous media demonstrates clearly that this area of fluid mechanics is studied extensively. Recent books by Ingham and Pop (1998), Nield and Bejan (1999), Vafai (2000), Pop and Ingham (2001) and the review papers by Hadim and Vafai (1999) and Vafai and Hadim (1999) present a comprehensive account of the information presently available on these flows and, in particular, stress the importance of the many extensions to Darcy's law which are required in various practical applications.

Transformation processes, such as phase changes and chemical reactions, can occur either through the whole volume or just at interfaces in the system under investigation. When fluids are involved in these processes then some of the fluid properties, such as the fluid density or viscosity, can be affected by the transformation. By analysing just the effects of changes in the fluid density, a whole new spectrum of phenomena can arise within the general framework of chemical engineering research. Changes in the density gradients in fluid-saturated porous media have been reported in many physical situations, in particular, when a fluid is unstably stratified and when both thermal and concentration effects are present, see Trevisan and Bejan (1990) and Mojtabi and Charrier-Mojtabi (2000). As changes in fluid density gradients induce natural convection, chemical reactions can provide a distributed driving force for the onset of secondary flows. When a non-isothermal chemical reaction takes place in the system, the heat generated (consumed) by the exothermic (endothermic) reaction, as well as the difference in molecular weight between the products and reactants, will determine density gradients.

As mentioned above, a large amount of information has been published regarding the onset of convection in non-reactive porous media. However, it is well known that chemical reactions can greatly affect buoyancy driven flows, and, as a result, the interactions between reaction and convection play a decisive role in the development of the concentration and temperature fields. When studying the progress of a chemical reaction under conditions of natural convection, two phenomena are manifested. First, the introduction of natural convection due to the existence of a chemical reaction, e.g., the transition from conduction–reaction regimes to conduction–convection–reaction regimes in reactive fluid media, and secondly, the influence of natural convection on the development of the chemical reaction, e.g., the existence of non-uniformities in deposited films by chemical vapour deposition. Examples of the interaction of chemical reaction and free convection occur in tubular laboratory reactors, chemical vapour deposition systems, oxidation of solid materials in large containers, synthesis of ceramic materials by a self-propagation reaction. Many chemically reacting systems involve both homogeneous and heterogeneous reactions, with examples occurring in combustion, catalysis and biochemical systems. The interaction between the homogeneous reactions in the bulk of the fluid and the heterogeneous reactions occurring on some catalytic surface is generally very complex, involving the production and consumption of reactant species at different rates, both within the fluid and on the catalytic surface, as well as the feedback on these reaction rates through temperature variations within the reacting fluid which in turn modifies the fluid motion.

Chemically reactive flows in both viscous (non-porous) fluids and fluid-saturated porous media have received far less attention than those of non-reactive systems. Chemical reactions greatly affect the buoyancy driven flows and, as a result, the interactions between reaction and convection plays a decisive role in the development of the concentration and temperature fields. These processes are important in chemical engineering, as well as in many other branches of engineering and science. Examples are combustion *in situ* in underground reservoirs for enhanced oil recovery, see Gottfried (1968), ceramic radiant porous burners used by industrial firms as an efficient heat transfer device, see Sathe *et al.* (1991), and the reduction of hazardous combustion products using catalytic porous beds. It is well known that exhaust gases from internal combustion engines contain carbon monoxide, residual hydrocarbons and nitrogen oxides that are hazardous to living organisms, so their formation must be reduced to an acceptable level. A review by McDermott (1971) indicated that catalytic converters are the only available technology in the automobile industry to meet the most stringent emission control standards. A catalytic converter is essentially a porous bed which converts the residual hydrocarbons and carbon monoxide to carbon dioxide and water vapour at relatively low temperatures.

There are many chemical reactions with important practical applications which proceed only very slowly, or not at all, except in the presence of a catalyst. A common configuration for such reactions is for the reactants (usually, but not exclusively, in the gaseous phase) to be made to flow over the solid catalyst, with the reaction taking place on the surface of the catalyst. The reaction is maintained by a fresh supply of reactants being brought to the catalyst surface by the flow. The detailed modelling of these, often complex, reaction systems usually also includes the effect of the reaction in the bulk (homogeneous reaction). This effect can, in some cases, play a significant role in the overall combustion process but there are many operating conditions for which it plays only a minor role, with the response of the combustion system being dominated by the surface (or heterogeneous) reactions. Even in cases where the homogeneous reaction cannot be ignored, a catalytic surface reaction is required for bulk reaction to be sustained. An excellent review of the chemical aspects of surface or heterogeneous reactions has been given by Gray and Scott (1990) and Scott (1991). Also, a full discussion of catalysis and a description of many of its practical applications may be found in Bond (1987). Examples of catalytic surface reactions, which are of importance in the chemical industry, are provided by the work, both experimental and theoretical, on methane/ammonia and propane oxidation over platinum by Song *et al.* (1991a, 1991b) and Williams *et al.* (1991a, 1991b).

Chemically reactive flows in porous media have received increasing interest over the last few years. Steinberg and Brand (1983) investigated the instability of a binary mixture with fast chemical reactions when heated from below or from above. It was found that oscillatory modes can develop at the first bifurcation points, depending on the sign and the magnitude of the heat of reaction. Gatica *et al.* (1987, 1989) studied the stability of chemical reaction and free convection in a two-dimensional planar channel filled with a porous medium. Hsu *et al.* (1991) analyzed the problem of premixed combustion in a porous medium using detailed chemical kinetics, and the effects of porous material, combustor geometry and kinetic parameters were discussed. Chao *et al.* (1994) studied the non-premixed burning of a condensed fuel in a porous medium with a natural convection oxidizer flow adjacent to the wall and obtained a solution for the flame temperature, standoff distance and mass consumption rate. Other studies on chemical reacting flows in porous media at low temperatures are those by Kordylewski and Krajewski (1984), Viljonen and Hlavacek (1987), Malashetty *et al.* (1994) and Chao *et al.* (1996).

In order to gain some insight into the process of heat transfer in a porous medium, which is driven by heating from a surface on which there is a catalytic reaction, we review some of the existing work and up-to-date references on the subject. We consider a planar, or axisymmetric, body of arbitrary shape which is embedded in a fluid-saturated porous medium which contains a reactive species A that reacts to form some inert product B when in contact with the body surface. In particular, we study the following two types of reactions:

I   There is an exothermic catalytic reaction on the body surface whereby reactant A is converted to an inert product B via the single first-order Arrhenius kinetics as described by Toong (1983), i.e.,

$$A \to B, \quad \text{rate} = k_0 \bar{a} \exp\left(-\frac{E}{RT}\right), \tag{13.1}$$

where $\bar{a}$ and T are the concentration of the reactant A and the temperature on the body surface, respectively, k_0 is a rate constant (pre-exponential factor), E is the activation energy and R is the universal gas constant.

II   There is an isothermal cubic autocatalytic reaction within the porous medium which is given schematically by

$$A + 2B \to 3B, \quad \text{rate} = k_1 \bar{a} \bar{b}^2, \tag{13.2}$$

where $\bar{b}$ is the concentration of the reactant B and k_1 is a constant. We will assume in both cases that there is no autocatalyst B in any external flow and that the reactant A has a constant concentration a_∞.

These schemes guarantee, in a natural way, that the reaction rate will be zero in the external flows and thus zero at the outer edge of the boundary-layer. The reaction scheme (13.1), sometimes referred to as the FONI scheme, has been used extensively in modelling a wide variety of combustion processes, see, for example, Aris (1975), Gray and Scott (1990), Chaudhary and Merkin (1994, 1996), Merkin and Chaudhary (1994), Chaudhary *et al.* (1995) and Ingham *et al.* (1999) for a viscous (non-porous) medium and Mahmood and Merkin (1998), Merkin and Mahmood (1998) and Minto *et al.* (1998) for a porous medium. On the other hand, reaction scheme (13.2) has been used only by Chaudhary and Merkin (1995a, 1995b) in this context, although it is used extensively in modelling the chemical kinetics for a range of purely reaction–diffusion processes, as reviewed by Gray and Scott (1990) and Scott (1991). It is worth pointing out to this end that porous media combustion arises in many practical applications and in some of these situations the combusting material can be regarded as being homogeneous. However, there are practical applications where this is not so, for example, the localized wetting of a cellulosic material can lead to regions within the material where local reaction sites arise, see Gray and Wake (1990). This situation can be modelled as an exothermic surface reaction within an otherwise unreacting porous material. The important role that natural convection can play in porous material combustion is recognized in the so-called 'stacking problem', see Balakotaiah and Pourtalet (1990). Here blocks of porous material are stacked on top of each other and the heat released sets up a convective flow which can significantly alter the combustion characteristics of the material in the blocks.

We now present some flow models which are driven by heating from a chemical reaction on the surface of a body which is embedded in a fluid-saturated porous medium using the reaction schemes (13.1) and (13.2). However, we consider a much simplified model of this complex problem, isolating the free or forced convection aspects and the catalytic surface reaction. In particular, we assume that the flow takes place in the boundary-layer near the stagnation point of a cylindrical body, since this enables the governing equations to be substantially simplified. It is also assumed that the reaction takes place only on the catalytic surface. The simplicity of this model enables us to determine many of the important features analytically and also to identify clearly the basic mechanisms involved.

13.2 FREE CONVECTION NEAR A STAGNATION POINT OF A CYLINDRICAL BODY IN A POROUS MEDIUM DRIVEN BY THE CATALYTIC REACTION SCHEME I

We consider the two-dimensional steady free convection flow near the lower stagnation point of a cylindrical body which is embedded in a fluid-saturated porous medium of ambient temperature T_∞. If we assume that the porous medium is isotropic and homogeneous and that the Darcy–Boussinesq approximation is valid, then the equations governing the free convection boundary-layer flow in a fluid-saturated porous medium are, for the present model, given by

$$\frac{\partial \bar{u}}{\partial \bar{x}} + \frac{\partial \bar{v}}{\partial \bar{y}} = 0, \tag{13.3}$$

$$\bar{u} = \frac{gK\beta}{\nu}(T - T_\infty)\frac{\bar{x}}{l}, \tag{13.4}$$

$$\bar{u}\frac{\partial T}{\partial \bar{x}} + \bar{v}\frac{\partial T}{\partial \bar{y}} = \alpha_m \frac{\partial^2 T}{\partial \bar{y}^2}, \tag{13.5}$$

$$\bar{u}\frac{\partial \bar{a}}{\partial \bar{x}} + \bar{v}\frac{\partial \bar{a}}{\partial \bar{y}} = D_m \frac{\partial^2 \bar{a}}{\partial \bar{y}^2}, \tag{13.6}$$

see Merkin and Mahmood (1998). Here $\bar{u}$ and $\bar{v}$ are the velocity components, as given by Darcy's law, in the $\bar{x}$ and $\bar{y}$ directions, respectively, with $\bar{x}$ and $\bar{y}$ being coordinates measuring distance along and normal to the body surface, g is the magnitude of the acceleration due to gravity, K is the permeability of the porous medium, β is the coefficient of thermal expansion, ν is the kinematic viscosity, α_m and D_m are the equivalent thermal diffusivity and diffusion coefficients of the porous medium and l is some length scale for the impermeable surface. It should be noted that the present model assumes that the buoyancy forces arise only from differences in temperature. Buoyancy forces can also arise from differences in the reactant concentration. This additional effect requires a term, which is proportional to $\beta_c\,(\bar{a} - a_\infty)$, to be included in equation (13.4), where β_c is the coefficient of concentration expansion. However, here it is assumed that $|\beta_c| \ll |\beta|$.

Heat is released by the first-order surface reaction

$$A \to B, \quad \text{rate} = k_0 \bar{a} \exp\left(-\frac{E}{RT}\right), \tag{13.7}$$

with a heat of reaction $Q > 0$, which is taken from the body surface into the surrounding fluid–porous medium by conduction. The boundary conditions for equations (13.3) – (13.6) are then given by

$$\left.\begin{aligned}
\bar{v} &= 0 \\
k_m \frac{\partial T}{\partial \bar{y}} &= -k_0 Q \bar{a} \exp\left(-\frac{E}{RT}\right) \\
D_m \frac{\partial C}{\partial \bar{y}} &= k_0 \bar{a} \exp\left(-\frac{E}{RT}\right)
\end{aligned}\right\} \quad \text{on} \quad \bar{y} = 0, \quad \bar{x} \geqslant 0,$$

$$\bar{u} \to 0, \quad T \to T_\infty, \quad \bar{a} \to a_\infty. \quad \text{as} \quad \bar{y} \to \infty, \quad \bar{x} \geqslant 0, \tag{13.8}$$

where k_m is the thermal conductivity of the porous medium.

We introduce the stream function $\bar{\psi}$, defined so that $\bar{u} = \partial\bar{\psi}/\partial\bar{y}$ and $\bar{v} = -\partial\bar{\psi}/\partial\bar{x}$, and introduce the non-dimensional variables

$$x = \frac{\bar{x}}{l}, \quad y = Ra^{1/2}\left(\frac{\bar{y}}{l}\right), \quad \psi = Ra^{-1/2}\left(\frac{\bar{\psi}}{\alpha_m}\right), \quad \theta = \frac{T - T_\infty}{RT_\infty^2/E}, \quad \phi = \frac{\bar{a}}{a_\infty}, \tag{13.9}$$

where Ra is the modified Rayleigh number for the porous medium which is defined as

$$Ra = \frac{gK\beta RT_\infty^2 l}{\alpha_m \nu E}. \tag{13.10}$$

It is worth mentioning that the non-dimensional form for the temperature, given in expression (13.9), uses the standard Frank-Kamenskii (1969) variable since there is no imposed temperature difference scale in the present problem. Substituting expressions (13.9) into equations (13.4) – (13.6), we obtain

$$\frac{\partial\psi}{\partial y} = x\theta, \tag{13.11}$$

$$\frac{\partial\psi}{\partial y}\frac{\partial\theta}{\partial x} - \frac{\partial\psi}{\partial x}\frac{\partial\theta}{\partial y} = \frac{\partial^2\theta}{\partial y^2}, \tag{13.12}$$

$$\frac{\partial \psi}{\partial y}\frac{\partial \phi}{\partial x} - \frac{\partial \psi}{\partial x}\frac{\partial \phi}{\partial y} = \frac{1}{Le}\frac{\partial^2 \phi}{\partial y^2}, \qquad (13.13)$$

where $Le = \alpha_m/D_m$ is the Lewis number for the fluid-saturated porous medium. The boundary conditions (13.8) now become

$$\left.\begin{array}{l} \psi = 0 \\[4pt] \frac{\partial \theta}{\partial y} = -\delta\phi\exp\left(\frac{\theta}{1+\varepsilon\theta}\right) \\[4pt] \frac{\partial \phi}{\partial y} = \lambda\delta\phi\exp\left(\frac{\theta}{1+\varepsilon\theta}\right) \end{array}\right\} \quad \text{on}\quad y = 0, \quad x \geqslant 0,$$

$$\frac{\partial \psi}{\partial y} \to 0, \quad \theta \to 0, \quad \phi \to 1 \quad \text{as}\quad y \to \infty, \quad x \geqslant 0, \qquad (13.14)$$

where λ and δ are consumption parameters and ε is the activation energy parameter which are defined as follows:

$$\lambda = \frac{EQk_0a_\infty}{k_m RT_\infty^2}\left(\frac{\alpha_m l}{U_c}\right)^2, \quad \delta = \frac{RT_\infty^2 k_m}{D_m QEa_\infty}, \quad \varepsilon = \frac{RT_\infty}{E}, \qquad (13.15)$$

where $U_c = gK\beta RT_\infty^2/\nu E$ is a velocity scale.

The system of equations (13.11) – (13.13) can be reduced to a set of ordinary differential equations by the transformation

$$\psi = xf(y), \quad \theta = \theta(y), \quad \phi = \phi(y). \qquad (13.16)$$

This reduces equation (13.11) to the relationship $\theta = f'$ and equations (13.12) and (13.13) become

$$f''' + ff'' = 0, \qquad (13.17)$$
$$\phi'' + Lef\phi' = 0, \qquad (13.18)$$

along with the boundary conditions (13.14) which reduce to

$$\begin{array}{l} f(0) = 0, \\[4pt] f''(0) = -\lambda\phi_w\exp\left(\frac{\theta_w}{1+\varepsilon\theta_w}\right), \\[4pt] \phi'(0) = \lambda\delta\phi_w\exp\left(\frac{\theta_w}{1+\varepsilon\theta_w}\right), \\[4pt] f' \to 0, \quad \phi \to 1 \quad \text{as}\quad y \to \infty, \end{array} \qquad (13.19)$$

where $\theta_w = \theta(0)$, $\phi_w = \phi(0)$ and primes denote differentiation with respect to y.

Following Merkin and Mahmood (1998), the boundary-value problem defined by equations (13.17) – (13.19) can be reduced to the solution of a standard free convection problem. To do so, we take

$$f = \theta_w^{1/2}F(Y), \quad \phi = 1 - (1-\phi_w)\Phi(Y), \quad Y = \theta_w^{1/2}y, \qquad (13.20)$$

so that equations (13.17) and (13.18) become

$$F''' + FF'' = 0, \tag{13.21}$$

$$\Phi'' + LeF\Phi' = 0, \tag{13.22}$$

along with the boundary conditions

$$F(0) = 0, \quad F'(0) = 1, \quad \Phi(0) = 1,$$
$$F' \to 0, \quad \Phi \to 0 \quad \text{as} \quad Y \to \infty, \tag{13.23}$$

where primes now denote differentiation with respect to Y. Solving equations (13.21) – (13.23) numerically, Merkin and Mahmood (1998) have found that $-F''(0) = C_0 = 0.62756$ and $F(\infty) = d_0 = 1.14277$. Equation (13.22) can be solved in terms of $F(Y)$ and from this solution $C_1(Le) = -\Phi'(0)$ can be calculated as follows:

$$C_1(Le) = \left[\int_0^\infty \exp\left(-Le \int_0^Y F(s)\,\mathrm{d}s \right) \mathrm{d}Y \right]^{-1}. \tag{13.24}$$

From equation (13.24), $C_1 \sim \sqrt{2/\pi}\, Le^{1/2}$ as $Le \to \infty$ and $C_1 \sim d_0 Le$ as $Le \to 0$.

To determine the constants θ_w and ϕ_w, we use the boundary conditions (13.19) which can be written as follows:

$$C_0 \theta_w^{3/2} = \lambda \phi_w \exp\left(\frac{\theta_w}{1 + \varepsilon\theta_w} \right), \tag{13.25}$$

$$C_1(1 - \phi_w)\theta_w^{1/2} = \lambda\delta\phi_w \exp\left(\frac{\theta_w}{1 + \varepsilon\theta_w} \right). \tag{13.26}$$

Combining these relations we obtain

$$1 - \phi_w = \gamma(Le)\,\theta_w, \tag{13.27}$$

where $\gamma(Le) = \delta C_0 / C_1$. On using expressions (13.25) and (13.27), we obtain

$$\lambda = \frac{C_0\theta_w^{3/2} \exp\left[-\theta_w/(1 + \varepsilon\theta_w) \right]}{1 - \gamma\theta_w}, \quad \theta_w < \frac{1}{\gamma}. \tag{13.28}$$

If we assume that the reactant consumption can be neglected, which can be thought of as putting $\delta = 0$, then $\gamma = 0$ and therefore $\phi_w = 1$ or $\phi(y) \equiv 1$. In this case equation (13.28) becomes

$$\lambda = C_0\theta_w^{3/2} \exp\left(-\frac{\theta_w}{1 + \varepsilon\theta_w} \right). \tag{13.29}$$

This expression shows that $\lambda = 0$ when $\theta_w = 0$ and $\lambda \sim C_0\theta_w^{3/2} \exp(-1/\varepsilon)$ as $\theta_w \to \infty$ and $\varepsilon \neq 0$, whilst $\lambda = C_0\theta_w^{3/2} \exp(-\theta_w)$ for $\varepsilon = 0$. To gain some insight into the nature of the model when there is no reactant consumption ($\delta = 0$), bifurcation diagrams, i.e., plots of the wall temperature θ_w against λ as given by equation (13.29), are shown in Figure 13.1. There are two important features to note about these diagrams. First, multiple solutions appear for values of ε in the range $0 \leqslant \varepsilon \leqslant 1/6$. Second, there are turning points (critical points) on these graphs which are determined from the condition

$$\frac{\mathrm{d}\lambda}{\mathrm{d}\theta_w} = 0, \tag{13.30}$$

giving

$$\theta_w^{(1,2)} = \frac{1 - 3\varepsilon \pm \sqrt{1 - 6\varepsilon}}{3\varepsilon^2}. \tag{13.31}$$

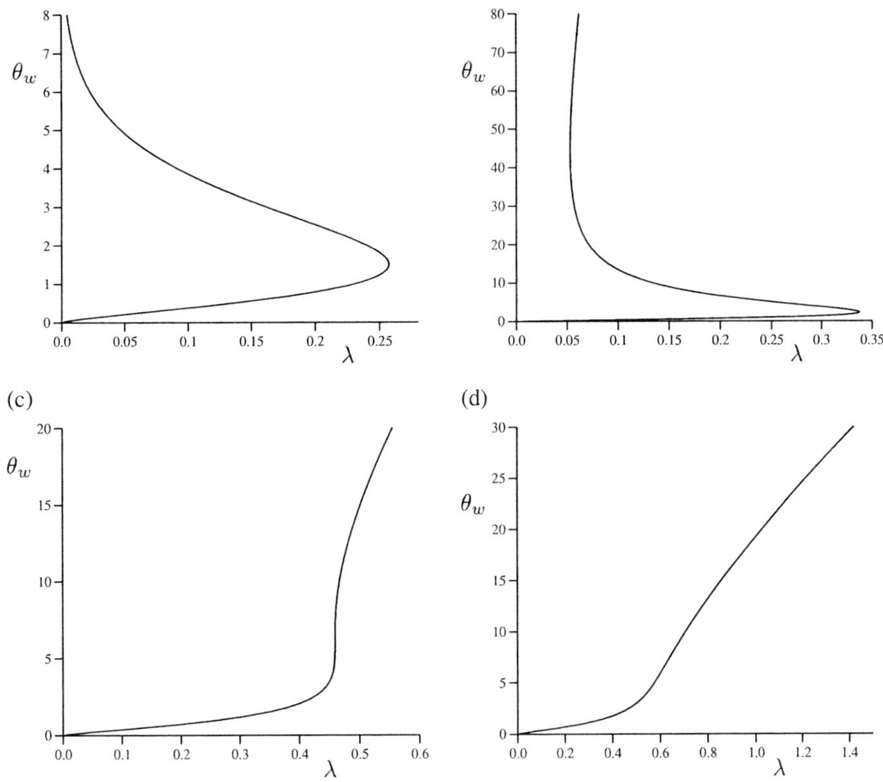

Figure 13.1 *Bifurcation diagrams, plots of θ_w as a function of λ, obtained from equation (13.28) for the case without reactant consumption ($\gamma = 0$), for (a) $\varepsilon = 0$, (b) $\varepsilon = 0.1$, (c) $\varepsilon = 1/6$ (hysteresis point) and (d) $\varepsilon = 0.2$*

This expression shows, for ε in the range $0 < \varepsilon < 1/6$, that there are two critical points at $\lambda_1 = \lambda\left(\theta_w^{(1)}\right)$ and $\lambda_2 = \lambda\left(\theta_w^{(2)}\right)$. Then, for $\varepsilon = 1/6$ there is a hysteresis bifurcation, where the slope becomes vertical, at $\theta_{w,\mathrm{hyst}} = 6$ and $\lambda_{\mathrm{hyst}} = C_0 6^{3/2} \mathrm{e}^{-3} = 0.4592$. For $\varepsilon > 1/6$ the θ_w as a function of λ curve is monotonic increasing.

Figure 13.1 shows the progression of the wall temperature, θ_w, from the exponential curve (for $\varepsilon = 0$, Figure 13.1a) with one critical point at $\theta_w = 3/2$ and $\lambda = C_0 (3/2)^{3/2} \mathrm{e}^{-3/2} = 0.2572$ to the typical S-shaped bifurcation diagram for $0 < \varepsilon < 1/6$ as shown in Figure 13.1(b). Figure 13.1(c) shows the hysteresis bifurcation for $\varepsilon = 1/6$ with the curve becoming vertical at $\lambda = \lambda_{\mathrm{hyst}} = 0.4592$. For $\varepsilon > 1/6$ the curve increases monotonically, see Figure 13.1(d). Finally, equation (13.31) gives, for $\varepsilon \ll 1$,

$$\theta_w^{(1)} \sim \frac{3}{2}\left(1 + 3\varepsilon + \ldots\right), \quad \theta_w^{(2)} \sim \frac{2}{3\varepsilon^2}\left(1 - 3\varepsilon + \ldots\right), \tag{13.32}$$

with $\theta_w^{(1)} \to 3/2$ and $\theta_w^{(2)} \to \infty$ as $\varepsilon \to 0$.

For the case when reactant consumption is included ($\delta \neq 0$ or $\gamma \neq 0$) we start by considering the exponential approximation ($\varepsilon = 0$). In this case equation (13.28) becomes

$$\lambda = \frac{C_0 \theta_w^{3/2} \mathrm{e}^{-\theta_w}}{1 - \gamma \theta_w}, \quad \theta_w < \gamma^{-1} \tag{13.33}$$

The bifurcation diagram (i.e., plots of θ_w against λ), have two critical points, and hence a range of multiple solutions, where equation (13.30) holds, gives

$$2\gamma\theta_w^2 - (2 + \gamma)\,\theta_w + 3 = 0, \quad \theta_w = \frac{2 + \gamma \pm \sqrt{4 - 20\gamma + \gamma^2}}{4\gamma}. \tag{13.34}$$

This holds provided $\gamma < \gamma_{\mathrm{hyst}} = 10 - 4\sqrt{6} = 0.20204$. There is a hysteresis point, i.e., equal roots in equation (13.34), at $\gamma = \gamma_{\mathrm{hyst}}$ ($\theta_{w,\mathrm{hyst}} = 2.7247$ and $\lambda_{\mathrm{hyst}} = 0.4117$). It should be noted that any solution of equation (13.34) has to satisfy $\theta_w < 1/\gamma$. For $\gamma > \gamma_{\mathrm{hyst}}$ the curve is monotonic increasing. This is illustrated in Figure 13.2 where we plot θ_w as a function of λ for $\gamma = 0.1$ and $\gamma_{\mathrm{hyst}} = 0.3$. An important feature to note about these bifurcation diagrams, compared to those shown in Figure 13.1, is that there is an upper bound on θ_w of $1/\gamma$ for large values of λ. Previously, θ_w increased indefinitely as λ increased and in dimensional terms this corresponds to a maximum possible surface temperature $T_{\max}$ of

$$T_{\max} = T_\infty + \frac{D_m Q a_\infty}{k_m}\left(\frac{C_0}{C_1 (Le)}\right). \tag{13.35}$$

The general case ($\varepsilon \neq 0$) has been treated by Merkin and Mahmood (1998). Differentiating equation (13.28) with respect to θ_w gives the equation

$$\gamma\varepsilon^2\theta_w^3 + \left(2\gamma\varepsilon - 3\varepsilon^2 - 2\gamma\right)\theta_w^2 + (\gamma + 2 - 6\varepsilon)\,\theta_w - 3 = 0 \tag{13.36}$$

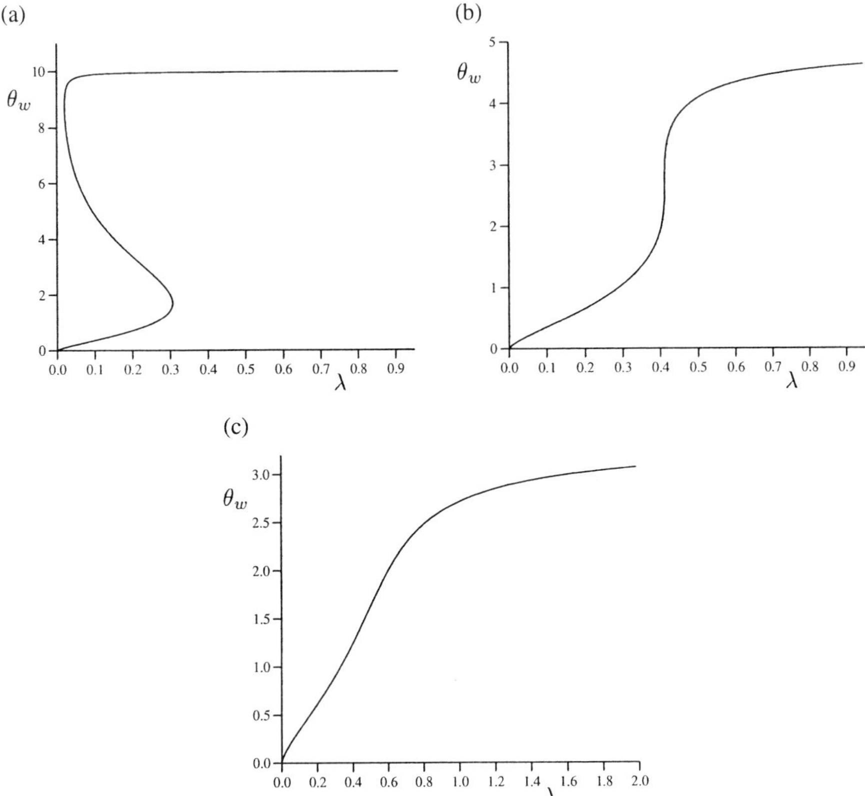

Figure 13.2 *Bifurcation diagrams, plots of θ_w as a function of λ, obtained from equation (13.33) for exponential approximation ($\varepsilon = 0$) with reactant consumption ($\gamma \neq 0$), for (a) $\gamma = 0.1$, (b) $\gamma = 0.20204$ (hysteresis point) and (c) $\gamma = 0.3$*

for determining the critical points. The hysteresis bifurcation curve in (ε, γ) parameter space, which separates regions where there is, and there is not, possible multiple solutions is determined by solving equation (13.36) simultaneously with the equation

$$3\gamma\varepsilon^2\theta_w^2 + 2\left(2\gamma\varepsilon - 3\varepsilon^2 - 2\gamma\right)\theta_w + \gamma + 2 - 6\varepsilon = 0. \tag{13.37}$$

Equations (13.36) and (13.37) were solved numerically by Merkin and Mahmood (1998) and the result is shown in Figure 13.3. This figure gives the hysteresis bifurcation curves which divides the (ε, γ) plane into a region where multiple solutions are possible (below the curve) and where the bifurcation diagrams are monotonic (above the curve). The plots

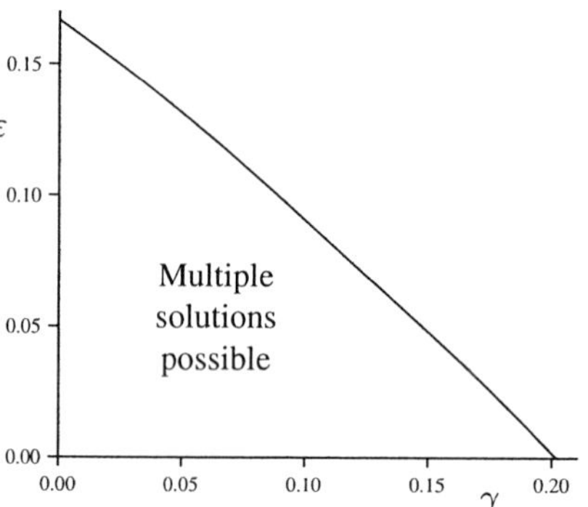

Figure 13.3 *The hysteresis bifurcation curve in the (ε, γ) plane. The region where multiple solutions are possible is indicated*

of θ_w against λ for $\varepsilon \neq 0$ are very similar to those shown in Figure 13.2, see also Merkin and Mahmood (1998).

13.3   FORCED CONVECTION FLOW NEAR A STAGNATION POINT OF A CYLINDRICAL BODY IN A POROUS MEDIUM DRIVEN BY THE CATALYTIC REACTION SCHEME II

Here we consider the steady homogeneous–heterogeneous reaction near a stagnation point of a cylindrical body which is embedded in a fluid-saturated porous medium. Within the porous medium we have the isothermal cubic autocatalytic reaction (13.2), while on the catalytic surface we have the isothermal, first-order reaction

$$A \to B, \quad \text{rate} = k_s \bar{a}, \tag{13.38}$$

where k_s is a constant.

The equations for $\bar{a}$ and $\bar{b}$, the concentrations of the reactants A and B, are, with the reaction scheme (13.2), see Chaudhary and Merkin (1995a, 1995b), given by

$$\bar{u}\frac{\partial \bar{a}}{\partial \bar{x}} + \bar{v}\frac{\partial \bar{a}}{\partial \bar{y}} = D_A \frac{\partial^2 \bar{a}}{\partial \bar{y}^2} - k_1 \bar{a}\bar{b}^2, \tag{13.39}$$

$$\bar{u}\frac{\partial \bar{b}}{\partial \bar{x}} + \bar{v}\frac{\partial \bar{b}}{\partial \bar{y}} = D_B \frac{\partial^2 \bar{b}}{\partial \bar{y}^2} + k_1 \bar{a}\bar{b}^2, \tag{13.40}$$

where D_A and D_B are the respective diffusion coefficients. The boundary conditions to be applied are

$$\left. \begin{array}{l} D_A \frac{\partial \bar{a}}{\partial \bar{y}} = k_s \bar{a} \\[2mm] D_B \frac{\partial \bar{b}}{\partial \bar{y}} = -k_s \bar{a} \end{array} \right\} \quad \text{on} \quad \bar{y} = 0, \quad \bar{x} \geqslant 0,$$

$$\bar{a} \to a_\infty, \quad \bar{b} \to 0 \quad \text{as} \quad \bar{y} \to \infty, \quad \bar{x} \geqslant 0. \tag{13.41}$$

The model, equations (13.39) – (13.41), assumes a large Rayleigh number, allowing the usual boundary-layer approximations to be made.

We consider an outer stagnation point flow $U_\infty (\bar{x}/l)$, where l is a measure of the body size, and then introduce the dimensionless variables

$$x = \frac{\bar{x}}{l}, \quad y = Ra^{1/2} \left(\frac{\bar{y}}{l} \right), \quad \bar{\psi} = \alpha_m Ra^{1/2} \left(\frac{\bar{x}}{l} \right) f(y),$$

$$a(y) = \frac{\bar{a}}{a_\infty}, \quad b(y) = \frac{\bar{b}}{a_\infty}. \tag{13.42}$$

The velocity field is given simply by $\psi = xy$ (forced convection flow). Near the stagnation point $a = a(y)$ and $b = b(y)$ so that equations (13.39) and (13.40) become

$$a'' + ya' - \Lambda ab^2 = 0, \tag{13.43}$$

$$\sigma b'' + yb' + \Lambda ab^2 = 0, \tag{13.44}$$

while the boundary conditions (13.41) become

$$a'(0) = \Lambda_s a_w, \quad b'(0) = -\Lambda_s a_w,$$

$$a \to 1, \quad b \to 0 \quad \text{as} \quad y \to \infty, \tag{13.45}$$

where $a_w = a(0)$. The parameter Λ gives a measure of the strength of the homogeneous reaction, Λ_s measures the strength of the heterogeneous surface reaction and σ is the ratio of the diffusion coefficients and they are defined as follows:

$$\Lambda = \frac{k_1 a_\infty^2 l^2}{\alpha_m Ra}, \quad \Lambda_s = \frac{k_s l}{D_A Ra^{1/2}}, \quad \sigma = \frac{D_B}{D_A}. \tag{13.46}$$

The Rayleigh number Ra is now defined by

$$Ra = \frac{U_\infty l}{D_A}. \tag{13.47}$$

Following Chaudhary and Merkin (1995a), we deal with the case which is encountered in many practical applications, namely the diffusion coefficients D_A and D_B of the chemical species A and B are of comparable size. This suggests we set $D_A = D_B$, giving $\sigma = 1$ and then from equations (13.43) – (13.45) we have

$$a(y) + b(y) = 1. \tag{13.48}$$

Equations (13.43) and (13.44) reduce to the single equation

$$a'' + ya' - \Lambda a \left(1 - a\right)^2 = 0, \tag{13.49}$$

which has to be solved subject to the boundary conditions

$$a'\left(0\right) = \Lambda_s a_w, \quad a \to 1 \quad \text{as} \quad y \to \infty. \tag{13.50}$$

We start by considering the two limiting cases, namely $\Lambda = 0$ and $\Lambda_s \to \infty$. With $\Lambda = 0$, equations (13.49) and (13.50) have the solution

$$a\left(y\right) = \frac{1}{1 + \Lambda_s \sqrt{\pi/2}} \left(1 + \Lambda_s \int_0^y e^{-t^2/2}\, \mathrm{d}t\right), \quad a_w = \frac{1}{1 + \Lambda_s \sqrt{\pi/2}}. \tag{13.51}$$

Equation (13.51) shows that, in this limit, a_w is monotonic decreasing in Λ_s and approaches zero as $\Lambda_s \to \infty$.

For the limit $\Lambda_s \to \infty$, equations (13.49) and (13.50) have to be solved subject to the boundary condition $a\left(0\right) = 0$. This has to be obtained numerically and the results are shown in Figure 13.4, with a plot of $a'\left(0\right)$ as a function of Λ. The figure shows that the curve is monotonic decreasing in Λ, starting with $a'\left(0\right) = \sqrt{\pi/2}$ at $\Lambda = 0$ obtained from expression (13.51) by letting $\Lambda_s \to \infty$, and decreasing to zero as Λ increases.

Both these limiting forms do not have multiple solutions. This is not the case in general, as can be seen in Figure 13.5(a), where we plot a_w as a function of Λ for $\Lambda_s = 0.05$. This case has a range of values of Λ over which there are three solutions. These arise at the two critical points, or saddle-node bifurcations, at $\lambda = 8.817$, $a_w = 0.890$ and $\lambda = 4.493$, $a_w = 0.403$ in this case. These saddle-node points can be followed in the (Λ, Λ_s) plane by

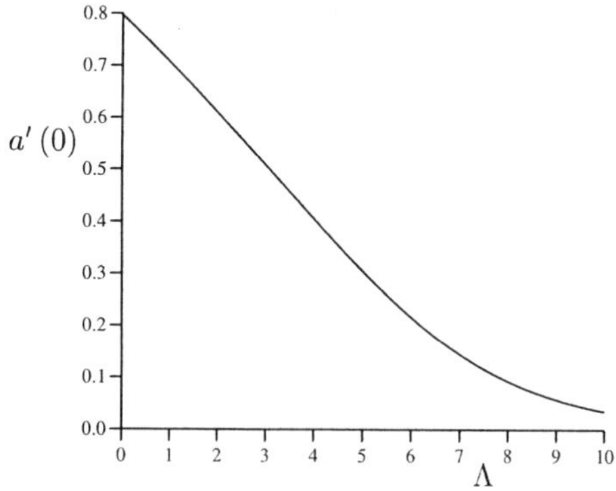

Figure 13.4 *Variation of the concentration flux $a'\left(0\right)$ of the reactant A as a function of Λ, obtained from the numerical solution to equations (13.49) and (13.50) for the limiting case $\Lambda_s \to \infty$*

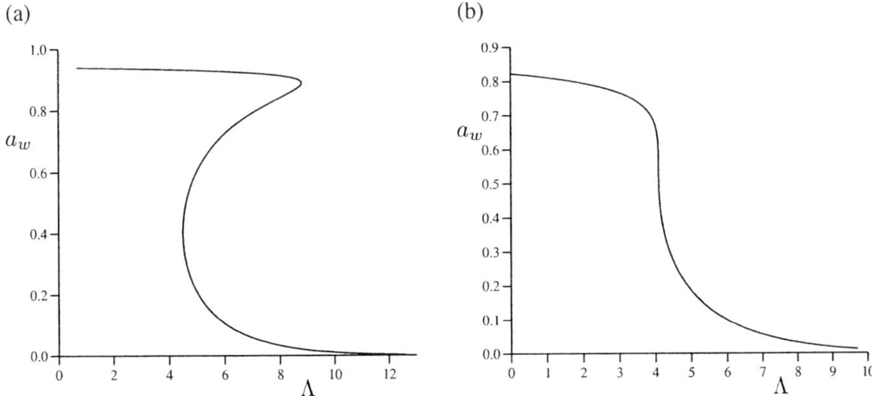

Figure 13.5 *Variation of the wall concentration a_w of the reactant A as a function of Λ, obtained from the numerical solution to equations (13.49) and (13.50) for (a) $\Lambda_s = 0.05$ and (b) $\Lambda_s = 0.172$ (hysteresis point)*

noting that they are determined by considering a perturbation a_1 to the equations (13.49) and (13.50). Thus we have to solve equations (13.49) and (13.50) for a_0 (say) together with the linear perturbed equation

$$a_1'' + ya_1' - \Lambda \left(1 - a_0\right)\left(1 - 3a_0\right) a_1 = 0,$$
$$a_1'\left(0\right) = \Lambda_s a_1\left(0\right), \quad a_1 \to 0 \quad \text{as} \quad y \to \infty \tag{13.52}$$

having a non-trivial solution. Graphs of the critical points obtained in this way are shown in Figure 13.6. This figure shows that there is a hysteresis point at $\Lambda_H = 4.094$ and $\Lambda_{s,H} = 0.172$. Multiple solutions are possible for $\Lambda_s < \Lambda_{s,H}$ with ranges of Λ over which there are three solutions, see Figure 13.5(a), given by the critical points shown in Figure 13.6. The hysteresis point is shown in Figure 13.5(b), where we plot a_w against Λ for $\Lambda_s = 0.172$. For $\Lambda_s > \Lambda_{s,H}$ the bifurcation diagrams are monotonic (as in Figure 13.4, for example).

13.4 CHEMICALLY REACTIVE FLOW NEAR THE STAGNATION POINT OF A CATALYTIC POROUS BED

We consider the steady two-dimensional laminar, premixed chemically reactive stagnation point flow in a catalytic porous bed of finite thickness $\bar{h}$, with the origin of the Cartesian coordinate system placed at the stagnation point as shown in Figure 13.7. The flow is bounded by an impermeable surface at the other side of the catalytic bed. It is assumed that the flow can be divided into two regions: a homogeneous gas phase region before the flow enters the porous bed and a two phase, solid–gas region within the bed. It is also

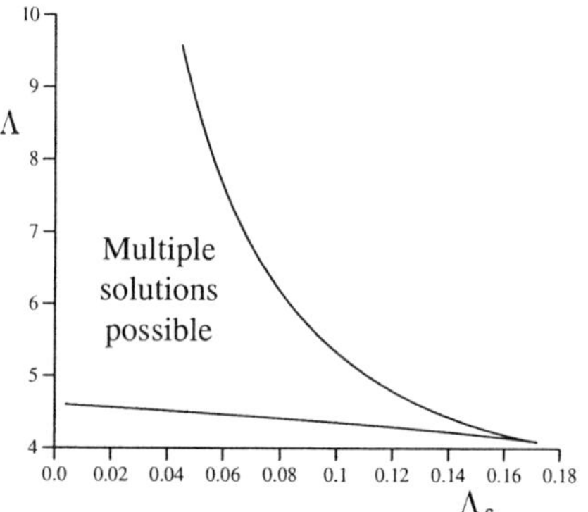

Figure 13.6 *Critical points in the (Λ, Λ_s) parameter plane. The region where multiple solutions are possible is indicated*

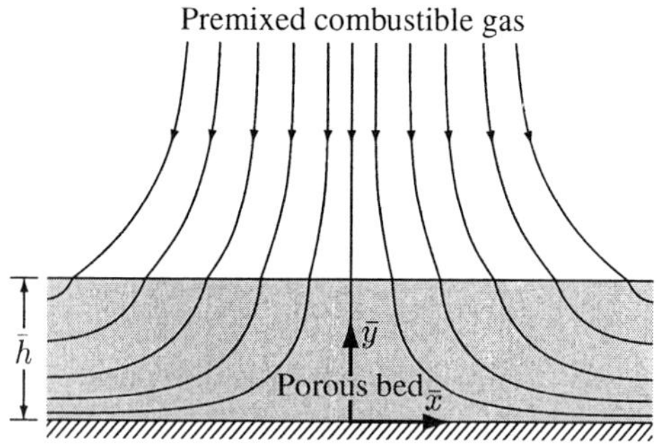

Figure 13.7 *Physical model and coordinate system*

assumed that the reaction rate follows a single-reactant, first-order, one-step Arrhenius kinetics. We notice that this flow model has been developed first by Chao *et al.* (1996).

Since the two regions described are controlled by different transport characteristics, the conservation equations need to be formulated separately. In the gas phase region bounded by $\bar{h} < \bar{y} < \infty$, there is a potential flow so that the velocity field is given by

$$\bar{u} = \bar{c}\bar{x}, \quad \bar{v} = -\bar{c}\bar{y}, \tag{13.53}$$

where the constant $\bar{c}$ is the parameter describing the flow strain rate. Due to the high activation energy and low flow temperature in this region, chemical reaction is negligible so that under the assumption of constant physical quantities, the energy and species (concentration) equations are given by

$$\frac{k_g}{\rho c_p}\frac{\mathrm{d}^2 T^+}{\mathrm{d}\bar{y}^2} + \bar{c}\bar{y}\frac{\mathrm{d}T^+}{\mathrm{d}\bar{y}} = 0, \tag{13.54}$$

$$D_g\frac{\mathrm{d}^2 C^+}{\mathrm{d}\bar{y}^2} + \bar{c}\bar{y}\frac{\mathrm{d}C^+}{\mathrm{d}\bar{y}} = 0, \tag{13.55}$$

subject to the far boundary conditions

$$T^+ \to T_\infty, \quad C^+ \to C_\infty \quad \text{as} \quad \bar{y} \to \infty, \tag{13.56}$$

where D_g and k_g are the diffusivity and thermal conductivity of the gas phase, respectively, T_∞ and C_∞ are prescribed quantities and $+$ denotes quantities in the gas phase region.

In the porous bed, $0 < \bar{y} < \bar{h}$, Darcy's law is applicable, so that the forced convection flow is again a potential flow and the velocity components are given by

$$K\bar{u} = \bar{c}\bar{x}, \quad K\bar{v} = -\bar{c}\bar{y}, \tag{13.57}$$

where K is the permeability of the porous medium. The energy and concentration (species) equations can now be written as

$$\frac{k_m}{\rho c_p}\frac{\mathrm{d}^2 T^-}{\mathrm{d}\bar{y}^2} + \bar{c}\bar{y}\frac{\mathrm{d}T^-}{\mathrm{d}\bar{y}} = -\frac{k_0 Q}{\rho c_p}C^- \exp\left(-\frac{E}{RT^-}\right), \tag{13.58}$$

$$D_m\frac{\mathrm{d}^2 C^-}{\mathrm{d}\bar{y}^2} + \bar{c}\bar{y}\frac{\mathrm{d}C^-}{\mathrm{d}\bar{y}} = \frac{k_0}{\rho}C^- \exp\left(-\frac{E}{RT^-}\right), \tag{13.59}$$

where the superscript $-$ denotes the quantities in the porous region. We assume that the impermeable wall ($\bar{y} = 0$) is isothermal and that there is no net species concentration flux.

It is also assumed that at the interface between these two regions, $\bar{y} = \bar{h}$, the temperature and the species concentration must be continuous. Thus, the wall and interface conditions for equations (13.54), (13.55), (13.58) and (13.59) are

$$T^- = T_w, \quad \frac{\mathrm{d}C^-}{\mathrm{d}\bar{y}} = 0 \qquad \text{on} \quad \bar{y} = 0,$$

$$\left.\begin{array}{l} T^+ = T^-, \quad k_g\frac{\mathrm{d}T^+}{\mathrm{d}\bar{y}} = k_m\frac{\mathrm{d}T^-}{\mathrm{d}\bar{y}} \\[2ex] C^+ = C^-, \quad D_g\frac{\mathrm{d}C^+}{\mathrm{d}\bar{y}} = D_m\frac{\mathrm{d}C^-}{\mathrm{d}\bar{y}} \end{array}\right\} \quad \text{on} \quad \bar{y} = \bar{h}. \tag{13.60}$$

The following non-dimensional quantities are defined:

$$y = \bar{y}\left(\frac{k_g}{\bar{c}\rho c_p}\right)^{-1/2}, \quad \theta = \frac{T - T_\infty}{QC_\infty/c_p}, \quad \phi = \frac{C}{C_\infty}, \quad \theta_\infty = \frac{c_p T_\infty}{QC_\infty},$$

$$\Delta = \frac{E/R}{QC_\infty/c_p}, \quad k = \frac{k_m}{k_g}, \quad D = \frac{D_m}{D_g}, \quad Le = \frac{k_g/\rho c_p}{D_m}, \quad Da = \frac{k_0}{\rho\bar{c}}, \tag{13.61}$$

where Da is the Damköhler number representing the chemical reactivity. Substituting expressions (13.61) into equations (13.54), (13.55), (13.58) and (13.59), we obtain

$$\frac{\mathrm{d}^2\theta^+}{\mathrm{d}y^2} + y\frac{\mathrm{d}\theta^+}{\mathrm{d}y} = 0, \tag{13.62}$$

$$\frac{1}{Le}\frac{\mathrm{d}^2\phi^+}{\mathrm{d}y^2} + y\frac{\mathrm{d}\phi^+}{\mathrm{d}y} = 0, \tag{13.63}$$

$$k\frac{\mathrm{d}^2\theta^-}{\mathrm{d}y^2} + y\frac{\mathrm{d}\theta^-}{\mathrm{d}y} = -D_a\phi^- \exp\left(-\frac{\Delta}{\theta^- + \theta_\infty}\right), \tag{13.64}$$

$$\frac{1}{Le}\frac{\mathrm{d}^2\phi^-}{\mathrm{d}y^2} + y\frac{\mathrm{d}\phi^-}{\mathrm{d}y} = D_a\phi^- \exp\left(-\frac{\Delta}{\theta^- + \theta_\infty}\right). \tag{13.65}$$

The boundary conditions (13.56) and (13.60) for equations (13.62) – (13.65) become

$$
\begin{aligned}
&\theta^- = \theta_w, \quad \frac{\mathrm{d}\phi^-}{\mathrm{d}y} = 0 && \text{on} \quad y = 0, \\[2mm]
&\left.\begin{aligned}
\theta^+ &= \theta^-, \quad \frac{\mathrm{d}\theta^+}{\mathrm{d}y} = k\frac{\mathrm{d}\theta^-}{\mathrm{d}y} \\[1mm]
\phi^+ &= \phi^-, \quad \frac{\mathrm{d}\phi^+}{\mathrm{d}y} = \frac{\mathrm{d}\phi^-}{\mathrm{d}y}
\end{aligned}\right\} && \text{on} \quad y = h, \\[2mm]
&\theta^+ \to 0, \quad \phi^+ \to 1 && \text{as} \quad y \to \infty,
\end{aligned}
\tag{13.66}
$$

where $h = \bar{h}\left(k_g/\bar{c}\rho c_p\right)^{-1/2}$.

Chao *et al.* (1996) have solved equations (13.62) – (13.65), subject to the boundary conditions (13.66), analytically and numerically assuming the cases of a thin bed and a thick bed. We shall present further some results only for the case of a thin bed, while the results for a thick bed can be found in the paper by Chao *et al.* (1996). It was assumed that the reactants are supplied at 300 K and the adiabatic flame temperature is 2180 K so that $\theta_\infty = 0.1376$. Also, the wall temperature was assumed to be $T_w = 700$ K and we have $\theta_w = 0.3211$. Other values used for the computation are $D = 1, Le = 1, Da = 1, \Delta = 1$ and $k = 20$. The analytical (full lines) solutions for large k and the numerical (dashed lines) for the temperature $\theta(y)$ and mass fraction $\phi(y)$ are shown, for some selected values of h, in Figure 13.8. It is seen that the perturbation and numerical solutions for $\phi(y)$ agree well with each other, whilst the results for $\theta(y)$ are different. However, the agreement between these two approaches for $\theta(y)$ is satisfactory when the porous bed is thin, $h = 2$ (say). By increasing the bed thickness, the agreement becomes poorer, which means that the perturbation (analytical) analysis cannot properly describe the heat transfer process when the porous bed is thick. This is due to high flow velocities in the gaseous region and in most of the porous bed except near the wall. Figure 13.8 shows that in the porous bed, the temperature $\theta(y)$ remains a constant to the second-order approximation because of the large value of k. The heat generated through the reaction is completely transferred to the gas phase to preheat the gaseous flow and to the impermeable surface. The gas temperature then increases when approaching the interface. As expected, $\phi(y)$ decreases with increasing y in the porous bed due to the catalytic reaction. The value of $\phi(y)$ starts decreasing from unity before the flow enters the porous bed due to mass

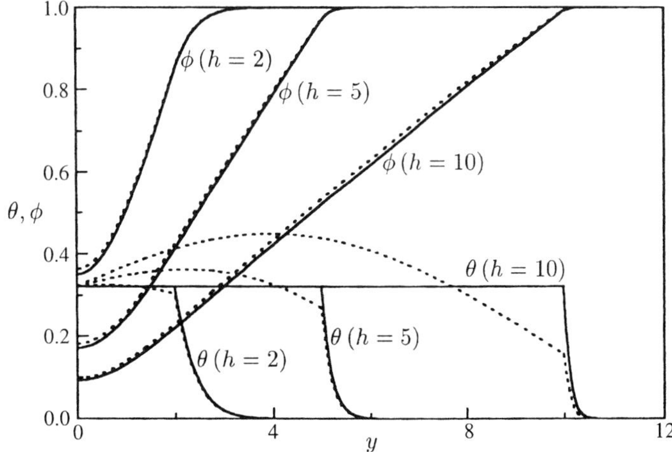

Figure 13.8 *Temperature θ and concentration ϕ profiles as a function of y for different values of the bed thickness h, where $Le = 1$, $Da = 1$, $\Delta = 1$ and $k = 20$. —— analytical solution, - - - numerical solution*

diffusion. It is also noticed from Figure 13.8 that for a thicker bed (larger value of h), the reactant concentration $\phi\,(y)$ is lower. This is because for a thicker bed then the residence time is larger and thus the conservation from reactants to products is higher.

The effect of Lewis number, Le, on $\theta\,(y)$ and $\phi\,(y)$ is illustrated in Figure 13.9. The values of Le considered are in the range 0.5 to 1.5 because, for a gaseous flow, Le is close to unity. Since Le is the ratio of gaseous thermal diffusivity to mass diffusivity, a lower value of Le means a higher mass diffusion rate when the thermal diffusion rate is held fixed. Thus, for small values of Le, the reactant concentration is lower in the region relatively far away from the wall because the reactant is diffused into the porous bed at a higher rate. However, because of the high effective thermal conductivity, the temperature remains the same. Although more reactants are converted to products, all the extra reactants cannot be consumed. The remaining product is accumulated in the region near the wall which results in a higher concentration. In the region extremely far away from the surface, the reactant concentration is not affected by Le because the mass transport is convection controlled.

It should be mentioned that Chao *et al.* (1996) have treated numerically the transient temperature and reactant concentration (species) and obtained the time required to reach steady state.

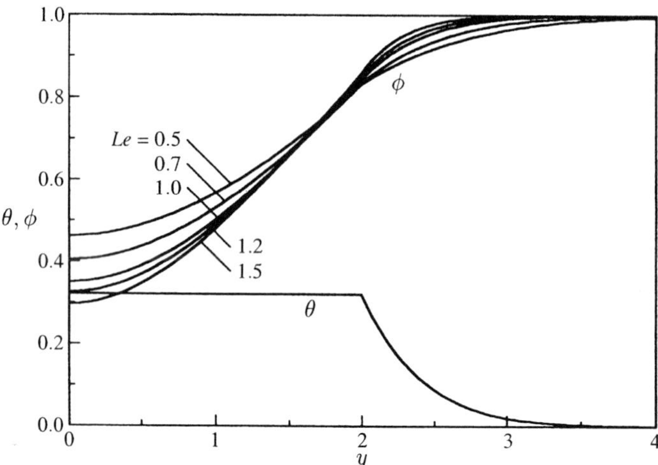

Figure 13.9 *Temperature θ and concentration ϕ profiles as a function of y for a thin bed ($h = 2$), where $Da = 1$, $\Delta = 1$ and $k = 20$*

13.5 CONCLUSION

In this chapter, we have discussed some models of catalytic surface reactions in porous media in which there are no external flows with heat generated by the homogeneous exothermic reactions setting up purely free convection boundary-layer flows on a cylindrical surface. In particular, the steady-state flow near the lower stagnation point of this surface is analysed in detail. The reaction has been modelled as a single first-order reaction with an Arrhenius temperature dependence or it is given by an isothermal cubic autocatalator kinetics. The situations considered for these models are combustion energies and these are highly exothermic. Critical points of the systems where the behaviour undergoes a change from a slow reactive state (low temperatures) to a highly reactive state (higher temperatures) over a relatively short distance have been identified. However, there are several features of the described models that need some comments.

- We have imposed the Boussinesq approximation. This has the advantage of giving a simplified model and is applicable when temperature differences are relatively small, so it could well be used to give a reliable indication of the critical points although it will be less applicable at the larger temperatures generated downstream.

- We have assumed that buoyancy forces arise only from differences in temperature. However, these forces can also arise from differences in reactant concentration and this additional effect requires a term proportional to chemical differences to be included into the Darcy equation. This aspect remains to be investigated further to see if any qualitatively different features are observed when the temperature and concentration components of the buoyancy forces differ substantially.

- We have ignored the effects of any imposed external flow. It would be interesting to analyse such flows which sometimes lead to boundary-layer separation.

- Initial-value (unsteady) problems of these chemical reactive modes can be also considered and analysed on the line proposed by Chao *et al.* (1996), Chaudhary and Merkin (1996) and Merkin and Mahmood (1998).

- The stability predictions of these chemically driven systems should also be investigated. Such an analysis proved to be useful to uncover many interesting features of the interaction between energy, concentration (species) and momentum transfer when the situation of reactive flow through porous media is considered.

- The finite element method to model and predict the convective flows in porous cavities driven by catalytic reactions is also an open topic which is worthy of being considered. In this context it is worth pointing out that the finite element method has been recently used by Zhao *et al.* (2000) to model and predict the dissipative structures of chemical species for a non-equilibrium chemical reaction system in fluid saturated porous medium. In particular, these authors have explored the conditions under which dissipative structures of the species (concentrations) may exist in the Brusselator model for chemical reaction. In the literature, the Brusselator, which is a trimolecular model and involves a third-order chemical reaction, has been widely used to study dissipative structures of far-from-equilibrium chemical reactions, see Prigogine (1980). Zhao *et al.* (1993) have shown that the agreement between the proposed numerical and analytical solutions demonstrate that the proposed finite element method is robust enough for dealing with the chemical instability problems in a fluid-saturated porous medium. Recently, the analysis of thermal instability has been extended to investigate the formation of giant ore deposits in hydrothermal systems of the Earth's crust, see Zhao *et al.* (1997, 1998).

Finally, we would like to stress the fact that models, such as those outlined in this chapter, can be used with great confidence to elucidate unknown features of several complex systems in porous media.

Acknowledgement

D.B. Ingham and I. Pop wish to thank The Royal Society for their financial support.

REFERENCES

Aris, R. (1975). *The Mathematical Theory of Diffusion and Reaction in Permeable Catalysts.* Clarendon Press, Oxford.

Balakotaiah, V. and Pourtalet, P. (1990). Natural convection effects on thermal ignition in a porous medium. I. Semenov model. *Proc. Roy. Soc. Lond.* **A429**, 533–554.

Bond, G. C. (1987). *Heterogeneous Catalysis, Principles and Applications.* Clarendon Press, Oxford.

Chao, B. H., Cheng, P., and Le, T. (1994). Free convective diffusion flame sheet in porous media. *Combustion Sci. Tech.* **99**, 221–234.

Chao, B. H., Wang, H., and Cheng, P. (1996). Stagnation point flow of a chemically reactive fluid in a catalytic porous bed. *Int. J. Heat Mass Transfer* **39**, 3003–3019.

Chaudhary, M. A. and Merkin, J. H. (1994). Free convection stagnation point boundary layers driven by catalytic surface reactions. I. The steady states. *J. Eng. Math.* **28**, 145–171.

Chaudhary, M. A. and Merkin, J. H. (1995a). A simple isothermal model for homogeneous–heterogeneous reactions in boundary-layer flow. I. Equal diffusivity. *Fluid Dyn. Res.* **16**, 311–333.

Chaudhary, M. A. and Merkin, J. H. (1995b). A simple isothermal model for homogeneous–heterogeneous reactions in boundary-layer flow. II. Different diffusivities for reactant and autocatalyst. *Fluid Dyn. Res.* **16**, 335–359.

Chaudhary, M. A. and Merkin, J. H. (1996). Free convection stagnation point boundary layers driven by catalytic surface reactions. II. Times to ignition. *J. Eng. Math.* **30**, 403–415.

Chaudhary, M. A., Liñan, A., and Merkin, J. H. (1995). Free convection boundary layers driven by exothermic surface reactions: critical ambient temperatures. *Math. Eng. Ind.* **5**, 129–145.

Frank-Kamenskii, D. A. (1969). *Diffusion and Heat Transfer in Chemical Kinetics* (2nd edn). Plenum Press, New York.

Gatica, J. E., Viljonen, H. J., and Hlavacek, H. (1987). Stability analysis of chemical reaction and free convection in porous media. *Int. Comm. Heat Mass Transfer* **14**, 391–403.

Gatica, J. E., Viljonen, H. J., and Hlavacek, H. (1989). Interaction between chemical reaction and natural convection in porous media. *Chem. Eng. Sci.* **44**, 1853–1870.

Gottfried, B. S. (1968). Combustion of crude oil in a porous medium. *Combustion and Flames* **12**, 5–13.

Gray, B. F. and Wake, G. C. (1990). The ignition of hygroscopic combustible material by water. *Combustion and Flames* **79**, 2–6.

Gray, P. and Scott, S. K. (1990). *Chemical Oscillations and Instabilities.* Clarendon Press, Oxford.

Hadim, H. and Vafai, K. (1999). Overview of current computational studies of heat transfer in porous media and their applications—forced convection and multiphase heat transfer. In *Advances in Numerical Heat Transfer* (eds W. J. Minkowycz and E. M. Sparrow), Vol. II, pp. 291–329. Taylor & Francis, Washington, DC.

Hsu, P. F., Howell, J. R., and Matthews, R. D. (1991). A numerical investigation on premixed combustion within porous inert media. In *ASME/ISME Conference*, Las Vegas, Nevada.

Ingham, D. B. and Pop, I. (eds) (1998). *Transport in Porous Media.* Pergamon, Oxford.

Ingham, D. B., Harris, S. D., and Pop, I. (1999). Free convection boundary layers at a three-dimensional stagnation point driven by exothermic surface reaction. *Hybrid Meth. Eng.* **1**, 401–417.

Kordylewski, W. and Krajewski, Z. (1984). Convection effects on thermal ignition in porous media. *Chem. Eng. Sci.* **39**, 610–612.

Mahmood, T. and Merkin, J. H. (1998). The convective boundary-layer flow on a reacting surface in a porous medium. *Transport in Porous Media* **32**, 285–298.

Malashetty, M. S., Cheng, P., and Chao, H. B. (1994). Convective instability in a horizontal porous layer saturated with a chemically reacting fluid. *Int. J. Heat Mass Transfer* **37**, 2901–2908.

McDermott, J. (1971). Catalytic conversion of automobile exhaust. *Pollution Control Rev.* **2**. Hoyes Data Corporation, Park Ridge, NJ.

Merkin, J. H. and Chaudhary, M. A. (1994). Free convection boundary layers on vertical surfaces driven by an exothermic surface reaction. *Quart. J. Mech. Appl. Math.* **47**, 405–428.

Merkin, J. H. and Mahmood, T. (1998). Convective flows on reactive surfaces in porous media. *Transport in Porous Media* **33**, 279–293.

Minto, B. J., Ingham, D. B., and Pop, I. (1998). Free convection driven by an exothermic reaction on a vertical surface embedded in porous media. *Int. J. Heat Mass Transfer* **41**, 11–23.

Mojtabi, A. and Charrier-Mojtabi, M. C. (2000). Double-diffusive convection in porous media. In *Handbook of Porous Media* (ed. K. Vafai), pp. 559–603. Marcel Dekker, New York.

Nield, D. A. and Bejan, A. (1999). *Convection in Porous Media* (2nd edn). Springer–Verlag, New York.

Pop, I. and Ingham, D. B. (2001). *Convective Heat Transfer: Mathematical and Computational Modelling of Viscous Fluids and Porous Media*. Pergamon, Oxford.

Prigogine, I. (1980). *From Being to Becoming*. Freeman, New York.

Sathe, S. B., Kulkarni, M. R., Peck, R. E., and Tong, T. W. (1991). An experimental and theoretical study of porous radiant burner performance. In *Proceedings of 23rd International Symposium on Combustion*, pp. 1011–1018.

Scott, S. K. (1991). *Chemical Chaos*. Clarendon Press, Oxford.

Song, X., Schmidt, L. D., and Aris, R. (1991a). Steady states and oscillations in homogeneous–heterogeneous reaction systems. *Chem. Eng. Sci.* **46**, 1203–1215.

Song, X., Williams, W. R., Schmidt, L. D., and Aris, R. (1991b). Bifurcation behavior in homogeneous–heterogeneous combustion. II. Computations for stagnation point flow. *Combustion and Flames* **84**, 292–311.

Steinberg, V. and Brand, H. (1983). Convective instabilities of binary mixtures with fast chemical reaction in a porous medium. *J. Chem. Phys.* **78**, 2655–2660.

Toong, T.-Y (1983). *Combustion Dynamics: The Dynamics of Chemically Reacting Fluids*. McGraw–Hill, New York.

Trevisan, O. V. and Bejan, A. (1990). Combined heat and mass transfer by natural convection in a porous medium. *Adv. Heat Transfer* **20**, 315–352.

Vafai, K. (ed.) (2000). *Handbook of Porous Media*. Marcel Dekker, New York.

Vafai, K. and Hadim, H. (1999). Overview of current computational studies of heat transfer in porous media and their applications—natural and mixed convection. In *Advances in Numerical Heat Transfer* (eds W. J. Minkowycz and E. M. Sparrow), Vol. II, pp. 331–369. Taylor & Francis, Washington, DC.

Viljonen, H. and Hlavacek, H. (1987). Chemically driven convection in a porous medium. *AIChE J.* **33**, 1344–1350.

Williams, W. R., Stenzel, M. T., Song, X., and Schmidt, L. D. (1991a). Bifurcation behaviour in homogeneous–heterogeneous combustion. I. Experimental results over platinum. *Combustion and Flames* **84**, 277–291.

Williams, W. R., Zhao, J., and Schmidt, L. D. (1991b). Ignition and extinction of surface and homogeneous oxidation of NH_3 and CH_4. *AIChE J.* **37**, 641–649.

Zhao, C., Hobbs, B. E., Muhlhaus, H. B., and Ord, A. (2000). Finite element modelling of dissipative structures for non-equilibrium chemical reactions in fluid-saturated porous media. *Comp. Meth. Appl. Mech. Eng.* **184**, 1–14.

Zhao, C., Muhlhaus, H. B., and Hobbs, B. E. (1997). Finite element analysis of steady-state natural convection problems in fluid-saturated porous media heated from below. *Int. J. Numer. Anal. Meth. Geomech.* **21**, 863–881.

Zhao, C., Muhlhaus, H. B., and Hobbs, B. E. (1998). Finite element modelling of temperature gradient driven rock alteration and mineralization in porous rock masses. *Comp. Meth. Appl. Mech. Eng.* **165**, 175–186.

Zhao, C., Xu, T. P., and Valliappan, S. (1993). Numerical modelling of mass transport problems in porous media: a review. *Comp. Struct.* **53**, 849–860.

14 METHANE HYDRATES IN POROUS LAYERS: GAS FORMATION AND CONVECTION

A. BEJAN*, L. A. O. ROCHA* and R. S. CHERRY[†]

*Department of Mechanical Engineering and Materials Science, Duke University, Durham, NC 27708-0300, USA

email: abejan@acpub.duke.edu

[†]Idaho National Engineering and Environmental Laboratory, P. O. Box 1625, Idaho Falls, ID 83415-2203, USA

email: chy@inel.gov

Abstract

This is a review of recent analytical and numerical work on the generation and flow of methane gas through a layer of porous medium impregnated with solid clathrate hydrates. The porous layer is depressurized suddenly on its lower plane and the phase-change front advances under the influence of heat conduction and convection. The first part of the chapter describes a simplified analytical solution based on a unidirectional phase-change model in which the conduction in the gas-filled region behind the front is neglected. The chapter continues with numerical results for the evolution of the unidirectional phase-change process. Both methods lead to the conclusion that the rate of gas flow through the depressurized (bottom) plane of the layer decreases approximately as $t^{-1/2}$. Further numerical modeling shows that the presence of a vertical geothermal gradient has a significant effect on the rate of gas generation. Numerical results for phase change and gas generation in a porous sediment with non-uniform porosity and permeability are also reported.

Keywords: porous media, methane, clathrate hydrates, convection, phase change, dissociation, geothermal gradient, energy sources, exergy, porosity, permeability

14.1 INTRODUCTION

The objective of this chapter is to draw attention to an important new area of fundamental and applied research on convection in porous media: the extraction of methane gas from clathrate hydrates. Vast deposits of methane hydrates have been found all over the globe, under the oceanic floor, and under permafrost. Clathrate hydrates are solid crystals of water and methane, which form and exist at sufficiently high pressures and low temperatures, see Figure 14.1. 'Clathrate' comes from the Latin verb *clathrate*, which means to endow with a lattice. In chemistry, this terminology refers to a mixture in which the molecules of one substance, e.g., methane, are completely entrapped in the crystal lattice or cage-like structure of another substance, e.g., water.

Figure 14.2 shows the geometric hydrate crystal unit structures. Different gases or gas mixtures form hydrates of three different structures that have repetitive crystal units composed of 'cages' (polyhedra) of hydrogen-bounded water molecules (polyhedron's vertices). Each cage (void) contains one guest molecule that is held in by Van der Waals forces. For example, Figure 14.2(a) shows the front face of structure I (sI) cubic cell, where two $5^{12}6^2$ polyhedra, each with 12 pentagonal and 2 hexagonal faces, are connected to four 5^{12} polyhedra, each with 12 pentagonal faces. The complete sI contains eight polyhedra within the cube. Each of the six cube faces contain two halves of a $5^{12}6^2$ polyhedron formed by 12 pentagonal and 2 hexagonal faces, for a total of the six $5^{12}6^2$ polyhedra within the cell. Each of the eight vertices of the cube contains one-eighth of a 5^{12} polyhedron formed by 12 pentagonal faces, which, added to the 5^{12} polyhedron in the center of the cube, gives a total of two 5^{12} polyhedra per cell. The ideal unit cell formula

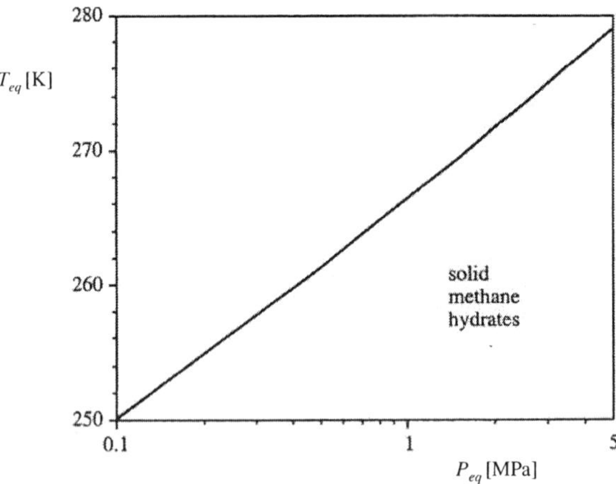

Figure 14.1   *The equilibrium pressure–temperature curve for clathrate hydrates of methane in water, see McKoy and Sinaglu (1963), Kamath and Holder (1987) and Lundgaard and Mollerup (1992)*

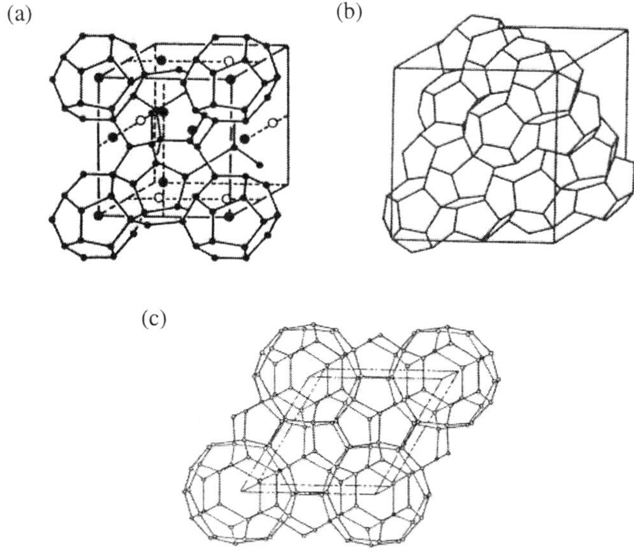

Figure 14.2 *Hydrate crystal unit structures: (a) sI, (b) sII, and (c) sH, see Sloan (1990)*

for sI is $6X \cdot 2Y \cdot 46H_2O$, where X represents the $5^{12}6^2$ polyhedron and Y represents the 5^{12} polyhedron. Pure methane forms structure I and when methane is filling up all the available cages (voids) that compose the hydrate sI, the formula that represents it is given by $8CH_4 \cdot 46H_2O$. Structure II (sII) and Structure H (sH) are also shown in Figures 14.2(b) and (c), respectively.

Today, there is a growing interest in understanding the formation and behavior of methane clathrates, because they represent a potentially huge source of gaseous fuel. Several methods are contemplated for capturing methane gas from hydrates: heating the hydrate, depressurizing the hydrate, mining the hydrate, and destabilizing the hydrate by using *in situ* combustion. The work reviewed in this chapter refers to the fundamentals of methane gas generation via depressurization, see Rocha *et al.* (2001). In addition, these phenomena are relevant to a wide spectrum of technologies, see Sloan (1990), as clathrate hydrate processes are involved in the plugging of pipe lines, see Behar (1994) and Lysne (1994), the sequestration of CO_2 in the ocean, see Makogon (1981) and Fontana and Mussumeci (1994), the endangering of the stability of foundations for offshore oil wells, see Briaud and Chaouch (1997), and the low-temperature storage of energy via clathrate formation, see Holder *et al.* (1994).

The fundamentals of the gas generation process have been studied under two scenarios, the heating of the sediment filled with solid hydrate, see Cherskii and Bondarev (1972), Selim and Sloan (1989), Islam (1994) and Briaud and Chaouch (1997), and the depressurization of the sediment, see Yousif *et al.* (1990). Rocha *et al.* (2001) considered the depressurization technique shown in Figure 14.3. The porous layer, which is filled with solid hydrate, lies immediately above sediment containing free gas that has not formed hydrate because the sediment temperature at that depth is too high. In the upper layer the solid hydrate occupies the pore spaces of a permeable porous medium. Here the temperature is sufficiently low and the pressure high such that the solid hydrate is stable. In the lower layer the generated gas flows through the pores of the same or another porous medium. The interface between the two layers marks the level where the initial temperature T_i and pressure P_i correspond to equilibrium (dissociation, phase change, T_{eq} and P_{eq}, see Figure 14.3).

The basic scales of the process characterize the rate at which the dissociated lens advances into the porous medium filled with solid hydrate, and the rate at which gas is being produced and captured in an unsteady, time-dependent fashion. There are two problems— two phenomena—that join hands during this process. One problem is the dissociation and advancement of the phase-change front into the hydrate-filled medium, see Figure 14.4, left. This problem can be studied first as a one-dimensional, time-dependent heat and fluid flow phenomenon, and is the subject of this chapter. The generated gas is driven downward through the dissociated layer of thickness $X(t)$ and the interface between the original two layers ($x = 0$) also serves as an interface between the two problems. The second problem is the nearly radial gas flow through the lower porous layer and into the well, see Figure 14.4, right. The gas enters vertically through the $x = 0$ interface, and flows radially towards the well.

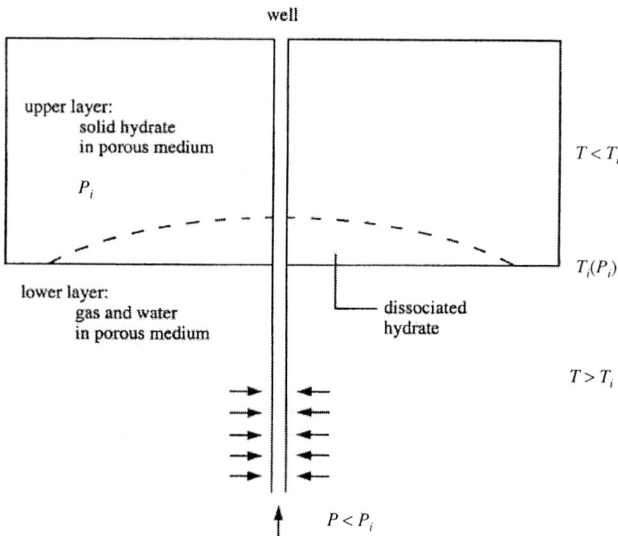

Figure 14.3 *The production of gas from a hydrate layer via depressurization, see Rocha et al. (2001)*

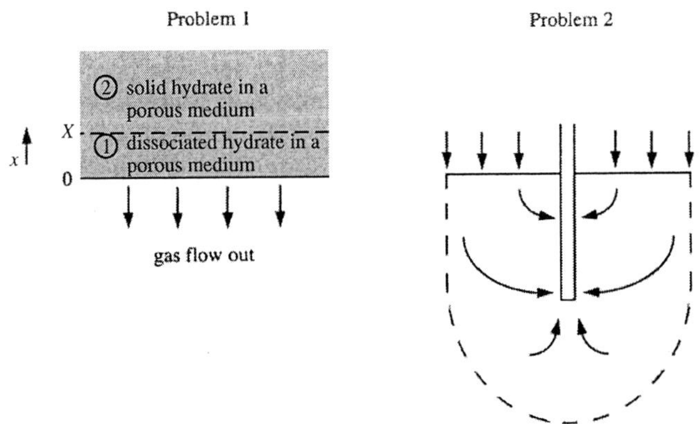

Figure 14.4 *The two heat and fluid flow fields separated by the interface between the two layers of Figure 14.3, see Rocha* et al. *(2001)*

14.2 PHASE CHANGE AND GAS FLOW

The one-dimensional model of Figure 14.4 (left) is shown in greater detail in Figure 14.5. The dissociated sublayer—region 1—contains porous medium filled with a combination of gas and liquid water. Region 2 is the original porous medium in which the pores are filled completely or incompletely with a solid hydrate. In the beginning, $t = 0$, region 1 is absent ($X = 0$) and region 2 is initially isothermal at the equilibrium temperature $T_i\,(P_i)$ that characterizes the near-interface regions of Figure 14.3.

The dissociation starts when the well begins flowing. In Rocha *et al.* (2001), this event is modeled as the sudden lowering of the pressure at the $x = 0$ interface: the new pressure P_0 is considerably lower than P_i. Dissociation advances into region 2, on a front ($x = X$) of

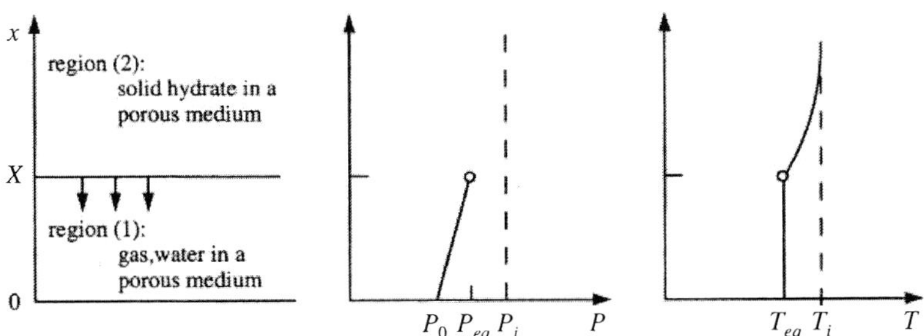

Figure 14.5 *One-dimensional model for phase change, heat transfer and fluid flow through the upper layer of Figure 14.3, see Rocha* et al. *(2001)*

Mollerup (1992), namely

$$P_{eq} = c\exp\left(a - \frac{b}{T_{eq}}\right) \quad \text{at} \quad x = X(t), \tag{14.12}$$

where a, b and c are three constants listed in Table 14.1. Equation (14.12) is presented in Figure 14.1.

14.3 SIMILARITY SOLUTION

The solution to this problem was developed by Rocha *et al.* (2001) in closed form, by introducing the similarity variables

$$\eta = \frac{x}{(4\alpha_2 t)^{1/2}} \quad \text{and} \quad \xi = \frac{X}{(4\alpha_2 t)^{1/2}}, \tag{14.13}$$

such that $\eta = \xi$ represents the position of the dissociation front. Integrating equation (14.4) once and combining it with equations (14.1) and (14.2), we obtain $(MP/ZRT_1)\,dP/dx =$ constant, or

$$\frac{M}{ZRT_1} \frac{P}{2(\alpha t)^{1/2}} \frac{dP}{d\eta} = \text{constant}. \tag{14.14}$$

The constant in equation (14.14) is supplied by equation (14.11): constant $= \rho v = -\omega\phi\rho_H \partial X/\partial t$, which also shows the origin of ξ in the subsequent equations, cf. the

Table 14.1   *Physical constants and properties used in the numerical work (Rocha et al., 2001)*

$a = 49.3185$	$T_{eq}(P_{eq} = P_0) = 280\,\text{K}$
$b = 9459\,\text{K}$	$T_{1,b,\max} = 300\,\text{K}$
$c = 1\,\text{Pa}$	$T_{sb} = 277\,\text{K}$
$c_p = 2162\,\text{J/kg}$	$\tilde{X}_0 = 0.01$
$c_1 = 2500\,\text{J/kg K}$	$Z = 1$
$c_2 = 2500\,\text{J/kg K}$	$\alpha_1 = 2.9 \times 10^{-6}\,\text{m}^2/\text{s}$
$k_1 = 5.6\,\text{W/m K}$	$\alpha_2 = 7 \times 10^{-7}\,\text{m}^2/\text{s}$
$k_2 = 2.7\,\text{W/m K}$	$\gamma_{\max} = 46\,\text{K/km}$
$K_1 = 1.4 \times 10^{-13}\,\text{m}^2$	$\Delta H = 4.1 \times 10^5\,\text{J/kg}$
$L_{\max} = 500\,\text{m}$	$\mu = 10^{-5}\,\text{kg/s m}$
$L_{\min} = 10\,\text{m}$	$\rho_1 = 1000\,\text{kg/m}^3$
$M = 16\,\text{kg/kmol}$	$\rho_2 = 1000\,\text{kg/m}^3$
$P_{eq,\max} = 5600.5\,\text{kPa}$	$\rho_H = 913\,\text{kg/m}^3$
$P_0 = 5600\,\text{kPa}$	$\phi = 0.3$
$R = 8314\,\text{J/kmol K}$	$\omega = 0.127\,\text{kg methane/kg hydrate}$
$T_i = 300\,\text{K}$	

second of equations (14.13). Integrating equation (14.14) from $P = P_0$ (at $\eta = 0$) to $P(\eta)$, and using equation (14.9), leads to

$$P^2 = P_0^2 + \frac{4\eta\xi T_{eq}\alpha_2\omega\phi\rho_H\mu ZR}{K_1 M}. \tag{14.15}$$

Finally, $P = P_{eq}$ at $\eta = \xi$ and the equilibrium condition (14.12) yields the following relation between T_{eq} and ξ:

$$c^2\left[\exp\left(a - \frac{b}{T_{eq}}\right)\right]^2 = P_0^2 + \frac{4\xi^2 T_{eq}\alpha_2\omega\phi\rho_H\mu ZR}{K_1 M}. \tag{14.16}$$

A second relation is supplied by the solution to the transient conduction problem in region 2. By using the classical method of dimensionless unidirectional time-dependent conduction in similarity formulation, see Carslaw and Jaeger (1959) and Bejan (1995), it can be shown that the similarity solution to the problem stated in equations (14.5) – (14.8) is given by

$$\frac{T_2(x,t) - T_i}{T_{eq} - T_i} = \frac{\operatorname{erfc}\eta}{\operatorname{erfc}\xi}. \tag{14.17}$$

Combining this expression with the energy conservation condition (14.10) and the ξ definition (14.13) yields a second relation between T_{eq} and ξ:

$$(T_i - T_{eq})\frac{\exp\left(-\xi^2\right)}{\xi\operatorname{erfc}\xi} = \frac{\pi^{1/2}\phi\rho_H\Delta H\alpha_2}{k_2}. \tag{14.18}$$

Equations (14.16) and (14.18) determine T_{eq} and ξ uniquely, as functions of the imposed boundary pressure P_0, the initial temperature T_i, and the thermophysical properties of regions 1 and 2. Rocha *et al.* (2001) showed that these equations establish ξ and T_{eq}/b as functions of five dimensionless parameters, namely

$$a, \quad \frac{T_i}{b}, \quad \frac{P_0}{c}, \quad G_2 \quad \text{and} \quad G_1, \tag{14.19}$$

where G_1 and G_2 are dimensionless groups that emerge in equations (14.16) and (14.18), respectively, namely

$$G_1 = \frac{b\alpha_2\omega\phi\rho_H\mu ZR}{c^2 K_1 M}, \tag{14.20}$$

$$G_2 = \frac{\phi\rho_H\Delta H\alpha_2}{k_2 b}. \tag{14.21}$$

The five parameters listed in equation (14.19) are pinpointed by using the representative values of the thermophysical properties listed in Table 14.1. Not every parameter is a constant: in equation (14.19), the parameters are listed sequentially starting with the parameter that is the least likely to vary (a), and ending with the ones that are the most likely. Specifically, in a field with methane hydrate sediments the group G_2 may vary (or may be difficult to be described by a single value) because of variations or uncertainties in ϕ. The group G_1 may vary on account of changes in both K_1 and ϕ. Changes in ϕ can also affect α_2 and k_2 if they are large enough, but α_2 and k_2 have been assumed to be constant here.

Figures 14.6 and 14.7 document the sensitivity of the solution (ξ, T_{eq}) obtained from equations (14.16) and (14.18), with respect to possible variations in four of the five parameters listed in equation (14.19). The changes in G_1 and G_2 are attributed to variations in permeability and hydrate saturation, while the remaining properties have the values listed in Table 14.1. Figure 14.6(a) shows that the equilibrium temperature at the interface is relatively insensitive to changes in permeability and hydrate saturation in the range $10^{-14} < K_1 < 10^{-11}$ and $0.1 < \phi < 0.3$. The hydrate saturation begins to show an effect in the limit of small K_1 and small ϕ. The corresponding chart for the dimensionless thickness of the dissociated layer, in Figure 14.6(b), shows that ξ decreases as the hydrate saturation increases. The effect of the permeability on ξ is insignificant.

The effect of variations in the initial temperature and the imposed low pressure is reported in Figure 14.7. The variation, or uncertainty, in the T_i value may be attributed to field conditions such as geographical position and geothermal gradient. The effect of the geothermal gradient is documented later in this chapter. Figure 14.7(a) shows that T_{eq} is relatively insensitive to the value of T_i in the expected range. The effect of the imposed low pressure P_0 is more noticeable: note that T_{eq} increases by about 10 K while P_0 increases by a factor of 10. Figure 14.7(b) shows that the thickness of the dissociated front decreases significantly as P_0 increases, and as T_i decreases, because in both cases

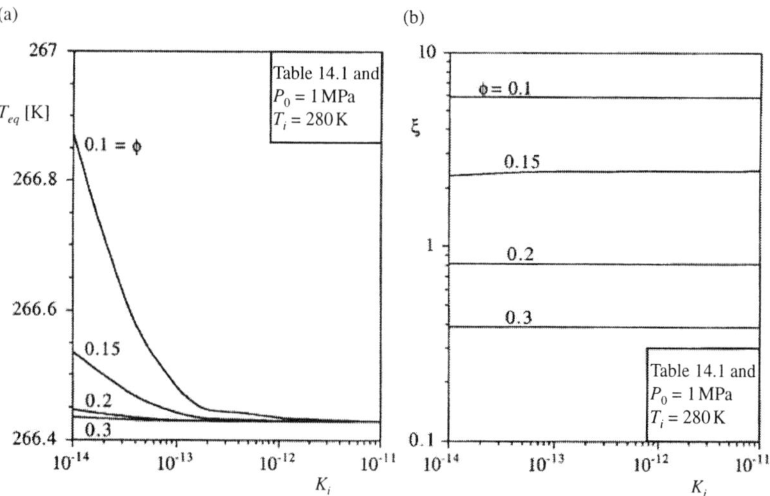

Figure 14.6 *The effect of changes in permeability and hydrate saturation on the position and temperature of the dissociation front, see Rocha et al. (2001)*

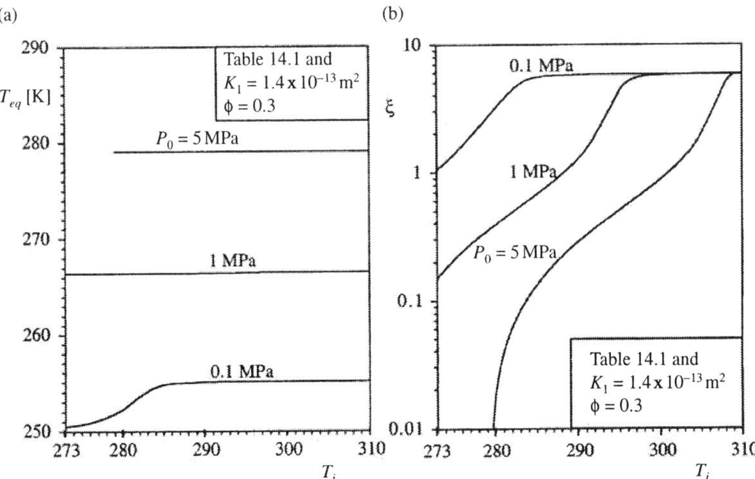

Figure 14.7 *The effect of changes in initial temperature and imposed (low) pressure on the position and temperature of the dissociation front, see Rocha* et al. *(2001)*

the temperature difference $(T_i - T_{eq})$ that drives heat to the dissociation front becomes smaller.

The rate of hydrate dissociation is obtained from equation (14.11) as follows:

$$\rho\,(-v) = \omega\phi\rho_H\xi \left(\frac{\alpha_2}{t}\right)^{1/2}. \tag{14.22}$$

Since the group ξ is time-independent, see Figures 14.6 and 14.7, the flow rate in equation (14.22) decreases in time as $t^{-1/2}$. This flow rate has the same value at any value of x in the dissociated zone (region 1), cf. equation (14.4). Equation (14.22) may be rearranged in dimensionless terms as follows:

$$\frac{\rho}{\rho_H} \frac{(-v)}{(\alpha_2/t)^{1/2}} = \phi\omega\xi \tag{14.23}$$

to show that the rate of gas formation increases not only with ϕ, ω and ξ, but also with changes in the remaining parameters that lead to increases in ξ, e.g., Figures 14.6 and 14.7.

14.4 NUMERICAL SOLUTION FOR A PLANE-SHAPED DISSOCIATION FRONT

The phase-change convection problem of Figure 14.5 was solved numerically by relaxing the simplifying assumption made in equation (14.9). Details are provided in Rocha *et al.* (2001). The one-dimensional domain was assumed finite, $0 \leqslant x \leqslant L$, where L is considerably larger than the thickness X of the developing dissociated layer. The dissociated layer (region 1) is no longer isothermal. The heat transfer through this region is governed by the energy equation for unidirectional convection in the x direction, see Nield and Bejan (1999), namely

$$\frac{\partial T_1}{\partial t} + \frac{\partial}{\partial x}\left(\frac{\rho c_P}{\rho_1 c_1} v T_1\right) = \alpha_1 \frac{\partial^2 T_1}{\partial x^2}, \tag{14.24}$$

where α_1, (ρc_p) and $(\rho_1 c_1)$ are the thermal diffusivity of the saturated porous medium (region 1), the heat capacity of the gas phase alone, and the heat capacity of a volume sample of saturated medium from region 1, respectively. Because of the temperature gradient in region 1, the movement of the dissociation front is driven by conduction from both sides of the interface. This means that instead of equation (14.10) we use

$$k_2 \frac{\partial T_2}{\partial x} - k_1 \frac{\partial T_1}{\partial x} = \phi \rho_H \Delta H \frac{\mathrm{d}X}{\mathrm{d}t}, \tag{14.25}$$

and the governing equations are equations (14.24) and (14.25) in combination with equations (14.1) – (14.8). This formulation was non-dimensionalized by using L as length scale, L^2/α_2 as time scale, and $[T_i - T_{eq}\,(P_{eq} = P_0)]$ as temperature scale, where $T_{eq}\,(P_{eq} = P_0)$ and L are constants introduced solely for the purpose of setting up the numerical scheme in a finite domain. An important feature of the numerical solution is to show that L is sufficiently large so that it has little effect on the dissociation-driven flow. To test this behavior, Rocha *et al.* (2001) repeated the numerical solution for several values of L in the range $L_{\min} \leqslant L \leqslant L_{\max}$. The dimensionless version of equations (14.5), (14.24) and (14.25) are given by

$$\frac{\partial \theta_2}{\partial \tau} = \frac{\partial^2 \theta_2}{\partial \tilde{x}^2}, \tag{14.26}$$

$$\frac{\partial \theta_1}{\partial \tau} = \tilde{\alpha}\frac{\partial^2 \theta_1}{\partial \tilde{x}^2} + G\frac{\partial \tilde{X}}{\partial \tau}\frac{\partial \theta_1}{\partial \tilde{x}}, \tag{14.27}$$

$$\frac{\partial \tilde{X}}{\partial \tau} = \frac{Ste}{\tilde{\rho}_H \phi}\left(\frac{\partial \theta_2}{\partial \tilde{x}} - \tilde{k}\frac{\partial \theta_1}{\partial \tilde{x}}\right), \tag{14.28}$$

where

$$\left(\tilde{x}, \tilde{X}\right) = \frac{(x, X)}{L}, \quad \tau = \frac{t}{L^2/\alpha_2}, \tag{14.29}$$

$$\theta_{1,2} = \frac{T_{1,2} - T_{eq}\left(P_{eq} = P_0\right)}{T_i - T_{eq}\left(P_{eq} = P_0\right)}, \quad \tilde{P} = \frac{P_1 - P_0}{P_{eq,\max} - P_0}, \tag{14.30}$$

$$\tilde{k} = \frac{k_1}{k_2}, \quad \tilde{\alpha} = \frac{\alpha_1}{\alpha_2}, \quad \tilde{\rho}_H = \frac{\rho_H}{\rho_2}, \quad \tilde{\rho}_1 = \frac{\rho_1}{\rho_2}, \tag{14.31}$$

$$G = \frac{\omega\phi\tilde{\rho}_H c_p}{\tilde{\rho}_1 c_1}, \quad Ste = \frac{c_2\left[T_i - T_{eq}\left(P_{eq} = P_0\right)\right]}{\Delta H}. \tag{14.32}$$

The non-dimensionalized pressure is based on the maximum difference $(P_{eq,\max} - P_0)$, where $P_{eq,\max}$ was calculated by solving equations (14.1), (14.2), (14.11), (14.12) and (14.28). The dimensionless initial and boundary conditions for regions 1 and 2 are shown in Figure 14.8 and given by

$$\theta_1 = \theta_2 = 1 \quad \text{and} \quad \tilde{X} = \tilde{X}_0 \quad \text{at} \quad \tau = 0, \tag{14.33}$$

$$\frac{\partial \theta_1}{\partial \tilde{x}} = 0 \qquad \text{at} \quad \tilde{x} = 0, \tag{14.34}$$

$$\theta_1 = \theta_{eq} \quad \text{and} \quad \theta_2 = \theta_{eq} \quad \text{at} \quad \tilde{x} = \tilde{X}(\tau), \tag{14.35}$$

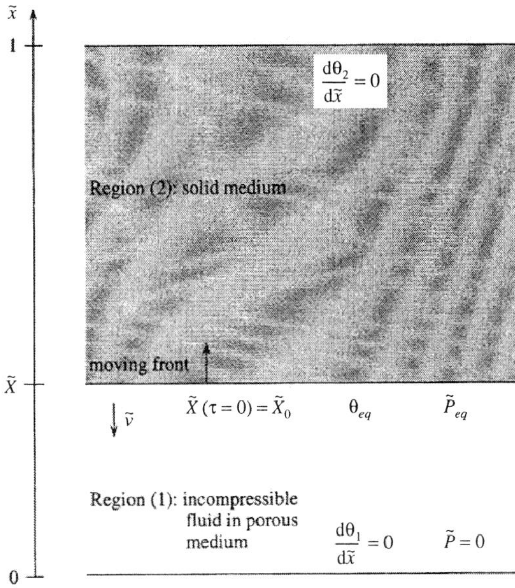

Figure 14.8 *The boundary conditions for the numerical simulation of the plane-front phase-change phenomenon, see Rocha et al. (2001)*

$$\frac{\partial \theta_2}{\partial \tilde{x}} = 0 \qquad \text{at} \quad \tilde{x} = 1. \qquad (14.36)$$

It was assumed that the initial thickness of the dissociated layer (X_0) is considerably smaller than the overall dimension of the computational domain (L), i.e., $\tilde{X}_0 \ll 1$. In addition, the dimension L is sufficiently large, or the dimensionless time domain τ sufficiently short, so that the heat transfer effect of the dissociation process is not felt at the far boundary of the domain $(\tilde{x} = 1)$. In this plane the temperature remains close to the initial temperature $(\theta_2 \cong 1)$, even though the invoked boundary condition is that of zero heat flux, see equation (14.36).

Equations (14.26) – (14.28) and conditions (14.33) – (14.36) are sufficient for determining the unknowns $\theta_1 (\tau), \theta_2 (\tau)$ and $\tilde{X} (\tau)$. The pressure and gas velocity distribution in region 1 follow from equations (14.1), (14.2) and (14.4) and equations (14.26) – (14.28) were solved numerically based on a finite-difference scheme. The algorithm and accuracy tests are described in Rocha *et al.* (2001) and the numerical results were generated by assuming the physical properties listed in Table 14.1. Each numerical run begins with assuming an initial (small) thickness for the lower layer $\tilde{X}_0$ and the behavior of the temperature distribution in region 2 is illustrated in Figure 14.9. In this case the phase-change interface is marked by the equilibrium temperature $\theta \cong 0$, which remains almost constant as the time increases. The $\theta (\tilde{x})$ variation has the characteristic shape of most one-dimensional transient diffusion problems. The far-field condition $(\partial \theta / \partial \tilde{x} = 0$ as $\tilde{x} \to \infty)$ is satisfied approximately at $\tilde{x} = 1$ when the time τ is sufficiently short and this is why in Figure 14.9 the numerical simulation was terminated at $\tau = 0.07$. The calculations ended when the boundary temperature dropped by more than 2.5% from the specified boundary condition $(\theta (\tilde{x} = 1) = 0.97$ at $\tau = 0.07)$.

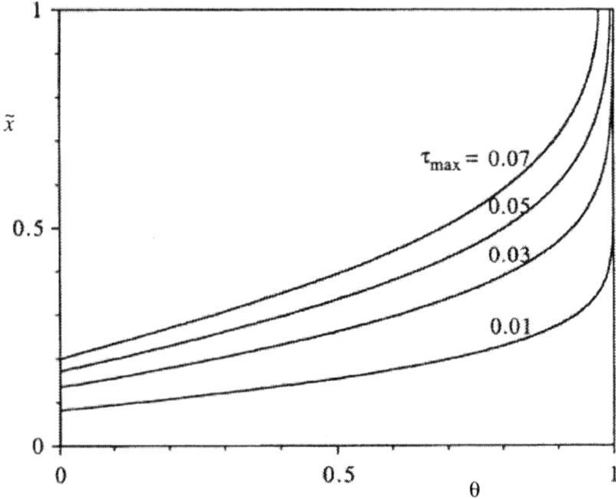

Figure 14.9 *Numerical solution for the temperature distribution in region 2 of Figure 14.8, see Rocha* et al. *(2001)*

The corresponding pressure distribution in region 1 is almost linear in $\tilde{x}$ for every value of τ and these aspects are documented in Rocha *et al.* (2001). The interface pressure decreases in time, but this decrease is accompanied by an almost imperceptible decrease in the interface temperature, see Figure 14.9.

Figure 14.10 summarizes the complete behavior of the phase-change front $\tilde{X}$, as a function of both τ and the assumed $\tilde{X}_0$ value. The hydrate dissociation rate history is represented by the dimensionless group $\tilde{\rho}\,(-\tilde{v})$ calculated from the numerical solution using the expression

$$[\tilde{\rho}\,(-\tilde{v})]_{\text{num}} = \frac{\rho\,(-v)}{\rho_H \alpha_2/L} = \phi\omega \frac{d\tilde{X}}{d\tau}. \tag{14.37}$$

According to the analytical solution, equation (14.23), the dissociation rate $\rho\,(-v)$ is expected to decrease as $t^{-1/2}$, such that the dimensionless group A, defined as

$$A = \frac{\rho\,(-v)}{\rho_H\,(\alpha_2/t)^{1/2}\,\phi\omega\xi}, \tag{14.38}$$

is equal to 1. On the other hand, the value produced by the numerical solution is, cf. equations (14.29) and (14.37), given by

$$A = [\tilde{\rho}\,(-\tilde{v})]_{\text{num}} \frac{\tau^{1/2}}{\phi\omega\xi} = \frac{2\tau}{\tilde{X}} \frac{d\tilde{X}}{d\tau}. \tag{14.39}$$

This numerical result is reported in Figure 14.11, where the reference level $A = 1$ represents the analytical result, i.e., equation (14.23). It should be noted that the A group is the same as the ratio $[\rho\,(-v)]_{\text{numerical}} / [\rho\,(-v)]_{\text{analytical}}$. The value of A is smaller

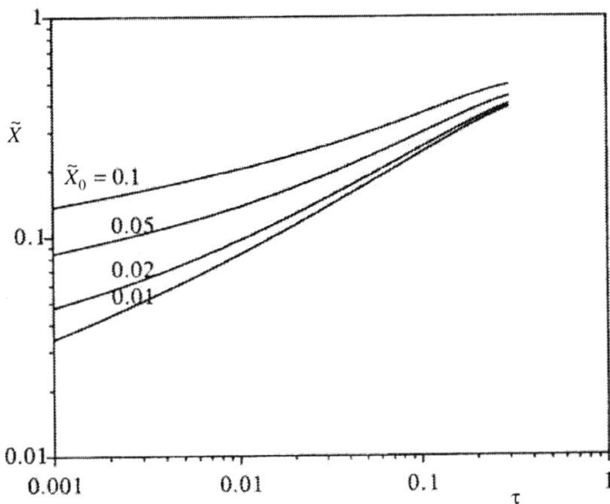

Figure 14.10 *The evolution of the plane phase-change front, see Rocha* et al. *(2001)*

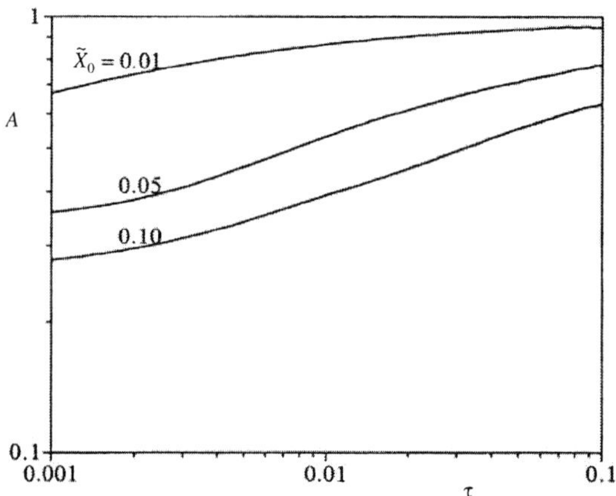

Figure 14.11 *The effect of the initial thickness of the plane dissociated layer on the agreement between the numerical and analytical solutions, see Rocha* et al. *(2001)*

than 1 because of the assumed finite initial thickness $\tilde{X}_0$ in the numerical solution. This means that the numerical solution underestimates the dissociation rate. The latter is theoretically infinite at $t = 0^+$, as in the analytical solution. Figure 14.11 shows that the numerical solution approaches the analytical solution as the assumed initial layer thickness $\tilde{X}_0$ vanishes. The smallest $\tilde{X}_0$ value that could be used by Rocha *et al.* (2001) was $\tilde{X}_0 = 0.005$, where a total of 1000 nodes were used to cover the $\tilde{x}$ domain. The discrepancy between the numerical and analytical solutions also decreases as the time τ increases. This effect is also due to the assumption made in equation (14.9), which becomes more accurate at larger times.

Rocha *et al.* (2001) also showed the effect of hydrate saturation and initial temperature on the agreement between the numerical and analytical solutions. The agreement is better when ϕ is small and T_i is large and, in general, the agreement improves as the time increases. The two solutions are in better agreement when the imposed low pressure P_0 is of the order of 1 MPa or lower.

An important feature of the numerical formulation is the assumed overall size of the system, L. This dimension was varied while repeating the numerical procedure for several values of the dimensionless parameter $\tilde{L} = L/L_{\max}$. For example, a test conducted for a relatively small time ($\tau = 10^{-4}$) showed that L is large enough to be irrelevant in the final numerical results when $\tilde{L}$ is of the order of 0.04, or larger, and physically this case means that $L \geqslant 20\,\mathrm{m}$.

14.5 THE EFFECT OF A GEOTHERMAL GRADIENT

In this section we document the effect of the vertical geothermal gradient on the phase-change and gas convection process. This is a continuation of the numerical work based on the plane model of Figure 14.8. This time the isothermal initial temperature is replaced by the assumption of a uniform initial temperature gradient in the vertical direction (x). The geothermal gradient represents the rate of increase of the temperature in the earth with depth. It is generated by the continuous flow of heat outward through the crust of the earth and its value varies from place to place depending on the heat flow in the region and on the thermal conductivity of the rock.

The geothermal gradient is one of the important parameters that controls the thickness and the stability of the hydrate zone in marine environment, see Holder *et al.* (1987), Roadifer *et al.* (1987), Hanumantha Rao *et al.* (1998) and Singh and Singh (1999). The depth, extent and stability of the hydrate zone are governed by the phase diagram for mixtures of methane and hydrate, and is determined by the ambient pressure and temperature. At sea depths greater than about 300 m, the pressure is high enough and the temperature is low enough for hydrates to occur at the sea floor. The base of the hydrate zone is a phase boundary between solid hydrate and free gas and water and its depth is determined principally by the value of the geothermal gradient, see Roadifer *et al.* (1987) and Willoughby and Edwards (1997). The most favorable conditions under which gas hydrates are likely to occur are the normal range of geothermal gradients, which are below 60 K/km, see Subrahmanyam *et al.* (1998). The geothermal gradient has also been reported to vary in the range $20 - 50$ K/km, see Briaud and Chaouch (1997), Win and Rik (1999) and Gering *et al.* (2000).

To the numerical model of Figure 14.8, we add the initial temperature gradient, which covers region 2, namely

$$\gamma = -\frac{\partial T}{\partial x} = \frac{q''}{k_s}, \tag{14.40}$$

where q'' is the local heat flux generated by the earth, and k_s is the thermal conductivity of the sediments. The initial and boundary conditions are given by

$$T_2 = \gamma \left(L - x \right) + T_{sb} \quad \text{at} \quad t = 0, \tag{14.41}$$

and $T_2 = T_{eq}$ at $x = X$ and $T_2 = T_{sb}$ at $x = L$, where T_{sb} is the temperature at the sea bottom. The dimensionless governing equations are the same as equations (14.26) – (14.32), except that

$$\theta_1 = \frac{T_1 - T_{sb}}{T_{1,b,\max} - T_{sb}}, \quad \theta_2 = \frac{T_2 - T_{sb}}{T_{1,b,\max} - T_{sb}}, \tag{14.42}$$

$$\theta_1 = \tilde{\gamma}\tilde{L}(1 - \tilde{x}), \quad \theta_2 = \tilde{\gamma}\tilde{L}(1 - \tilde{x}) \quad \text{and} \quad \tilde{X} = \tilde{X}_0 \quad \text{at} \quad \tau = 0, \tag{14.43}$$

$$\frac{\partial \theta_1}{\partial \tilde{x}} = -\tilde{\gamma} \qquad\qquad\qquad \text{at} \quad \tilde{x} = 0, \tag{14.44}$$

$$\theta_1 = \theta_{eq} \quad \text{and} \quad \theta_2 = \theta_{eq} \qquad \text{at} \quad \tilde{x} = \tilde{X}(\tau), \tag{14.45}$$

$$\theta_2 = 0 \qquad\qquad\qquad \text{at} \quad \tilde{x} = 1 \tag{14.46}$$

and $\tilde{\gamma}$ is the dimensionless temperature gradient

$$\tilde{\gamma} = \frac{\gamma}{\gamma_{\max}}, \quad \gamma_{\max} = \frac{T_{1,b,\max} - T_{sb}}{L_{\max}}. \tag{14.47}$$

The temperature at the bottom of region 1 is given by

$$T_{1,b} = T_{sb} + \gamma L \tag{14.48}$$

and its maximum value is reached when the geothermal gradient is a maximum, namely

$$T_{1,b,\max} = T_{sb} + \gamma_{\max} L_{\max}. \tag{14.49}$$

The size of the computational domain (L) is now defined as the distance from the bottom of region 1 to the sea bottom.

The dimensionless temperature at the bottom of region 1, $\theta_{1,b}$, was varied in the range $0.4 - 1$ and in this way we could study the effect of varying of the dimensionless geothermal gradient. Figure 14.12 shows how the geothermal temperature distribution varies. It is assumed that the initial thickness of the dissociated layer (X_0) is considerably smaller than the overall dimension of the computational domain (L), i.e., $\tilde{X}_0 \ll 1$.

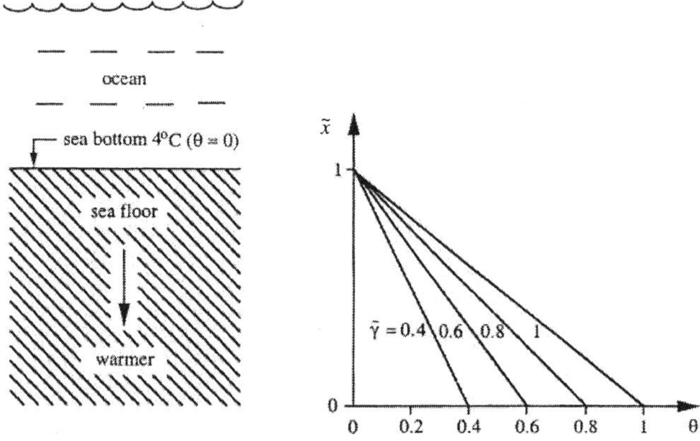

Figure 14.12 *Initial temperature distributions showing the assumed geothermal gradient*

The hydrate dissociation rate history is represented by the dimensionless group $\tilde{\rho}\,(-\tilde{v})$, which is calculated from equation (14.37) and details of the numerical work are provided by Rocha *et al.* (2001). Numerical results were generated by assuming the physical properties listed in Table 14.1 and the value $\tilde{X}_0 = 0.01$ was selected based on numerical tests such as those shown in Figure 14.11. Each numerical run began with assuming a small enough initial thickness for the lower layer $\tilde{X}_0$ and the behavior of the temperature distribution is illustrated in Figure 14.13. In this case, the phase-change interface is marked by the equilibrium temperature $\theta \cong 0.12$, which remains almost constant as the time increases. The $\theta\,(x)$ distribution gradually returns to the initial distribution because of the continuous heat flux generated by the earth.

The temperature at the bottom of region 1 is shown in Figure 14.14. This temperature drops quickly and reaches a minimum value close to the equilibrium temperature ($\theta \cong 0.13$, when $\tau = 10^{-4}$). This result is in excellent agreement with the results of Rocha *et al.* (2001) for phase change following an initial isothermal field. At greater times, the bottom temperature increases monotonically, in agreement with the behavior illustrated in Figure 14.13.

The evolution of the pressure distribution in region 1 is illustrated in Figure 14.15. The gradient is nearly uniform in x, and the pressure decreases in time at every point in the medium. The upper end of each curve indicates the position and pressure of the phase-change interface. The interface pressure decreases in time, and this decrease is accompanied by a small decrease in the interface temperature. This temperature decrease is too small to be visible in Figure 14.13. The high values of dP/dx at short times could trigger mechanical failure in the sediments, and, subsequently, the gas may flow through a porous medium that experiences erosion.

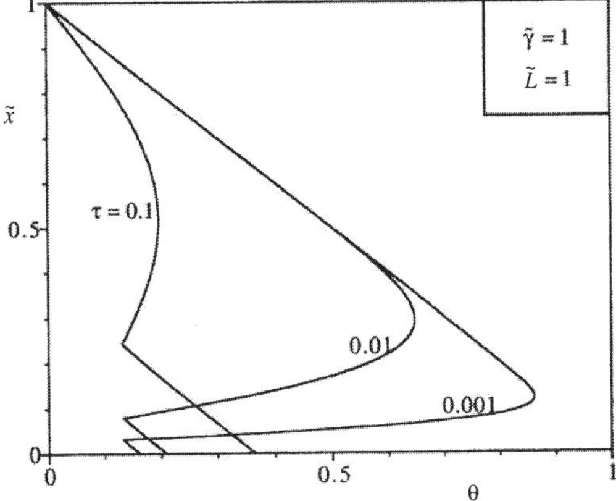

Figure 14.13 *The evolution of the temperature distribution following an initial geothermal gradient*

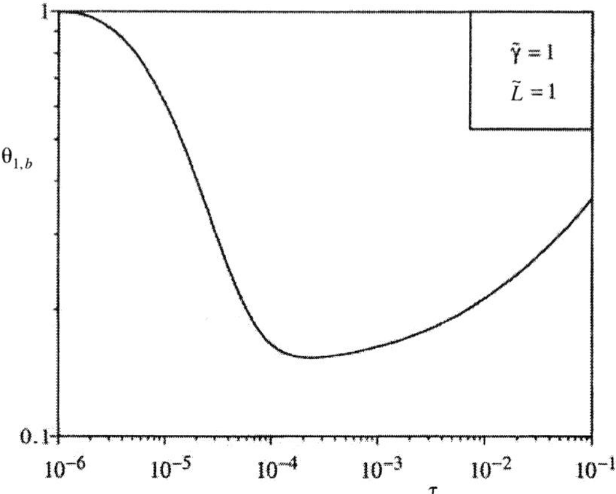

Figure 14.14 *The evolution of the temperature of the bottom of region 1*

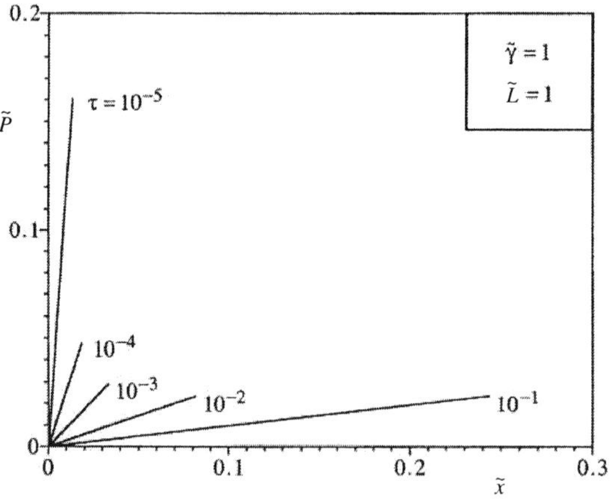

Figure 14.15 *The evolution of the distribution of pressure in region 1*

Figure 14.16 shows the effect of the initial temperature gradient on the position of the phase-change front. The thickness of the dissociated layer is larger when the gradient is larger. The corresponding rate of gas generation is reported in Figure 14.17, which confirms that the gas flow rate increases as $\tilde{\gamma}$ increases. The dashed line in Figure 14.17 is related to, but is not the same as, the scenario described in the preceding section, where the initial temperature distribution was isothermal. In the present case, the isothermal initial condition was $\theta = 1$, but the bottom temperature of region 1 was not fixed at $\theta_{1,b} = 1$. The bottom condition was constant heat flux. Figure 14.17 shows that the dashed curve agrees very well with the solid curve for $\tilde{\gamma} = 1$ at small times and this is because in this limit the temperature at the bottom of region 1 is essentially the same for both scenarios, $\theta_{1,b} = 1$. As the time increases, the dashed-line curve is lower than that obtained using $\tilde{\gamma} = 1$, and this effect is stronger as time increases. Again, the disagreement between these two curves is due to the fact that the curve $\tilde{\gamma} = 1$ was obtained using a heat flux boundary condition at the bottom of region 1, to account for the continuous heat flux generated by the earth.

To understand better the disagreement between the dashed curve and the solid curve in Figure 14.17, we compared the rate of the heat flux at the dissociation front (on the hydrate side, region 2), and the heat flux at the dissociation front (on the dissociated side, region 1), by calculating the ratio

$$\frac{\tilde{q}_2''}{\tilde{q}_1''} = \frac{(\partial\theta_2/\partial\tilde{x})|_{\tilde{x}=\tilde{X}}}{\tilde{k}\,(\partial\theta_1/\partial\tilde{x})|_{\tilde{x}=\tilde{X}}}. \tag{14.50}$$

Figure 14.18 shows that $\tilde{q}_1''$ is the most important portion of the total heat flux for maintaining the dissociation when $\tau > \tau_c$, where τ_c is the time when the ratio $\tilde{q}_2''/\tilde{q}_1'' = 1$. This means that, when $\tau > \tau_c$, the heat flux generated by the earth plays the main role, and this

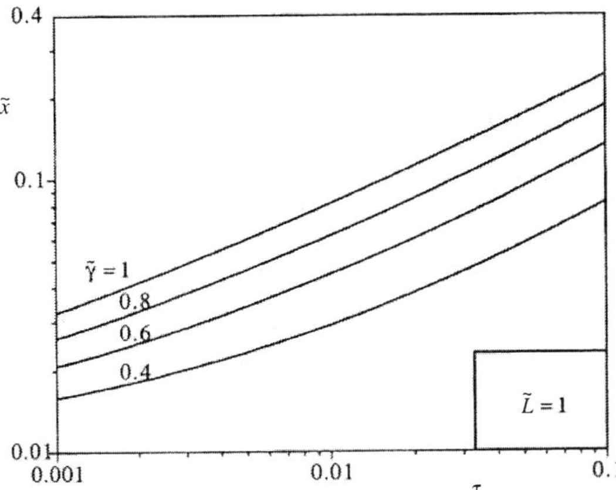

Figure 14.16 *The effect of the geothermal gradient on the evolution of the phase-change front*

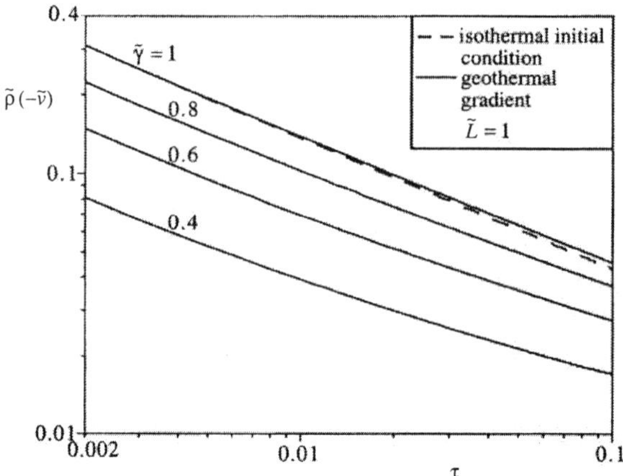

Figure 14.17   *The effect of the geothermal gradient on the rate of gas generation*

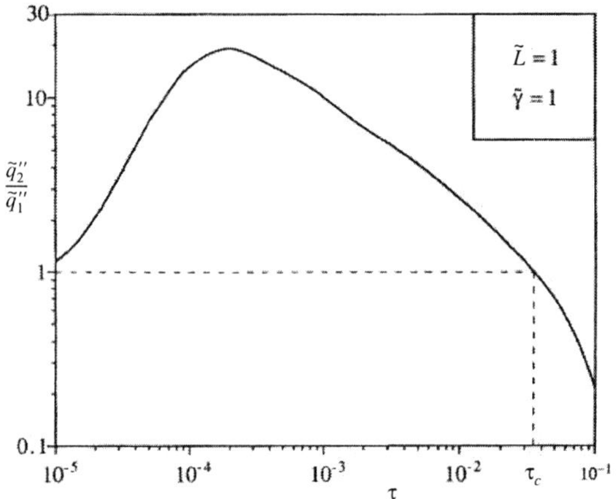

Figure 14.18   *The evolution of the ratio between the upper and lower heat fluxes, $\tilde{q}_2'' / \tilde{q}_1''$*

causes the difference observed in Figure 14.17. The geothermal gradient does not make a difference in the numerical model until the time that the bottom $\tilde{q}_1''$ dominates the upper $\tilde{q}_2''$.

Figure 14.19 shows that the effect of the hydrate saturation (ϕ) on τ_c is insignificant when the geothermal gradient is kept constant ($\tilde{\gamma} = 1$). Further, it should be noted that τ_c decreases when the geothermal gradient decreases and ϕ is kept constant. This effect occurs because the temperature variation in region 2 is smaller when the geothermal

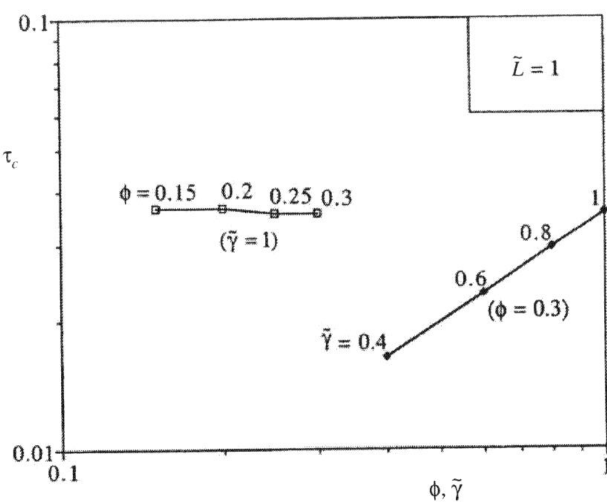

Figure 14.19 *The effect of porosity and geothermal gradient on the τ_c value*

gradient decreases. Therefore, the available $\tilde{q}_2''$ is smaller, and $\tilde{q}_1''$ dominates earlier the heat flux used in the dissociation process.

The effect caused by the low imposed pressure (P_0) on the τ_c value is documented in Figure 14.20. The critical time τ_c increases when the pressure decreases. These results show that the heat transfer generated by the geothermal gradient dominates earlier the dissociation process when the dissociation pressure is larger. This makes sense because a larger dissociation pressure corresponds to a larger equilibrium temperature and smaller available heat flux in region 2.

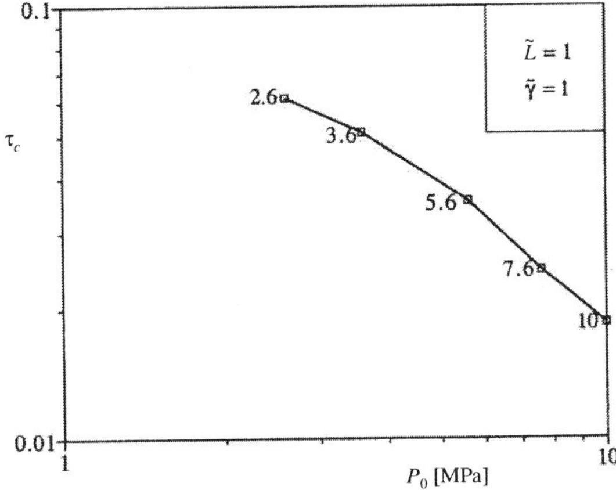

Figure 14.20 *The effect of the imposed low pressure (P_0) on the τ_c value*

The rate of gas generation depends on other parameters, such as the hydrate saturation (ϕ) and the imposed low pressure (P_0) and these effects are documented in Figures 14.21 and 14.22. Gas generation is enhanced when ϕ and P_0 decrease. The same behavior is exhibited by the numerical solution for the model with imposed constant temperature at the bottom of region 1, see Rocha *et al.* (2001). It is also interesting to note that $P_0 = 2.6\,\mathrm{MPa}$ was the smallest value simulated in Figure 14.22. The fundamental reason for choosing this limit value is that it corresponds to a $T_{eq} = 274\,\mathrm{K}$, and this equilibrium temperature rules out the possibility that the water from the hydrates freezes as ice and blocks the gas flow.

The rate of gas generation also depends on the size of the physical domain being modeled. Figure 14.23 shows dimensional results obtained for $L = 500, 400$, and $300\,\mathrm{m}$ using the geothermal gradient fixed on $\gamma = 0.046\,\mathrm{K/m}$. We can observe in this figure that the hydrate dissociation rate decreases when L also decreases.

How important is it to incorporate the geothermal gradient as a feature in the model of Figure 14.8? In dimensional terms, based on the properties listed in Table 14.1, the results of Figure 14.17 indicate that when $L = 500\,\mathrm{m}$ and $\gamma = 0.046\,\mathrm{K/m}$ the gas flow rate drops to the level $\rho\,(-v) = 3.26\,\mathrm{kg/}\left(\mathrm{m}^2 \times \mathrm{years}\right)$ in 317 years but the same dissociation rate drops to the same level in only 22 years when $\gamma = 0.0184\,\mathrm{K/m}$. These figures emphasize the strong role of the geothermal gradient in the dissociation process and shows that this geophysical feature must be accounted for in more accurate models.

In addition, the numerical simulations showed that almost the same results are obtained with isothermal and geothermal gradient conditions, when both have the same temperature at the bottom of region 1, and the upper $\tilde{q}_2''$ dominates the bottom $\tilde{q}_1''$, i.e., $\tau < \tau_c$. In this case the geothermal gradient does not make a difference during the numerical simulation.

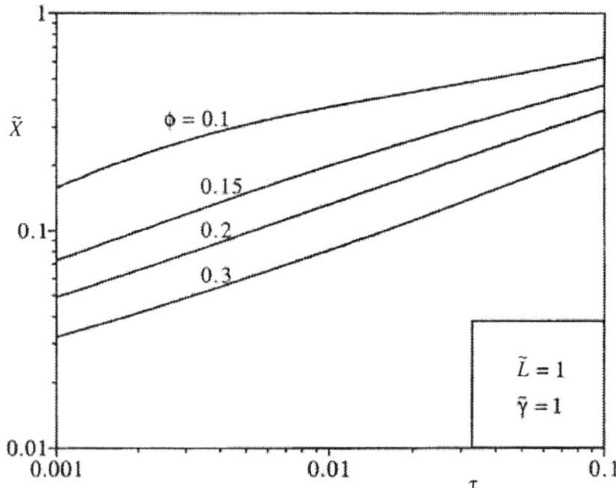

Figure 14.21   *The effect of hydrate saturation on the thickness of the dissociated layer*

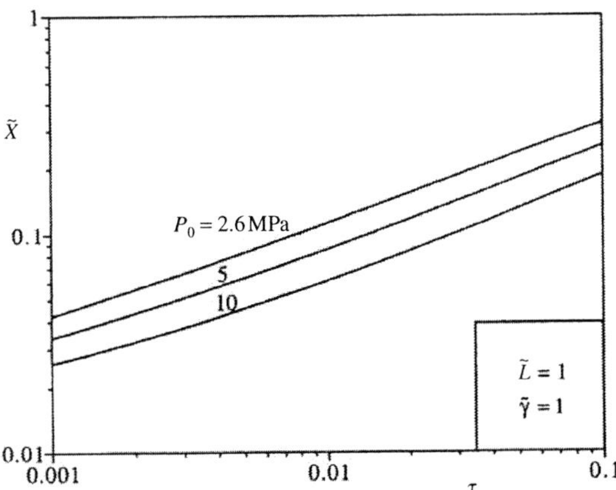

Figure 14.22 *The effect of the imposed low pressure on the thickness of the dissociated layer*

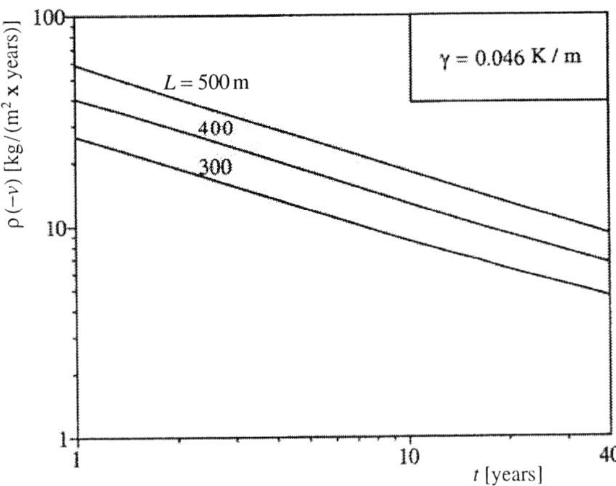

Figure 14.23 *The effect of the domain size (L) on the gas flow rate*

It seems that the superposition principle applies here, and the thermal effects of hydrate dissociation and the effect of geothermal gradient do not interfere with each other when they are based on the same temperature at the dissociation front. This explains the good agreement found at short times. The eventual movement of the dissociation front to a region of different initial temperature explains why the results for the cases with and without geothermal gradient diverge over time.

14.6   THE EFFECT OF POROSITY AND PERMEABILITY NON-UNIFORMITIES

The porosity of sediments under the sea is not uniform, for example, the porosity varies in the vertical direction because of sedimentation and compacting in time. According to the Athy model the porosity decreases with depth exponentially, see Athy (1930), Rubey and Hubbert (1959), Magara (1971), Shi and Wang (1986), Yuan *et al.* (1994), Hart *et al.* (1995) and Gering (2001), as follows:

$$\phi\left(x\right) = \phi_0 \exp\left(-\frac{L-x}{\lambda}\right), \tag{14.51}$$

where λ is an empirical constant, for example, $\lambda = 500\,\mathrm{m}$, and ϕ_0 is the porosity at the sea bottom. The corresponding permeability decreases with depth as well. If we use the Carman–Kozeny model for the permeability of beds of particles of fixed size, e.g., Nield and Bejan (1999), then the permeability K varies with x via the porosity function $\phi\left(x\right)$, as follows:

$$K\left(x\right) = K_0 \left(\frac{\phi}{\phi_0}\right)^3 \left(\frac{1-\phi_0}{1-\phi}\right)^2, \tag{14.52}$$

where the value of K_0 is the permeability at the sea bottom, $K\left(1\right)$. We investigated the effect of non-uniform porosity and permeability by substituting $\phi\left(x\right)$ and $K\left(x\right)$ into the numerical formulation presented starting with equation (14.24). For example, in equation (14.28), ϕ is replaced by $\phi\left(x\right)$ and in place of equation (14.27) we have

$$\frac{\partial\theta_1}{\partial\tau} = G\frac{\partial\tilde{X}}{\partial\tau}\left\{\frac{\partial\left[\phi_1\left(\tilde{x}\right)\right]}{\partial\tilde{x}}\left[\theta_1 + \frac{T_{sb}}{T_{1,b,\max} - T_{sb}}\right] + \phi_1\left(\tilde{x}\right)\frac{\partial\theta_1}{\partial\tilde{x}}\right\} + \tilde{\alpha}\frac{\partial^2\theta_1}{\partial\tilde{x}^2}, \tag{14.53}$$

where $G = \omega\tilde{\rho}_H c_p \phi_0/\left(\tilde{\rho}_1 c_1\right)$ and $\phi_1 = \exp\left[-\left(L-x\right)/\lambda\right]$. The initial and boundary conditions are the same as in equations (14.43) – (14.46).

The numerical results shown in Figures 14.24 to 14.28 correspond to a computational domain with $L = 500\,\mathrm{m}$, $\tilde{X}_0 = 0.01$, $\tilde{\gamma} = 1$ and were obtained by assuming the physical properties listed in Table 14.1. Figures 14.24 and 14.25 document the evolution of the phase-change front and the effect of changing the sea-bottom values for porosity (assumed 100% saturated with hydrate) and permeability. Contrary to the trend seen in Figure 14.10, the function $\tilde{X}\left(\tau\right)$ departs more visibly from $\tilde{X} \sim \tau^{1/2}$. This departure is illustrated further in Figures 14.26 and 14.27, which show the evolution of the rate of gas

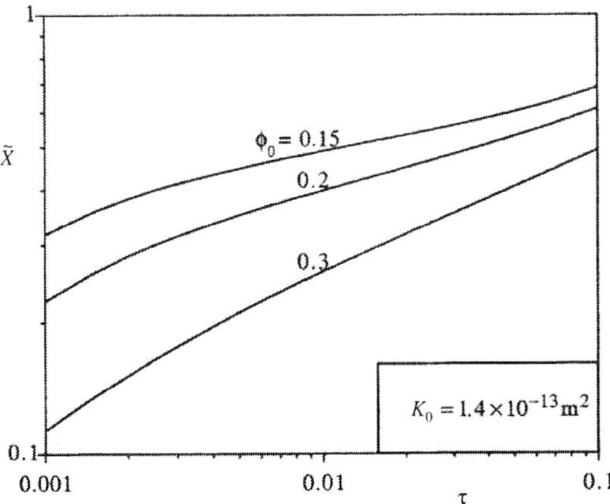

Figure 14.24 *The evolution of the phase-change front when the porosity and permeability decrease with depth; the effect of changing the porosity at the sea bottom*

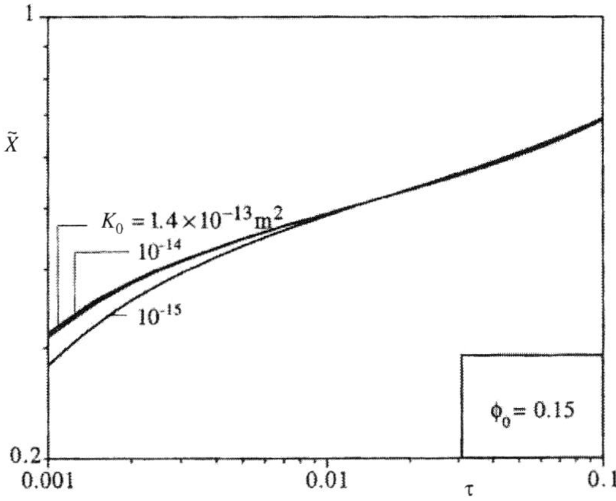

Figure 14.25 *The evolution of the phase-change front when the porosity and permeability decrease with depth; the effect of changing the permeability at the sea bottom*

generation. Further, the group $\tilde{\rho}\left(-\tilde{v}\right)$ decays as τ^{-n}, where n decreases monotonically in time.

The main conclusion is that the variations in porosity and permeability have a significant effect on global performance indicators, such as $\tilde{\rho}\left(-\tilde{v}\right)$. We see this more clearly in Figure

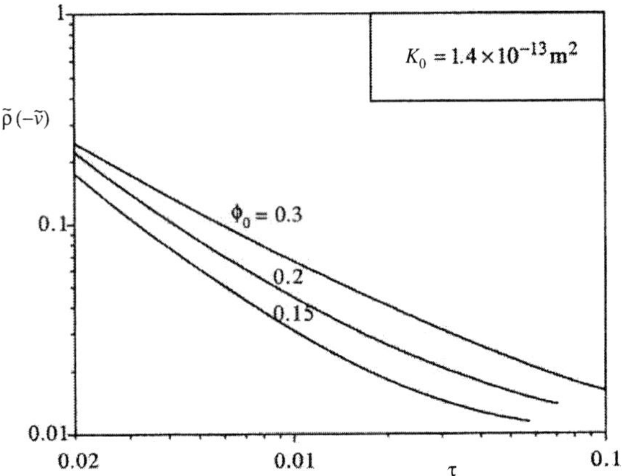

Figure 14.26 *The evolution of the gas flow rate when the porosity and permeability decrease with depth; the effect of changing the porosity at the sea bottom*

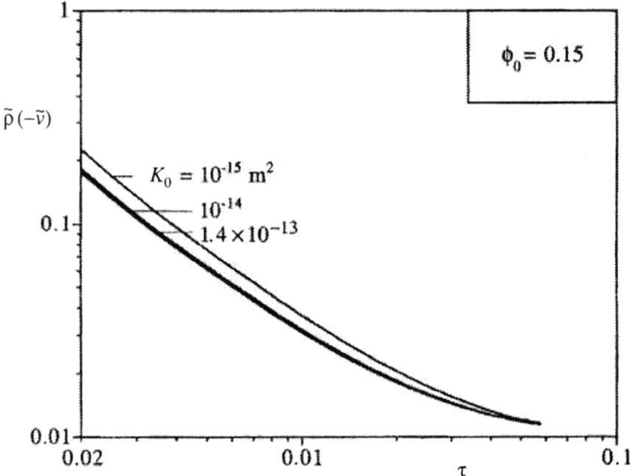

Figure 14.27 *The evolution of the gas flow rate when the porosity and permeability decrease with depth; the effect of changing the permeability at the sea bottom*

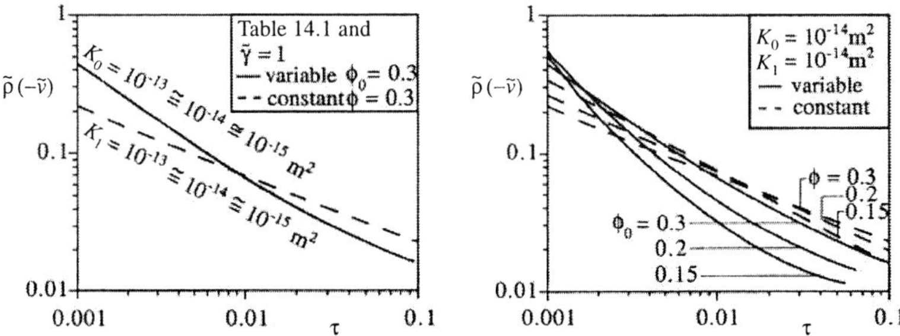

Figure 14.28 *The effect of the porosity model on the evolution of the gas flow rate. Gas flow rates calculated with variable ϕ and K models, versus flow rates calculated with constant ϕ and K models*

14.28, which shows a comparison between the gas flow rates calculated by assuming constant porosity and permeability (dashed lines), and the flow rates based on variable properties (solid lines). The decrease in ϕ and K with depth has the effect of decreasing the gas flow rate, and the effect is more noticeable at longer times.

14.7 CONCLUDING REMARKS

In this chapter we have reviewed an analytical and numerical study of the generation of gas (methane) through the depressurization of clathrate hydrates embedded in a porous medium. The model was simplified based on the assumption that the resistance to gas flow through the lowest region, Figure 14.4, right, is negligible. This permitted the one-dimensional modeling of the phase-change and convection processes in regions 1 and 2, see Figures 14.5 and 14.8.

The analytical solution was simplified further by the assumption that conduction in region 1 can be neglected, i.e., the movement of the dissociation front is driven mainly by conduction through region 2, ahead of the advancing front. The same one-dimensional configuration was subjected to a direct numerical simulation which confirmed, in an order of magnitude sense, the behavior and results based on the simplified analytical solution. The effects of geothermal gradient and non-uniform porosity and permeability expanded the coverage of the numerical results developed, based on the time-dependent phase-change and convection model.

The central conclusion provided by both methods is that the hydrate dissociation rate decreases monotonically in time. This decrease is proportional to $t^{-1/2}$ in the analytical solution, confirming a characteristic of one-dimensional phase-change processes in the media with constant properties, see Carslaw and Jaeger (1959). The numerical study showed that the decay of the hydrate dissociation rate is somewhat more complicated, as it depends on the initial thickness of region 1. If we regard the numerical model as more realistic and the numerical solution as more exact, then the numerical tests reported in detail by Rocha *et al.* (2001) show that the analytical solution is more accurate in the limit of long time (τ), small hydrate saturation (ϕ), high initial temperature (T_i), and low imposed pressure (P_0).

The hydrate dissociation rate—its level and behavior in time—is the central question in the modeling of the phase-change and depressurization process. With this information, one can assess the exergy-production potential of a known deposit, and its lifetime. The decaying production rate is, in this sense, similar to the rate of exergy extraction from hot dry rock deposits, see Lim *et al.* (1992). The depressurization process is considerably more complicated because of the interaction between region 1 and the well embedded in the lowest porous region.

This chapter also documented the effect of the geothermal gradient in the dissociation process. We have showed that this physical feature plays an important role and must be accounted for in more accurate models, see Figures 14.16 and 14.17. However, if $\tau < \tau_c$, the numerical simulation using isothermal initial temperature is a good approximation when the temperature at the bottom of region 1 is the same as in the numerical simulation in the presence of a geothermal gradient, see Figure 14.17.

Porosity and permeability of sediments under the sea are not uniform. The variations in porosity and permeability have a significant effect on global performance indicators, such as the flow rate $\tilde{\rho}\,(-\tilde{v})$, see Figures 14.26 to 14.28. The group $\tilde{\rho}\,(-\tilde{v})$ decreases in time as τ^{-n}, where n decreases monotonically in time. The decrease in ϕ and K with the depth has the effect of decreasing the gas flow rate.

Acknowledgements

This work was supported by the Laboratory Directed Research and Development program at the INEEL, which is managed and operated by Bechtel BWXT Idaho Inc. under DOE Idaho Operations Office Contract DE-AC07-99ID13727. Luiz A. O. Rocha's work was also supported by CAPES–Brasília, Brazil.

REFERENCES

Athy, L. F. (1930). Density, porosity, and compaction of sedimentary rocks. *Amer. Assoc. Petroleum Geol. Bull.* **14**, 1–24.

Behar, E. (1994). Plugging control of production facilities by hydrates. In *International Conference on Natural Gas Hydrates*, pp. 94–105. Annals New York Academy of Sciences.

Bejan, A. (1995). *Heat Transfer*. Wiley, New York.

Briaud, J.-L. and Chaouch, A. (1997). Hydrate melting in soil around hot conductor. *J. Geotech. Geoenviron. Eng.* **123**, 645–653.

Carslaw, H. S. and Jaeger, J. C. (1959). *Conduction of Heat in Solids* (2nd edn). Oxford University Press.

Cherskii, N. V. and Bondarev, E. A. (1972). Thermal method of exploiting gas-hydrated strata. *Soviet Phys. Dokl.* **17**, 211–213.

Fontana, R. L. and Mussumeci, A. (1994). Hydrates offshore Brazil. In *International Conference on Natural Gas Hydrates*, pp. 106–113. Annals New York Academy of Sciences.

Gering, K. L. (2001). Private communication. Idaho National Engineering and Environmental Laboratory, Idaho Falls, ID.

Gering, K. L., Cherry, R. S., and Weinberg, D. M. (2000). Mechanisms for methane gas accumulation under hydrate deposits in sediment. *Annals New York Acad. Sci.* **912**, 623–632.

Hanumantha Rao, Y., Reddy, S. I., Khanna, R., Rao, T. G., Thakur, N. K., and Subrahmanyam, C. (1998). Potential distribution of methane hydrates along the Indian continental margins. *Current Sci.* **74**, 466–468.

Hart, B. S., Flemings, P. B., and Deshpande, A. (1995). Porosity and pressure: Role of compaction disequilibrium in the development of geopressures in a Gulf Coast pleistocene basin. *Geology* **23**, 45–48.

Holder, G. D., Kamath, V. A., and Godbole, S. P. (1994). The potential of natural gas hydrates as an energy resource. In *International Conference on Natural Gas Hydrates*, pp. 427–445. Annals New York Academy of Sciences.

Holder, G. D., Malone, R. D., and Lawson, W. F. (1987). Effects of gas composition and geothermal properties on the thickness and depth of natural-gas-hydrate zones. *J. Petroleum Tech.* **39**, 1147–1152.

Islam, M. R. (1994). A new recovery technique for gas production from Alaskan gas hydrates. *J. Petroleum Sci. Eng.* **11**, 267–281.

Kamath, V. A. and Holder, G. D. (1987). Dissociation heat transfer characteristics of methane hydrates. *AIChE J.* **33**, 347–350.

Lim, J. S., Bejan, A., and Kim, J. H. (1992). Thermodynamics of energy extraction from fractured hot dry rock. *Int. J. Heat Fluid Flow* **13**, 71–77.

Lundgaard, L. and Mollerup, J. (1992). Calculation of phase diagrams of gas-hydrates. *Fluid Phase Equilibria* **76**, 141–149.

Lysne, D. (1994). Hydrate plug dissociation by pressure reduction. In *International Conference on Natural Gas Hydrates*, pp. 514–517. Annals New York Academy of Sciences.

Magara, K. (1971). Permeability considerations in generation of abnormal pressures. *Soc. Petroleum Eng.* **11**, 236–242.

Makogon, Y. F. (1981). *Hydrates of Natural Gas*. Pennwell Books, Tulsa, Oklahoma.

McKoy, V. and Sinanglu, O. (1963). Theory of dissociation pressure of some gas hydrates. *J. Chem. Phys.* **38**, 2946–2956.

Nield, D. A. and Bejan, A. (1999). *Convection in Porous Media* (2nd edn). Springer–Verlag, New York.

Roadifer, R. D., Godbole, S. P., and Kamath, V. A. (1987). Thermal model for establishing guidelines for drilling in the Arctic in the presence of hydrates. SPE California Regional Meeting, Ventura, California. SPE 16361.

Rocha, L. A. O., Neagu, M., Bejan, A., and Cherry, R. S. (2001). Convection with phase change during gas formation from methane hydrates via depressurization of porous layers. *J. Porous Media* **4**. To appear.

Rubey, W. W. and Hubbert, M. K. (1959). Role of fluid pressure in mechanics of overthrust faulting, ii. *Bull. Geol. Soc. Amer.* **70**, 167–206.

Selim, M. S. and Sloan, E. D. (1989). Heat and mass transfer during dissociation of hydrates in porous media. *AIChE J.* **35**, 1049–1052.

Shi, Y. and Wang, C. (1986). Pore pressure generation in sedimentary basins: Overloading versus aquathermal. *J. Geophys. Res.* **91**, 2153–2162.

Singh, A. and Singh, B. D. (1999). Methane gas: An unconventional energy resource. *Current Sci.* **76**, 1546–1551.

Sloan, E. D. (1990). *Clathrate Hydrates of Natural Gases.* Marcel Dekker, New York.

Subrahmanyam, C., Reddy, S. I., Thakur, N. K., Gangadhara Rao, T., and Sain, K. (1998). Gas-hydrates—a synoptic view. *J. Geol. Soc. India* **52**, 497–512.

Willoughby, E. C. and Edwards, R. N. (1997). On the resource evaluation of marine gas-hydrate deposits using sea floor compliance methods. *Geophys. J. Int.* **131**, 751–766.

Win, J. A. M. and Rik, J. J. (1999). Thermal reservoir simulation model of production from naturally occurring gas hydrate accumulations. SPE Annual Technical Conference, Houston, Texas. SPE 56550.

Yousif, M. H., Li, P. M., Selim, M. S., and Sloan, E. D. (1990). Depressurization of natural gas hydrates in Berea sandstone cores. *J. Inclusion Phenomena Molec. Recognition Chem.* **8**, 71–88.

Yuan, T., Spence, G. D., and Hyndman, R. D. (1994). Seismic velocities and inferred porosities in the accretionary wedge sediments at the Cascadia Margin. *J. Geophys. Res.* **99**, 4413–4427.

15 GRAVITY DRIVEN FLOWS IN POROUS ROCKS: EFFECTS OF LAYERING, REACTION, BOILING AND DOUBLE ADVECTION

A. W. WOODS

BP Institute, University of Cambridge, Madingley Rise, Madingley Road, Cambridge, CB3 OEZ, UK

email: andy@bpi.cam.ac.uk

Abstract

We develop a hierarchy of models to describe gravity driven flows in porous rocks and the models account for a series of phenomena which can change the structure and spreading rate of the current. These include draining through the lower boundary of the formation or into a fracture intersecting the channel; reaction of the injected liquid with the formation which changes the permeability; boiling of the injected fluid as occurs in a superheated system; and the effects of thermal inertia in producing a range of double advective phenomena in currents driven by both temperature and salinity. Where possible, theoretical models are compared with laboratory data and we include some discussion about the implications of the modelling for field scale processes.

Keywords: gravity, reactions, boiling, layering, draining, double advection, density

15.1 INTRODUCTION

Flow through porous media arises in many natural and industrial situations, see Bear (1988), Lake (1989) and Phillips (1991), and in many such cases, fluid of one density invades a porous layer saturated with fluid of different density, and this can lead to a gravity controlled flow, see Muskat (1949), Phillips (1991) and Woods (1999). Of the many natural or geological flows which are controlled by gravity, important examples include the migration of hydrocarbons from the site of their genesis to a shallow reservoir, the percolation of saline fluid into a rock saturated with relatively fresh water, following a period of sea-level rise, or as a result of fracturing during an earthquake which may connect two compositionally distinct aquifers; the percolation of rain water through the shallow sub-surface; and natural convective flows associated with the geothermal gradient, see Phillips (1991). Engineered flows include water injection into oil reservoirs, with the objective of displacing oil towards the producing well, see Lake (1989); water injection into superheated geothermal reservoirs, with the objective of producing vapour, see Woods and Fitzgerald (1993, 1997); and well-stimulation flows, in which acid is injected into porous layers with the purpose of dissolving some of the porous matrix and thereby enhancing permeability, see Lake (1989) and Raw and Woods (2000). More complex situations arise in heterogeneous formations in which the fluid may drain as it spreads along a high permeability layer, see Pritchard *et al.* (2001). In some waste management processes, fluid is injected to displace or enclose a second fluid already in the porous formation; this may lead to a gravity current composed of two fluids with different density and viscosity, see Woods and Mason (2000).

In this chapter, we aim to describe some of the physical controls on the gravity driven spreading of fluid through porous layers, accounting for the extra complications to the flow associated with rock–fluid interactions. Our approach is to build from simple self-similar models, in order to expose the fundamental controls on a variety of complex flow phenomena, rather than through numerical simulation. This provides a powerful means of describing the advance of a range of nonlinear flow processes and nonlinear interface development by reducing the problem to ordinary differential equations in the appropriate asymptotic limits, see Barenblatt (1996). As a building block for the subsequent analysis, we start by developing a model and then derive some simple similarity solutions to describe the motion of a finite release of dense fluid as it spreads through a porous layer. We also present a solution to describe the effects of draining from the end of the formation, as a model of draining into a fracture which bounds the porous layer. Also we expand the model and associated solutions to account for the motion of a gravity current moving through a high permeability layer, bounded below by a thin layer of much smaller permeability, but through which the current can eventually drain. In the remainder of the chapter, we investigate the effect of rock–fluid interactions, focusing on the role of

(i) reactions between the injected liquid and the porous matrix,

(ii) the thermal inertia of flow in a porous layer which introduces a lag between the mass and thermal signal associated with the injected liquid,

(iii) the effects of capillarity or vaporisation on the leading edge of the current, both of which may reduce the mass in the current, and

(iv) the effect of an overlying impermeable boundary on the development of the flow.

15.2 FUNDAMENTAL MODELS

We first examine the situation in which liquid of density $\rho + \Delta\rho$ spreads through a porous layer of permeability K and porosity ϕ and which is initially saturated with liquid of density ρ. As the current spreads along the lower impermeable boundary of the formation, the current becomes long and thin. If we denote the length and depth scales by L and H, with $L \gg H$, then from mass conservation, the ratio between the horizontal, u, and vertical, v, velocity has magnitude $u/v \sim L/H \gg 1$. For a one-dimensional channel flow, the flow may therefore be modelled as a unidirectional flow along the boundary, see Figure 15.1. The vertical velocity and any associated thinning or thickening of the current may then be described by an equation governing the depth of the current, $h\,(x, t)$, according to the vertically integrated continuity equation

$$\phi\frac{\partial h}{\partial t} = -\frac{\partial}{\partial x}\left(\int_0^h u\,dy\right). \tag{15.1}$$

We now adopt Darcy's law, see Bear (1988), to describe the slow viscous flow through the porous layer, namely

$$\boldsymbol{u} = -\frac{K}{\mu}\left(\nabla p - \rho\boldsymbol{g}\right), \tag{15.2}$$

where p is the pressure, μ the viscosity and g is the magnitude of the gravity acceleration vector. Owing to the small vertical velocity, the pressure in the current may be approximated as being hydrostatic, see Huppert and Woods (1995), so that the pressure in the current relative to that in the far field at the same level, p_r say, is given by

$$p - p_r = \Delta\rho g h. \tag{15.3}$$

This leads to the governing equation for the shape of the current, namely

$$\phi\frac{\partial h}{\partial t} = \frac{K\Delta\rho g}{\mu}\frac{\partial}{\partial x}\left(h\frac{\partial h}{\partial x}\right). \tag{15.4}$$

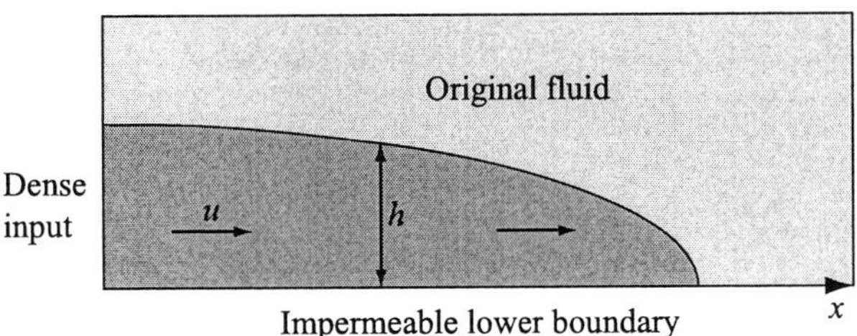

Figure 15.1 *Schematic of a gravity current propagating along an impermeable boundary in a porous layer of permeability K and porosity ϕ*

Equation (15.4) can be solved by application of suitable boundary conditions. In the present work, we describe flow in the region $x > 0$. The boundary conditions we apply at $x = 0$ may then be interpreted as either representing flow in a semi-infinite domain, with a rigid boundary at $x = 0$, or representing the conditions at the mid-point of a symmetrical flow $-\infty < x < \infty$. The flux condition at the source is given by

$$-\frac{K\Delta\rho g}{\mu} h\frac{\partial h}{\partial x} = Q\left(t\right) \tag{15.5}$$

and represents the flux supplied to the current from an injector at $x = 0$. The global mass conservation condition requires that

$$\int_0^t Q\left(t\right)\,\mathrm{d}t = \phi \int_0^{L(t)} h\,\mathrm{d}x, \tag{15.6}$$

where $L\left(t\right)$ is the length of the current. If the volume of fluid supplied after time t, per unit width of the cell, can be expressed in the form $V\left(t\right) = V_0 t^c$, then the above mathematical system admits similarity solutions of the form

$$h\left(x,t\right) = H\left(\omega t\right)^a f\left(\eta\right), \tag{15.7}$$

where $\eta = x/H\left(\omega t\right)^b$ and the shape factor f satisfies the ordinary differential equation

$$af - b\eta\frac{\mathrm{d}f}{\mathrm{d}\eta} = \frac{\mathrm{d}}{\mathrm{d}\eta}\left(f\frac{\mathrm{d}f}{\mathrm{d}\eta}\right). \tag{15.8}$$

Here the characteristic velocity is given by $S = Kg\Delta\rho/\mu\phi$, while the time scale over which the length of the flow evolves is given by $1/\omega$, where $\omega = \left(S^2/V_0\right)^{1/(c-2)}$ and the typical thickness of the current $H = \left(V_0/S^c\right)^{1/(2-c)}$. Here S is the gravitational descent velocity as described above. There are two special situations which admit analytical solutions to equation (15.8).

15.2.1   Finite release

For the case of a finite release at $t = 0$, the current has the shape of a parabola,

$$f = \frac{1}{6}\left(\eta_0^2 - \eta^2\right), \tag{15.9}$$

where $a = -b = -1/3$ and $\eta_0 = \left(9/\phi\right)^{1/3}$. In this solution, the depth decays with time according to the law

$$h\left(0,t\right) = h\left(0,t_0\right)\left(\frac{t}{t_0}\right)^{-1/3}, \tag{15.10a}$$

while the length of the current increases with time according to

$$L(t) = L(t_0)\left(\frac{t}{t_0}\right)^{1/3}.$$ (15.10b)

This fundamental solution of the diffusion equation was discovered by Pattle (1959) and has been confirmed by laboratory experiments using a Hele–Shaw cell, see Huppert and Woods (1995). In the experiments, syrup was placed behind a lock gate at one end of a Hele–Shaw cell, consisting of two parallel plates oriented so that the normal to the plates was horizontal. The syrup was then released from behind the lock gate and observed as it spread along the impermeable base of the cell. Figure 15.2 shows the very good agreement between the observed shape of the current and the model solutions. It should be noted that this experimental model of equation (15.4) is valid as long as the current is deeper than the width of the cell, so that the dominant friction is that due to the no-slip condition on the two vertical walls of the cell. Existence of this fundamental analytic solution is especially useful since it provides a reference for more complex flows which we will examine later in the chapter.

15.2.2 Flow with draining at $x = 0$

It is worth noting that there is a second analytic solution to the nonlinear diffusion equation, as described by Barenblatt (1996). This second solution corresponds to the case of a flow which is able to drain at $x = 0$, for example into a fracture, while spreading from $x = 0$ along the horizontal lower boundary of the porous layer, see Figure 15.3.

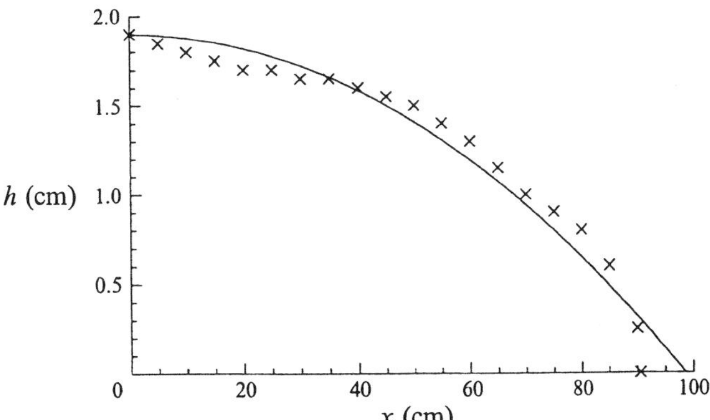

Figure 15.2 *Comparison of experimental measurements of the shape of a spreading gravity current of syrup, shown as crosses, with the theoretical prediction of equation (15.9), shown as the solid line. The experiment involved a finite release of fluid. After Huppert and Woods (1995)*

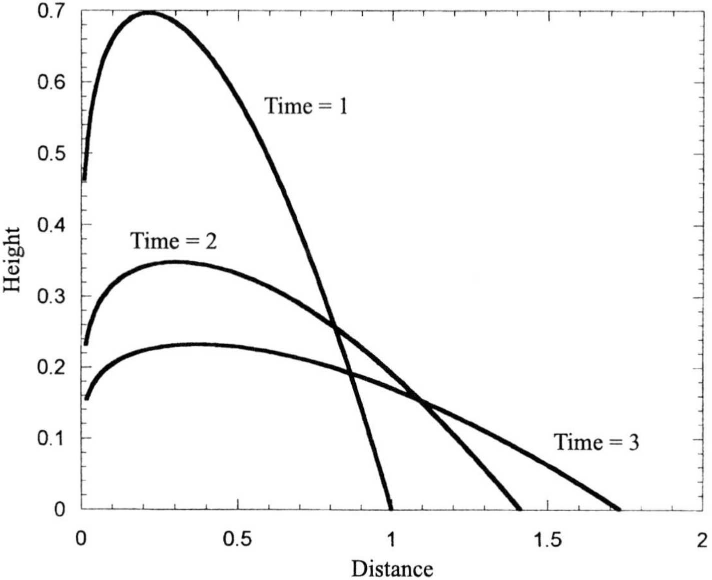

Figure 15.3 *Illustration of the distribution of fluid in a porous layer as it spreads under gravity while draining into a fracture at $x = 0$, see equation (15.11). Curves of the free surface are shown for three dimensionless times to illustrate the evolution of the flow*

The solution is given in terms of the similarity variable, $\eta = x/H\,(\omega t)^{1/4}$ and has form $h = H\,(\omega t)^{-1/2} f\,(\eta)$, where f is the shape factor which is given by

$$f = \frac{1}{6}\eta^{1/2}\left(\eta_0^{3/2} - \eta^{3/2}\right). \tag{15.11}$$

The dimensionless flux at $x = 0$, given by $-f\,\mathrm{d}f/\mathrm{d}\eta$, has value $\eta_0^3/12$. This solution corresponds to the situation in which the second, or dipole moment of the current shape, $\int_0^{L(t)} h^2\,\mathrm{d}x$, is constant. Owing to the loss of fluid at $x = 0$, the volume of the current decays gradually with time, according to a law of the form $V = V_0\,(t/t_0)^{-1/4}$. This solution is especially valuable for evaluating how far into a porous formation a release of fluid may advance while it drains away through a high permeability fracture. For example, in the engineering context of tracer tests, in which a discrete mass of tracer fluid is injected at one well and subsequently fluids are sampled at other wells, then even if the dominant flow path is the draining flow into the fracture at $x = 0$, some of the tracer may be recovered through gravity driven spreading through the porous layer.

15.2.3 Other injection rates

For power law injection rates, $Q(t) = Q_0 t^{c-1}$, with $c > 0$, there is a family of similarity solutions with $a = (2c - 1)/3$ and $b = (c + 1)/3$. The shape of these solutions is found by numerical integration of the ordinary differential equation (15.8), coupled with the global equation for mass conservation and the boundary condition at $x = 0$, see Huppert and Woods (1995). Although we do not know of any analytical solutions, power series approximations to these solutions may also be found, using a similar method to that described by Hatcher (2001). Perhaps it is worth noting that for a constant injection rate, $c = 1$, the current depth increases at a rate $h(0, t) = h(0, t_0)(t/t_0)^{1/3}$, while the current length increases at a rate $L(t) = L(t_0)(t/t_0)^{2/3}$.

A similar modelling approach may be used to describe radially symmetric gravity currents, for which the governing equation takes the form

$$r\frac{\partial h}{\partial t} = S\frac{\partial}{\partial r}\left(rh\frac{\partial h}{\partial r}\right) \tag{15.12}$$

in terms of the radial distance from the source, r, where $S = Kg\Delta\rho/\phi\Gamma$. In this case, the exponents which govern the spreading of the current are different due to the different geometry, although the modelling approach is analogous. For example, with a finite release of fluid at $t = 0$, the radius increases at a rate proportional to $t^{1/4}$, while the depth decreases at a rate proportional to $t^{1/2}$.

15.3 EFFECTS OF STRATIFICATION IN THE ROCK OR FLUID

15.3.1 Layered rock and the effects of draining

In many natural systems, the rock is highly layered, and this can have an important impact on gravity driven flows. Here, we consider two limiting cases to illustrate the potential impact on the motion of the current. First we examine the case in which there is a thin horizontal layer of low permeability, K_1, and thickness l, which separates two layers each with much greater permeability. The lower boundary of the upper layer is therefore permeable and a density current spreading through the upper layer gradually drains into the lower layer, see Pritchard *et al.* (2001). If the relatively dense current is draining under gravity, then, assuming a hydrostatic pressure distribution in the upper layer, the loss of mass through the intermediate low permeability channel is proportional to the depth of the current in the upper layer. For a permeability ratio $R = K_1/K \ll 1$, between the low permeability boundary and the upper layer, the equation of mass conservation for the current in the upper layer is given by the local balance

$$\phi\frac{\partial h}{\partial t} = -\frac{K\Delta\rho g}{\mu}\frac{\partial}{\partial x}\left(h\frac{\partial h}{\partial x}\right) - \frac{KR\Delta\rho gh}{\mu l}, \tag{15.13}$$

15.3.2   Continuously varying permeability

A further development of the model described earlier is to consider a layer with continuously stratified permeability. For example, we now examine the flow in a layer in which the permeability, $K(y)$, increases uniformly from the value $K(0) = 0$ to the form $K(y) = Gy$ for $y > 0$. This leads to a governing equation for the current depth, $h(x, t)$, which, after suitable reduction to dimensionless variables, has the form, see Huppert and Woods (1995),

$$\frac{\partial h}{\partial t} = \frac{\partial}{\partial x}\left(h^2 \frac{\partial h}{\partial x}\right).$$

(15.17)

By contrast with the earlier solutions for a finite release of fluid, this equation admits solutions with different exponents in the scaling for the length and depth of the current, namely

$$L(t) = L(t_0)\left(\frac{t}{t_0}\right)^{1/4} \quad \text{and} \quad h(t) \sim h(t_0)\left(\frac{t}{t_0}\right)^{-1/4}.$$

(15.18)

Since the mean permeability seen by the current decreases as the current thins, the spreading rate of the current is much more gradual than in a layer of uniform permeability in which the current length increases more rapidly, at a rate $L(t) = L(t_0)(t/t_0)^{1/3}$ and the depth decreases more rapidly, $h(t) = h(t_0)(t/t_0)^{-1/3}$. In contrast to these solutions, for a current supplied with a constant flux of fluid at the source, and hence in which the current length and depth both increase with time, the spreading rate is faster than in a layer of uniform permeability. Indeed, the length increases as $L(t) = L(t_0)(t/t_0)^{3/4}$ in comparison to the uniform layer in which $L(t) = L(t_0)(t/t_0)^{2/3}$. Again, although these solutions are highly idealised, they do reveal some of the complexities which are introduced as a result of layering or non-uniformity in the permeability structure of the rock. We now turn to effects of stratification in the fluid of which the current is composed.

15.3.3   Two-layer gravity currents

Woods and Mason (2000) extended the modelling approach to describe the motion of a two-layer or continuously stratified current. For a two-layer current, the flow of each layer depends on the depth of both layers, as may be seen by noting that, for each layer, the mass conservation equation satisfies the relation

$$\phi \frac{\partial h_i}{\partial t} = \frac{\partial}{\partial x}\left(h_i u_i\right),$$

(15.19)

where the speed u_i in each layer is given, to leading order, from Darcy's law and using the approximation of local hydrostatic pressure gradient

$$p_i = P_r - \Delta \rho_u g\left(h_l + h_u - y\right) - \rho g\left(H_r - y\right) \quad \text{for} \quad 0 < y < h_l,$$

(15.20a)

corresponding to the dense lower layer, and

$$p_i = P_r - \Delta\rho_l g \left(h_l - y\right) - \Delta\rho_u g h_u - \rho g \left(H_r - y\right) \quad \text{for} \quad h_l < y < h_l + h_u, \quad (15.20b)$$

corresponding to the less dense upper layer, where P_r is the reference pressure at height H_r above the boundary. Woods and Mason (2000) have shown that these equations combine to form the two nonlinear diffusion equations

$$\frac{\partial h_l}{\partial t} = RS\frac{\partial}{\partial x}\left(Bh_l\frac{\partial h_l}{\partial x} + h_l\frac{\partial h_u}{\partial x}\right), \qquad (15.21)$$

$$\frac{\partial h_u}{\partial t} = S\frac{\partial}{\partial x}\left(h_u\frac{\partial h_l}{\partial x} + h_u\frac{\partial h_u}{\partial x}\right), \qquad (15.22)$$

where $S = \Delta\rho_u g K/\phi\mu_u$, $R = \mu_u/\mu_l$ and $B = \Delta\rho_l/\Delta\rho_u$. Motivated by the fundamental solution (15.9), they showed that for a finite release of mass in each layer, then these coupled equations have similarity solutions which consist of piecewise continuous parabolae, with the morphology of the solution depending on the value of three parameters, namely

(i) the viscosity ratio of the fluids, $R = \mu_u/\mu_l$,

(ii) the buoyancy ratio of the fluids, $B = \Delta\rho_l/\Delta\rho_u$, and

(iii) the volume ratio of the two fluids, $V = \int_0^{L_u(t)} h_u \, dx \Big/ \int_0^{L_l(t)} h_l \, dx$.

The range of different solutions is sketched in Figure 15.5(a) and (b) following Woods and Mason (2000) and the different parameter regimes may be seen in Figure 15.6 in the case of a finite release of each layer. It may be seen that for a low viscosity and low density upper layer, the upper layer runs ahead of the lower layer. However, as the upper layer becomes progressively more viscous, the upper layer becomes attached to the source and the nose of the upper layer recedes relative to the lower layer. Eventually a critical condition is achieved for which the two currents behave as a single layer, even though they have different viscosity and density. As the upper layer becomes even more viscous, it begins to lag behind the lower layer and eventually the lower layer runs out ahead of the upper layer, separating from the source. Woods and Mason (2000) have considered these solutions in detail, establishing detailed conditions under which there are specific transitions in the flow structure, see Figure 15.5. They also showed how the different flow regimes for these two-layer currents can be realised in laboratory experiments by using a Hele–Shaw cell, and an example of these experiments is included in Figure 15.5(b).

The model may be extended to describe the case of a current which is continuously stratified in density or viscosity. Woods and Mason (2000) thereby established conditions under which such continuously stratified flows can behave in a self-similar fashion. Although this is a highly idealised problem, it identifies how a two-layer current can propagate through a porous layer. The insight associated with this is of great value for interpreting more complex flow structures, such as that associated with a reacting current, as considered in the next main section of the chapter.

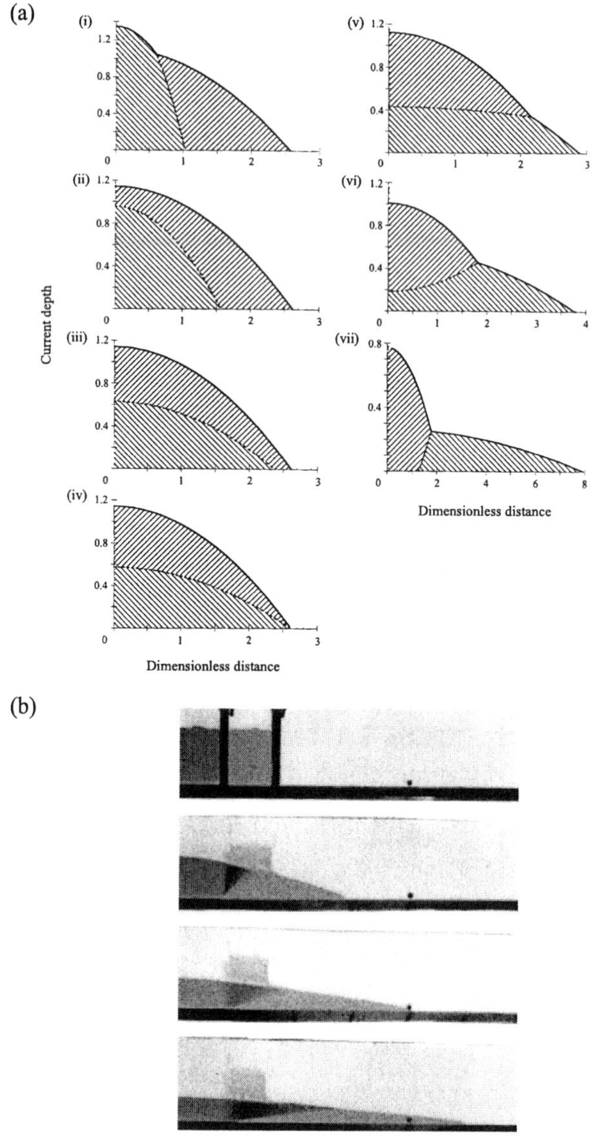

Figure 15.5 *(a) Diagram illustrating the range of flow regimes which may arise with a two-layer gravity current spreading through a porous layer. As the viscosity of the upper layer increases moving from figure (i) − (vii), the upper layer is progressively retarded and eventually the lower layer advances ahead of the upper layer. (b) Photograph of a laboratory experiment in which a finite release of low viscosity, dense fluid is displaced by higher viscosity, less dense fluid, see Woods and Mason (2000)*

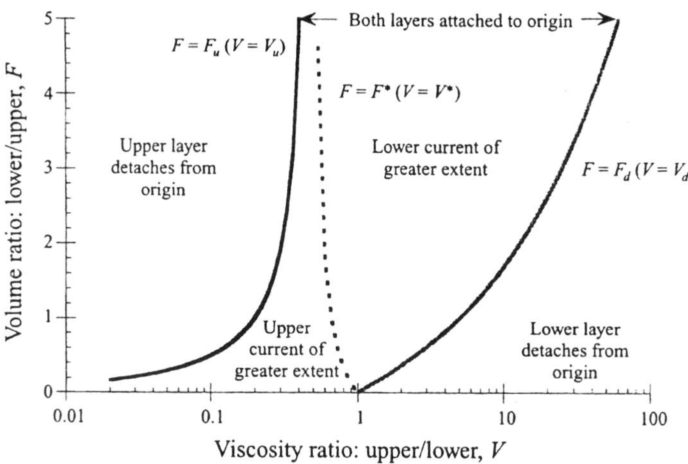

Figure 15.6 *Regime diagram illustrating the transitions in current structure, as seen in Figure 15.4, as a function of the volume ratio of the two layers and the viscosity ratio of the two layers, see Woods and Mason (2000)*

15.4   REACTING FLOWS

In some cases, fluid invading a reservoir reacts with the matrix leading to either precipitation or dissolution reactions, see Phillips (1991). This can change the permeability of the porous matrix and, in turn, this has a feedback on the morphology of the spreading current, see Raw and Woods (2000). Here we consider frontal reactions across which there is a change in permeability. In a typical frontal reaction between the matrix and the injected fluid, a volume of fluid ϕ is required to react with a volume $(1 - \phi)\,\lambda$ of the matrix where ϕ denotes the porosity of the matrix. As a result, the location of the reaction front lags a distance λ behind the front of the injected fluid. If the injected fluid is of different density from the fluid in the formation, then, as the injected fluid spreads through the formation and deepens, the flow will become progressively controlled by the buoyancy. In this limit, we can extend the model of gravity currents to account for the change in the permeability across the reaction front.

In general, the finite time for the reaction kinetics leads to the formation of a reaction zone between the injected fluid and the original rock of the formation. However, it follows from the two-layer gravity current solutions of the previous section, that as the flow develops, it migrates through the rock progressively more slowly. As a result, more time becomes available for the reaction per unit distance travelled. Thus the length of the reaction zone becomes a progressively smaller fraction of the domain occupied by the injected liquid. In the limit that this is a very small fraction, we can model the reaction zone as a sharp front dividing the region of reacted and unreacted rock within the current. If we denote the depth of the current in the unreacted region near the source by h_1 and the depth of the current in the reacted region ahead of the front by h_2, see Figure 15.7, then the equations for mass conservation in the lower, unreacted and upper, reacted zones become, to leading order, see Raw and Woods (2000),

$$\phi \frac{\partial h_1}{\partial t} = \lambda \frac{\partial}{\partial x} (h_1 u_1), \tag{15.23}$$

$$\phi \frac{\partial h_2}{\partial t} = (1 - \lambda) \frac{\partial}{\partial x} (h_1 u_1) + \frac{\partial}{\partial x} (h_2 u_2), \tag{15.24}$$

where the first term on the right-hand side of equation (15.24) denotes the flux of fluid across the reaction front as the fluid invades the reservoir. These equations are coupled with the approximation of hydrostatic pressure gradient for a long thin current, which takes the form

$$p(y) = P_r - \rho (H_r - y) - \Delta\rho (h_1 + h_2 - y). \tag{15.25}$$

Here, for simplicity, we assume that the density of the current does not change across the reaction front. This is consistent with a number of reactions in which the concentration of reactant in the fluid is small and has little impact on the density, compared to another dissolved, but unreacting phase, which controls the density. For example, the dolomitisation reaction sometimes involves transport of Mg ions by sea water invading a porous layer saturated with relatively fresh water, see Phillips (1991). However, effects of changes in density can be built into the model, see Raw and Woods (2001).

The resulting governing equations are as follows:

$$\frac{\partial h_1}{\partial t} = \lambda S \frac{\partial}{\partial x} \left(h_1 \frac{\partial h}{\partial x} \right), \tag{15.26}$$

$$\frac{\partial h}{\partial t} = S \frac{\partial}{\partial x} \left([K_R h + (1 - K_R) h_1] \frac{\partial h}{\partial x} \right), \tag{15.27}$$

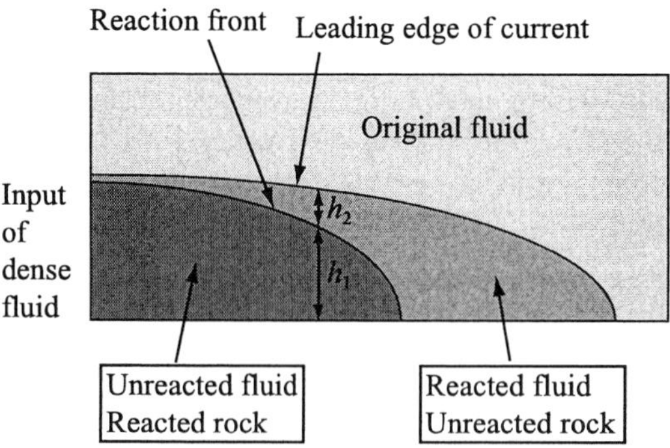

Figure 15.7 *Schematic of a double structure reacting gravity current, illustrating the region behind and ahead of the reaction front*

where $h = h_l + h_u$ is the full depth of the injected fluid zone, including that in contact with both the reacted and unreacted rock, and $K_r = K_u/K_l$ is the ratio of permeabilities in the upper and lower layers. This system of equations applies as long as the reaction front is ascending and hence migrating into the formation. For example, in the case of constant injection of liquid into the formation, the system admits (dimensionless) similarity solutions for a two-dimensional flow of the form

$$h_1 = H\,(\omega t)^{1/3}\, f_1\left(\frac{x}{H\,(\omega t)^{2/3}}\right) \quad \text{and} \quad h_2 = H\,(\omega t)^{1/3}\, f_2\left(\frac{x}{H\,(\omega t)^{2/3}}\right), \quad (15.28)$$

where the shape functions f_1 and f_2 satisfy the following ordinary differential equations, expressed in terms of f_1 and $f = f_1 + f_2$:

$$f_1 - 2\eta\frac{df_1}{d\eta} = 3\lambda S\frac{d}{d\eta}\left(f_1\frac{df}{d\eta}\right), \qquad (15.29a)$$

$$f - 2\eta\frac{df}{d\eta} = 3S\frac{d}{d\eta}\left[K_R f + (1 - K_R)\,f_1\right]\frac{df}{d\eta}, \qquad (15.29b)$$

and H and ω are scales for the height and inverse time scale of the flow.

The system is closed by imposing the boundary condition of constant flux at $x = 0$ given by

$$S h_1 \frac{\partial h_1}{\partial x} = Q, \qquad (15.30)$$

and the condition that the depth of the reacted zone at $x = 0$ is zero, $h_u = 0$. The solutions of the two ordinary differential equations depend on the permeability contrast across the reaction front, K_R, and the reaction parameter λ which denotes the volume of rock matrix which reacts with a volume $\phi/(1 - \phi)$ of liquid. As expected, the numerical solution of the equations identifies the very different structure of the flow for precipitation reactions in comparison with dissolution reactions. For precipitation reactions in which the rock permeability increases, see Figure 15.8(a), the reaction front remains very close to the source and forms a vertical front. As the injected fluid passes through the front, reacting with the rock, the reacted fluid is able to spread much more rapidly ahead of the front through the higher permeability original rock. The figure illustrates the variation of the morphology of the reaction front as a function of λ, the chemical constant for the reaction. If a large volume of liquid is required to react with unit volume of rock then the unreacted fluid runs far ahead of the reaction front, whereas when a smaller volume of liquid is required for the reaction then there is only a small reacted liquid zone ahead of the front.

In contrast, in dissolution reactions, see Figure 15.8(b), in which the permeability of the rock increases through reaction, then the gravitational force driving the flow tends to draw the oncoming fluid into the high permeability reacted zone. As a result, the reaction zone runs out along the base of the reservoir. As the oncoming injected fluid passes through this reaction front, it accumulates above the reaction zone, in the lower permeability original rock. Given the permeability contrast across the reaction front, there is a critical value

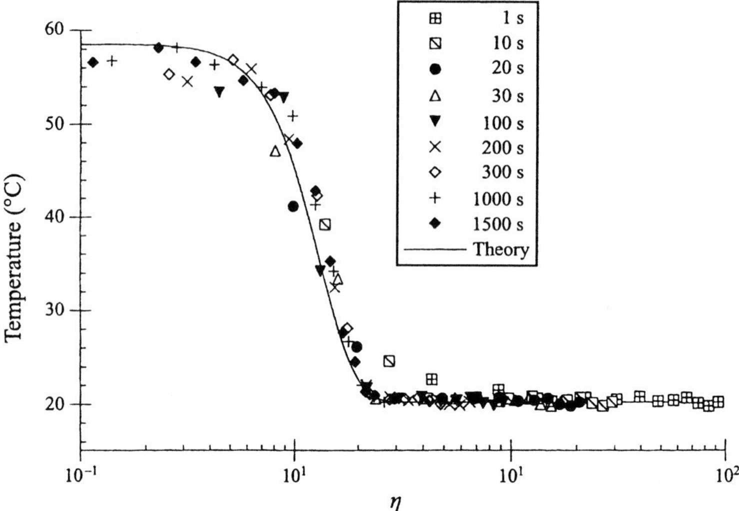

Figure 15.9 *Illustration of the thermal inertia effect in a porous rock as exhibited using an experiment in which cold liquid was injected from a central source into a two-dimensional hot porous layer. It is seen that a sharp transition in temperature occurs within the zone occupied by the injected liquid, see Woods and Fitzgerald (1997)*

consider currents which are dense relative to the water in the formation, both before and after the temperature has adjusted to the temperature of the rock, so that $\Delta\rho\,(\Delta T, \Delta S) > 0$ and also $\Delta\rho\,(0, \Delta S) > 0$. If the injected fluid is supplied at a constant rate then the current and thermal fronts will deepen as they spread into the reservoir. If the injected liquid is hotter than the reservoir water then as the injected fluid spreads through the rock and adjusts in temperature, and the density of the current will increase. Even though this leads to an increase in density across the thermal front, the interface may be neutrally stable; essentially, if a parcel of the cooled fluid moves back through the thermal front, it will be heated up again and becomes neutrally buoyant owing to the thermal inertia, see Phillips (1991). We now examine the motion of both cooling and heating currents, assuming that the depth of the thermal front decreases with distance from the source but increases with time.

The equation for heat transport through a porous layer has the form, see Phillips (1991),

$$\frac{\partial T}{\partial t} + \Gamma\,(\boldsymbol{u} \cdot \nabla)\,T = \kappa\nabla^2 T, \tag{15.32}$$

where

$$\Gamma = \frac{\rho_l C_{pl}}{\phi\rho_l C_{pl} + (1 - \phi)\,\rho_s C_{ps}}, \tag{15.33}$$

C_{pl} and C_{ps} are the specific heats for the liquid and solid parts, respectively, and typically $\Gamma = O(1)$. This identifies that the speed of the isotherms is a fraction $\Gamma\phi$ of the speed of the fluid which moves through the interstices. Typically, this has a value in the range $0.1-1.0$, e.g., in a number of North Sea oil reservoirs, see King (2001), it has value of about 0.3, while in laboratory experiments using glass ballotini it has a value in the range $0.4-0.5$, see Woods and Fitzgerald (1997). If we denote the fractional speed of the thermal front by f then the equations of motion of the current are given in terms of the depth of the region with the original temperature of the injected liquid, h_1, and the depth of the region with the formation temperature, h_2, see Figure 15.10. The mass conservation relations have the form

$$\phi \frac{\partial h_1}{\partial t} = f \frac{\partial}{\partial x} (h_1 u_1), \tag{15.34}$$

$$\phi \frac{\partial h_2}{\partial t} = \frac{\partial}{\partial x} (h_2 u_2) + (1 - f) \frac{\partial}{\partial x} (h_1 u_1), \tag{15.35}$$

where the horizontal speed of each layer is given by

$$u_i = -\frac{K}{\mu} \frac{\partial p_i}{\partial x} \tag{15.36}$$

and the second term on the right-hand side of equation (15.35) denotes the fluid passing through the thermal boundary-layer which is embedded within the current. In equation

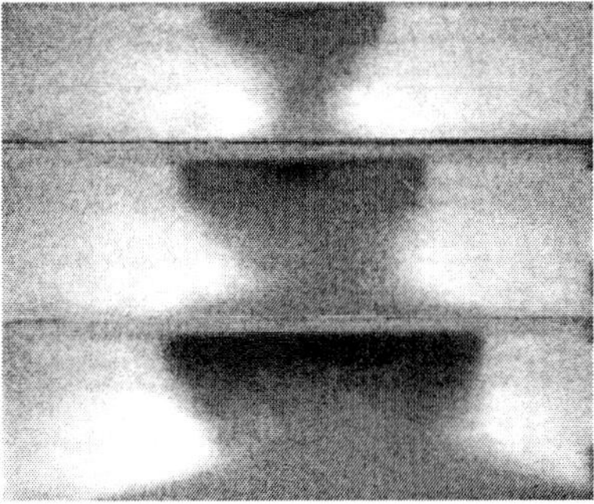

Figure 15.10 *Sequence of photographs illustrating the evolution of a current in which the buoyancy reverses on passing through the thermal front. Initially the injected liquid is less dense than the relatively cold but fresh fluid which is in the bead pack. However, on passing through the thermal front and cooling, the injected liquid becomes of comparable temperature, but is relatively saline and therefore detaches from the upper boundary and descends to the lower boundary. After Woods and Raw (2001)*

(15.36), the pressure takes the hydrostatic value

$$p\left(y\right) = p_r - \rho g\left(H_r - y\right) - \int_y^{h_1+h_2} \Delta\rho g \, \mathrm{d}y. \qquad (15.37)$$

The difference in pressure, Δp, from the original reservoir value, $p_r - \rho g\left(H_r - y\right)$, which is the key term for evaluating the horizontal pressure gradient, is given by

$$\Delta p\left(x,y,t\right) = \Delta\rho\left(\Delta T, \Delta S\right) g\left[h_1\left(x,t\right) - y\right] + \Delta\rho\left(0, \Delta S\right) gh_2\left(x,t\right), \qquad (15.38)$$

for the region $y < h_1$, in which the injected fluid retains the injection temperature, and

$$\Delta p\left(x,y,t\right) = \Delta\rho\left(0, \Delta S\right) g\left[h_1\left(x,t\right) + h_2\left(x,t\right) - y\right], \qquad (15.39)$$

for the region $h_1 < y < h_1 + h_2$ ahead of the thermal boundary-layer, in which the temperature of the injected fluid has adjusted to the temperature of the formation. The coupled equations then lead to the governing equations

$$\frac{\partial h_1}{\partial t} = Sf\frac{\partial}{\partial x}\left[h_1\left(\frac{\partial h_1}{\partial x} + R\frac{\partial h_2}{\partial x}\right)\right], \qquad (15.40)$$

$$\frac{\partial h_2}{\partial t} = RS\frac{\partial}{\partial x}\left[h_2\left(\frac{\partial h_1}{\partial x} + \frac{\partial h_2}{\partial x}\right)\right] + S\left(1 - f\right)\frac{\partial}{\partial x}\left[h_1\left(\frac{\partial h_1}{\partial x} + R\frac{\partial h_2}{\partial x}\right)\right], \quad (15.41)$$

where $R = \Delta\rho\left(0, \Delta S\right) / \Delta\rho\left(\Delta T, \Delta S\right)$. These equations have features in common with the system describing the propagation of a reaction front and the solutions are somewhat similar. However, the change in density across the front provides an important additional control on the flow, which inhibits the formation of shocks, see Figure 15.8(a).

An interesting additional case arises when the density of the input liquid relative to the formation fluid changes sign on passing through the thermal boundary-layer, see Woods and Raw (2001). Figure 15.10 illustrates an experimental investigation of this phenomena as modelled using a bead back. In the experiment, hot but saline fluid is injected at the top of the bead pack which is initially filled with cold but fresh liquid. It is seen that the gravity current initially spreads along the upper boundary of the formation, driven by its original buoyancy. However, on passing through the thermal boundary-layer, the fluid temperature evolves and the fluid becomes cold and salty. The salinity then controls the density and the injected liquid descends to the base of the system. In the final two images shown in this sequence of photographs, the injected fluid is dyed a darker colour. This injected fluid may be seen spreading along the upper boundary of the layer, through the zone which has been heated by the previous passage of injected liquid. The dark fluid then passes through the thermal boundary-layer and begins to descend to the base of the system.

In order to model the motion of such currents, we can appeal to the simple theory of gravity driven flows along impermeable boundaries described earlier in this chapter but extending the model to account for mass loss across the advancing front. This represents a very approximate method of modelling the effect of mass loss on such flows, since we are assuming that the gravitational separation of the fluid on passing through the thermal boundary-layer does not induce any significant pressure feedback on the gravity current proper. With this simplified model, a continuous input of fluid, we can model the advance of the thermal front simply according to the relation

$$\phi\frac{\partial h}{\partial t} = Sf\frac{\partial}{\partial x}\left(h\frac{\partial h}{\partial x}\right). \tag{15.42}$$

This equation is solved together with the boundary conditions that the volume of fluid in the current matches the volume of injected which remains dense, $f\int_0^t Q\left(t\right)\mathrm{d}t = \phi\int_0^{L(t)} h\,\mathrm{d}x$, and the source flux condition $Q = -Sh\,\partial h/\partial x$. For a finite release of fluid, the loss of fluid from the front of the current results in a progressively decreasing mass of fluid as a function of time. This process in fact yields similarity solutions of the second kind, and we consider this in the next section.

15.6   CURRENTS WITH MASS LOSS

A number of gravity current flows considered in this chapter have been described by similarity solutions. Each of these solutions has been based on the conservation of mass which has been imposed either through the injection of liquid at the source, or the requirement that the initial finite release of fluid remains fixed. One exception to this was the draining solution in which fluid leaked through the lower boundary of the flow domain. However, there are a number of different processes which lead to a loss of mass from the current as it evolves. In some such situations, the flow organises itself so that the evolution is self-similar. A classical problem, described by Barenblatt (1996), is the case of a finite release in an initially unsaturated porous layer, which slumps and spreads but also leaves a fraction of the fluid in the pore spaces owing to capillary retention. For an approximately homogeneous rock, the residual fluid in the formation occupies a fraction of the pore space, typically a volume fraction of the order $0.1 - 0.2$. Therefore as the current slumps, the receding front leaves a fraction f of fluid in the pore spaces. In contrast, at the leading edge of the current, the current invades the porous layer, and we assume that it fully saturates the porous layer. This leads to the following nonlinear diffusion equations describing the propagation of the current:

$$\frac{\partial h}{\partial t} = (1-f)S\frac{\partial}{\partial x}\left(h\frac{\partial h}{\partial x}\right)\quad\text{for}\quad\frac{\partial h}{\partial t} < 0, \tag{15.43}$$

$$\frac{\partial h}{\partial t} = S\frac{\partial}{\partial x}\left(h\frac{\partial h}{\partial x}\right)\quad\quad\text{for}\quad\frac{\partial h}{\partial t} > 0. \tag{15.44}$$

Since mass is lost in the receding front, the equation for conservation of mass may be written in the integral form

$$\frac{\mathrm{d}}{\mathrm{d}t} \int_0^{L(t)} h \, \mathrm{d}x = -f \int_0^{L_d(t)} \frac{\partial h}{\partial t} \, \mathrm{d}x, \tag{15.45}$$

where $L_d(t)$ is the location of the point in the current at which $\partial h/\partial t = 0$ and $L(t)$ is the full extent of the current in the porous layer. Furthermore, for this symmetrical slumping problem, the boundary condition at $x = 0$ takes the form $\partial h/\partial x = 0$. The system has a similarity solution of the form

$$h(x, t) = H(\omega t)^a \, \mathcal{H}\left(\frac{x}{H(\omega t)^b}\right), \tag{15.46}$$

where $\mathcal{H}$ is the shape function and now a and b depend on the value of the capillary retention constant, f, which determines the rate of loss of mass in the current. The mass of fluid in the current varies with time at a rate proportional to t^{a+b}. The exponent $a + b$ is shown in Figure 15.11 for four values of f and this identifies how the rate of loss of mass of fluid in the current decreases with f.

A second process in which a current loses mass occurs in superheated geothermal systems in which liquid migrating through the permeable rock is heated and boils, see Woods

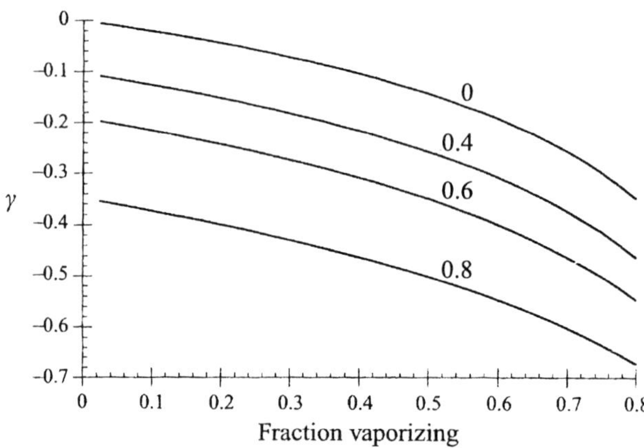

Figure 15.11 *Variation of the exponent controlling the rate of mass loss from the current, c, as a function of the fraction of the liquid which vaporizes as the liquid invades the porous layer, f. Curves are given for different values of the capillary retention fraction in the formation as indicated by the numbers on the curves. The vertical axis denotes the exponent of time which controls the mass of fluid in the current, see Woods (1998)*

(1998). At an advancing liquid front, the mass fraction which boils, b, is a function of the superheat of the system, the rate of thermal diffusion and the speed of the liquid. If the liquid propagates sufficiently rapidly that thermal diffusion is unimportant, but sufficiently slowly that the pressure gradient required to drive the liquid ahead of the front is negligible, then the mass fraction which vaporises is simply a function of the superheat, see Woods and Fitzgerald (1993). In that case, the governing equations for the structure of the current are very similar to equations (15.43) and (15.44), except that now it is the region of ascent $\partial h/\partial t > 0$ in which the mass is lost, so that the model equations take the form (15.44) in the region $\partial h/\partial t > 0$ and (15.43) in the region $\partial h/\partial t < 0$, while the global mass conservation equation (15.43) becomes

$$\phi \frac{\mathrm{d}}{\mathrm{d}t} \int_0^{L(t)} h \, \mathrm{d}x = -b \int_{L_d(t)}^{L(t)} \frac{\partial h}{\partial t} \, \mathrm{d}x. \tag{15.47}$$

The similarity solutions of the form (15.46) for this problem have been solved by Woods (1998), and the results are also shown in Figure 15.11. The solutions for the case of boiling liquid are of especial interest since they identify that all of the liquid injected into a geothermal system may boil off if the liquid is injected in discrete volumes at different wells. This contrasts with a more continuous input from a single well in which only a fraction of the injected liquid boils off and, instead, the remainder of the liquid is heated up, see Woods (1999).

We have discussed the dynamics of currents of reversing buoyancy in which the buoyancy changes sign on passing through the thermal front. If the fluid separates from the current on passing through the front then this represents a loss of mass which may be modelled in a fashion which is directly analogous to the boiling fronts in geothermal systems.

15.7 EFFECTS OF CONFINING GEOMETRY

In all the models considered herein, we have focussed on gravity spreading in an essentially unconfined layer, and therefore the current is free to deepen as it spreads. Here we illustrate how the modelling approach may be extended to account for the presence of a confining upper boundary, see Figure 15.12. The key principle that we adopt is that the length scale of the flow is long compared to the depth of the channel, so that at each point along the channel the vertical pressure gradient is hydrostatic. If we denote the pressure gradient along the base of the channel by $p_b(x, t)$ then the pressure is given by

$$P(x, y, t) = p_b(x, t) - \rho g y \quad \text{for} \quad 0 < y < h(x, t), \tag{15.48}$$

while in the upper less dense layer

$$P(x, y, t) = p_b(x, t) - \rho g y + \Delta \rho g(y - h) \quad \text{for} \quad h(x, t) < y < H. \tag{15.49}$$

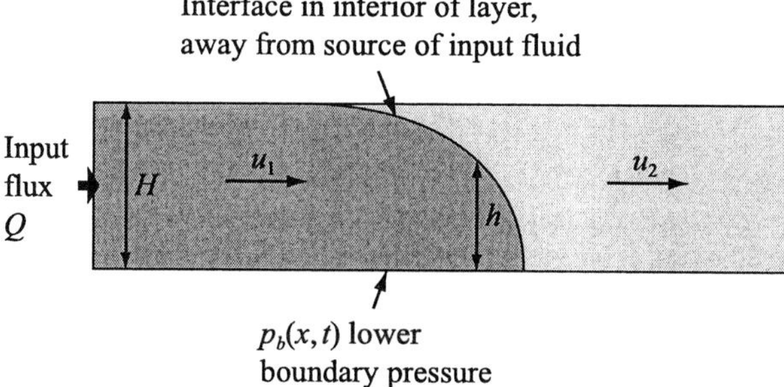

Figure 15.12 *Schematic diagram illustrating the geometry of a gravity current spreading through a layer of fixed vertical extent*

We now couple these relations with the Darcy law for each layer,

$$u_i = -\frac{K}{\mu_i}\frac{\partial P}{\partial x},$$ (15.50)

the mass conservation relations for each layer,

$$\phi\frac{\partial h_i}{\partial t} + \frac{\partial}{\partial x}\left(h_i u_i\right) = 0,$$ (15.51)

and the conservation of total flux, Q, at each point along the channel,

$$Q = u_1 h_1 + u_2 h_2,$$ (15.52)

to derive the equation for the depth of the interface, as a function of the position along the channel. For fluids of equal viscosity this is given by, see Huppert and Woods (1995),

$$\frac{\partial h}{\partial t} + \frac{2Q}{\phi H}\frac{\partial h}{\partial x} = \frac{2K\nabla\rho g}{H\mu\phi}\frac{\partial}{\partial x}\left[h\left(H - h\right)\frac{\partial h}{\partial x}\right].$$ (15.53)

This equation has the exact solution

$$h = \frac{H}{2}\left[1 + \left(\frac{2\mu\phi}{HK\nabla\rho g t}\right)^{1/2}\left(x - \frac{2Qt}{H\phi}\right)\right].$$ (15.54)

This has been compared to a series of laboratory experiments involving a lock exchange flow in a Hele–Shaw cell, see Hatcher (2001), and there is very good agreement for the case of saline water displacing fresh water, see Figure 15.13(a).

The model can be extended to account for fluids with different viscosities, and in the case of a pure exchange flow, $Q = 0$, the governing equation takes the form, see Hatcher

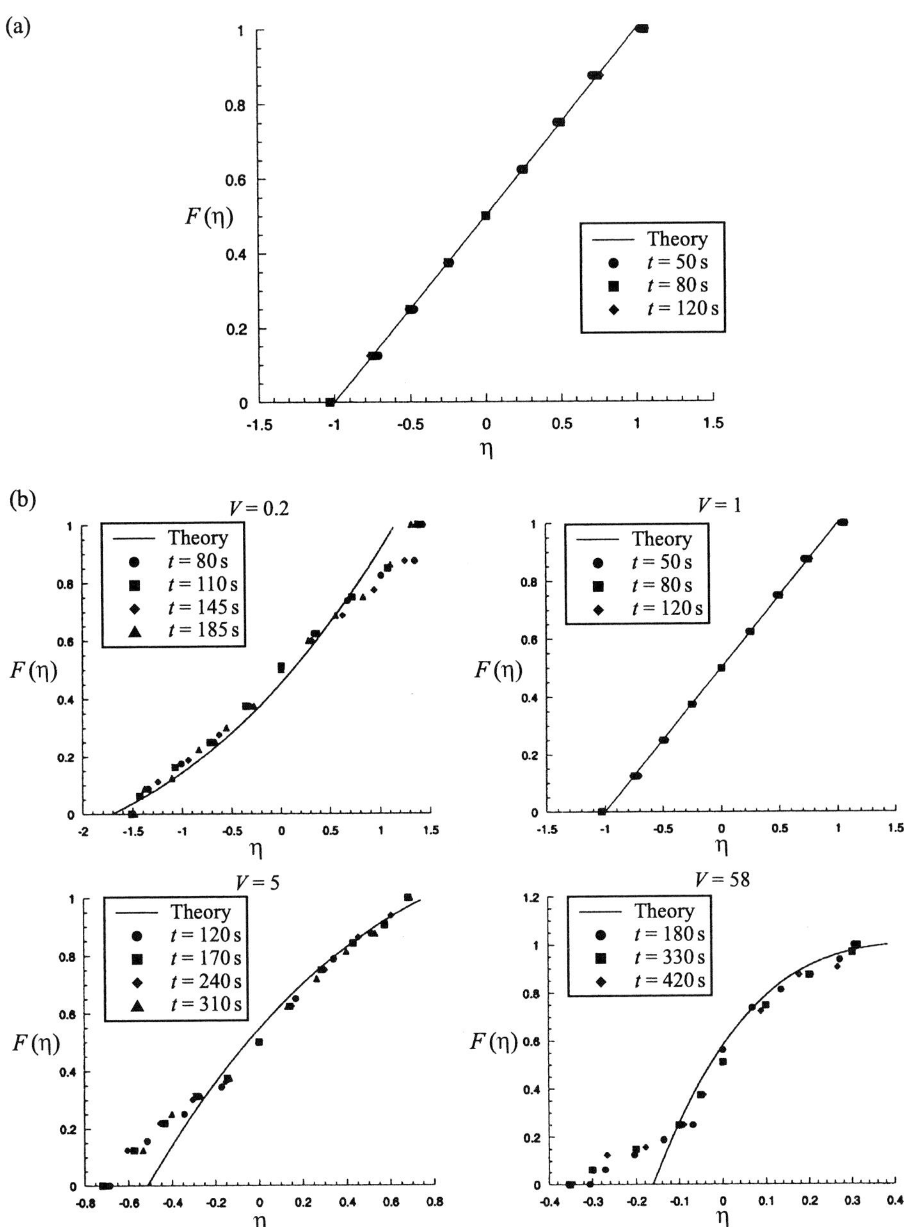

Figure 15.13 *Comparison of (a) the lock exchange self-similar flow solutions, equation (15.41), with a laboratory experiment, and (b) the lock exchange self-similar flow solutions, equation (15.43), for the case of fluids of different viscosity, see Hatcher (2001)*

(2001),

$$H\frac{\partial h}{\partial t} = \frac{HK\Delta\rho g}{\mu\phi}\frac{\partial}{\partial x}\left[\frac{h(H-h)}{h(1-V)+HV}\frac{\partial h}{\partial x}\right], \qquad (15.55)$$

where $V = \mu_{\text{lower}}/\mu_{\text{upper}}$ is the viscosity ratio of the upper to lower layers. This equation admits similarity solutions for the structure of the exchange flow given by $h(x,t) = Hf(\eta)$, where $\eta = x/(\omega t)^{1/2}$, with $\omega = HK\Delta\rho g/\phi\mu$. The distance the interface has spread after a time t increases as $t^{1/2}$ and the shape of the interface satisfies the governing equation

$$-\eta\frac{\mathrm{d}f}{\mathrm{d}\eta} = 2\frac{\mathrm{d}}{\mathrm{d}\eta}\left[\frac{f(1-f)}{f(1-V)+V}\frac{\mathrm{d}f}{\mathrm{d}\eta}\right]. \qquad (15.56)$$

This equation admits an exact nonlinear solution in the special case, $V = 1$, namely that $f = \eta - 1/2$, see Huppert and Woods (1995). However, for other viscosity ratios, the current tends to spread more rapidly in the direction of the advancing low viscosity fluid. Figure 15.13(b) compares the results of a series of laboratory experiments with the theoretical predictions of equation (15.55) for exchange flows in a confined channel, see Hatcher (2001). In the laboratory experiments, glycerol and syrup were used as the two working fluids and each of these was diluted with water in order to run experiments with different viscosity ratios. The general trend of the data is captured very accurately by the model, except in the case of large viscosity contrast, in which case a small finger of the viscous fluid appears to spread adjacent to the boundary ahead of the predicted current shape. This is thought to be a consequence of the effect that only a fraction of the low viscosity fluid is displaced by the higher viscosity fluid as it spreads through the Hele–Shaw cell, see Yang and Yortsos (1997). As a result, a small layer of low viscosity fluid on the walls of the cell lubricate the higher viscosity fluid, and allow it to spread through the centre of the cell more rapidly than predicted by the model. However, the overall predictions of the model, especially when the viscosities are comparable is quite accurate.

15.8 CONCLUSIONS

In this chapter we have introduced a model for the gravitational spreading of fluid through a porous layer. We have then developed these models to account for a number of real processes that occur in porous layers. These include reaction of the invading liquid with the porous matrix, which can lead to changes in the permeability; the thermal inertia of the liquid as it spreads through the layer, which can lead to changes in density in the current; and the loss of the fluid from the current either through a low permeability lower boundary of the flow channel or through capillary retention or boiling of the fluid. Finally, we examined how the modelling approach may be extended to include the effects of an impermeable upper boundary, so that the current is confined vertically and a porous exchange flow develops. In several parts of the chapter, we have presented data from laboratory experiments which have been used to test the modelling approach.

There are numerous possible extensions to the present work, including the effects of more complex geometry, coupling of the various phenomena and the study of immiscible displacements in which the two phases mix within the porous formation, in which case the effective permeability for the flow of each phase is reduced owing to the presence of the other phase within the pore space and which is sometimes known as relative permeability. However, the present work has revealed how some nonlinear and complex flow problems may be reduced to analytical form, in the limit that the interface between the phases remains sharp. Immiscible displacements can also be influenced by capillary forces. However, if the flow extends over sufficiently large scales, the gravitational pressure gradients are sufficient to overcome such forces. In this case, to good approximation the flow behaves as described herein with a well-defined interface, see Lake (1989). However, in such cases the effect of residual saturation of each phase needs to be included in the models. This will form the subject of future work.

REFERENCES

Barenblatt, G. I. (1996). *Scaling, Self-Similarity and Intermediate Asymptotics*. Cambridge University Press.

Bear, J. (1988). *Dynamics of Fluids in Porous Media*. Elsevier, Holland.

Hatcher, L. (2001). Ph.D. thesis. University of Bristol.

Hinch, J. and Bhatt, B. (1990). Stability of an acid front moving through porous rock. *J. Fluid Mech.* **212**, 279–288.

Huppert, H. E. and Woods, A. W. (1995). Gravity driven flows in porous layers. *J. Fluid Mech.* **292**, 55–69.

King, S. (2001). Private communication.

Lake, L. (1989). *Enhanced Oil Recovery*. Prentice Hall, New York.

Muskat, M. (1949). *Physical Principles of Oil Production*. McGraw–Hill, New York.

Pattle, R. E. (1959). Diffusion from an instantaneous point source with a concentration-dependent coefficient. *Quart. J. Mech. Appl. Math.* **12**, 407–409.

Phillips, O. M. (1991). *Reaction and Flow in Porous Rocks*. Cambridge University Press.

Pritchard, D., Woods, A. W., and Hogg, A. J. (2001). On the slow draining of a gravity current moving through a layered permeable medium. *J. Fluid Mech.* In press.

Raw, A. and Woods, A. W. (2000). Modelling well stimulation through acid injection. In *Proceedings of Stanford Geothermal Workshop*, Palo Alto, California. Stanford University Press.

Raw, A. and Woods, A. W. (2001). In preparation.

Woods, A. W. (1998). Vaporising gravity currents in a superheated porous media. *J. Fluid Mech.* **377**, 151–168.

Woods, A. W. (1999). Liquid and vapour flows in porous rocks. *Ann. Rev. Fluid Mech.* **31**, 171–199.

Woods, A. W. and Fitzgerald, S. D. (1993). The generation of vapour through injection of water into a hot rock. *J. Fluid Mech.* **251**, 563–579.

Woods, A. W. and Fitzgerald, S. D. (1997). Vaporisation of a liquid front migrating through a hot porous rock. Part II. Slow injection. *J. Fluid Mech.* **343**, 303–316.

Woods, A. W. and Mason, R. (2000). The dynamics of two-layer gravity-driven flows in permeable rock. *J. Fluid Mech.* **421**, 83–114.

Woods, A. W. and Raw, A. (2001). Double advective effects controlling liquid injection in geothermal reservoirs. In *Proceedings of Stanford Geothermal Workshop*, Palo Alto, California. Stanford University Press.

Yang, Y. and Yortsos, Y. (1997). Asymptotic solutions of miscible displacement in geometries of large aspect ratio. *Phys. Fluids* **9**, 286–298.

16 POROUS RIVERS: A NEW WAY OF CONCEPTUALISING AND MODELLING RIVER AND FLOODPLAIN FLOWS?

S. N. LANE and R. J. HARDY

School of Geography, University of Leeds, Leeds, LS2 9JT, UK

email: s.lane@geog.leeds.ac.uk

Abstract

This chapter aims to review the basic problems that we have in representing the complex bed geometry and complex surface characteristics associated with flows over rivers and floodplains. The mathematical basis of existing treatments is presented, and the over-reliance upon empirical analyses with a poor physical basis is noted, both for the roughness of complex river surfaces and situations where rivers contain vegetation. The problems of using existing treatments in fully three-dimensional models is especially important because of the numerical diffusion and potential numerical instability that may result. Analogies with atmospheric flows demonstrate the complexity of treating rough and vegetated surfaces in numerical models and lead to the recommendation of a new way of treating these surfaces using numerical porosity. Preliminary results from a high resolution study of fluid flow over a rough gravel surface are presented and the scope of this method is identified. The chapter concludes by noting that whilst porosity approaches require high resolution geometrical and vegetation data, and that such data is increasingly available, it is unlikely to be a particularly effective approach in practical terms. However, its core potential will be in allowing a much better physical justification of meaningful parameter values for practical predictions of flood conveyance and ecosystem habitat.

Keywords: rivers, conveyance, roughness, vegetation, drag coefficients, roughness heights, numerical porosity, atmospheric flows, canopy flows

16.1 INTRODUCTION

The traditional conception of a river channel is that it is a straight, trapezoidal channel with smooth boundaries. There are many examples of rivers that have been engineered so as to have this characteristic, see, for example, Figure 16.1(a). However, by far the majority of natural river channels have complex boundaries, see, for example, Figure 16.1(b) that

"

Figure 16.1 *(a) A classic engineered river channel with a trapezoidal section (tributary of the River Rhone, Martigny, Switzerland). (b) The ultimate in river channel complexity (a wide, gravel-bed river in South Island, New Zealand). (c) The rough bed and bank morphology of an upland gravel-bed river (the Upper Wharfe, North Yorkshire, UK). (d) A highly vegetated river (a tributary of the River Nene, Northamptonshire, UK). (e) The ultimate porous river, vegetation and shopping trolleys in an urban stream (Glasgow, UK, photograph supplied by JBA Consulting)*

are rough, see, for example, Figure 16.1(c), may contain vegetation, see, for example, Figure 16.1(d) and which may contain other obstacles, see, for example, Figure 16.1(e), especially when the channel flow through urban areas. The rivers in both Figures 16.1(d) and (e) represent an especially complex form of porous media: water can move within (see Figure 16.1d) and between (see Figures 16.1d and e) components on the surface. In addition to having a three-dimensional porosity structure that will be largely indeterminate, the structure of the porosity will evolve in relation to both the depth and the velocity of flow, and hence the river discharge. However, these sorts of channels are of immense practical importance. For instance, the porosity will control the relationship between water level and discharge. Figure 2 from Darby (1999) shows the simulated relationship between water level for a discharge with a five year return period in a river that is 6 m deep, with different percentages of bed vegetation cover and different grass heights up to 33% of water depth. This shows how

(i) higher percentages of bed vegetation, and

(ii) taller vegetation

result in higher water depths for a given discharge. This seems to imply that if we have more vegetation, the same discharge will cause more flooding. At the same time, it is now recognised that porosity in rivers is of immense ecological value. For instance, see, for example, Reiser (1998), egg nests (redds) are constructed by salmon and trout within river beds, commonly on the upstream end of riffles. This involves the construction of a small hole, laying of the eggs, and then covering of the eggs by gravels eroded by the mother fish from just upstream. Maintenance of a reasonable level of porosity in the buried eggs is crucial for the effective spawning of certain species.

This chapter is written with these issues in mind. If we are to develop more robust methods for predicting stage discharge relationships and habitat related parameters we need a much improved understanding of the nature of rivers which, in most cases, may be conceived as being porous.

16.2 RIVERS AS SOLID BOUNDARY PROBLEMS

16.2.1 Flows over complex boundaries

The conventional approach to modelling the interaction between river flows and bed topography is based upon well-established flow hydraulics for the case of straight, rectangular channels. In the UK, for flood routing and floodplain risk mapping purposes, the approach is predominantly one-dimensional, based upon the St Venant equations. The equation for mass balance is given by

$$\frac{\partial A}{\partial t} = -A\frac{\partial v}{\partial x} - v\frac{\partial A}{\partial x} + i \tag{16.1}$$

and for momentum by

$$\frac{\partial Av}{\partial t} + \frac{\partial Av^2}{\partial x} = \frac{\partial Agh}{\partial x} + gA\left(S_0 - S_f\right) = \frac{\partial Agh}{\partial x} + gAS_0 - gP\frac{f}{2g}v^2 \qquad (16.2)$$

$$\text{(a)} \;+\; \text{(b)} \;=\; \text{(c)} \;+\; \text{(d)} \;=\; \text{(c)} \;+\; \text{(e)} \;-\; \text{(f)},$$

where A is the cross-sectional area, t is time, v is the section-averaged downstream velocity, x is measured along the slope of the river in the downstream direction, i represents lateral inputs per unit length of channel, h is the average water depth, S_0 is the bed slope, S_f is the friction slope, P is the wetted perimeter, f is the Darcy–Weisbach friction factor, and g is the magnitude of the acceleration due to gravity. In equation (16.2), the temporal change in momentum (a) plus the downstream change in momentum flux (b) is equal to the net source of momentum, as determined by the downstream pressure gradient (c) and the net available energy (d). The latter has two components: available potential energy (e) and energy loss due to resistance to flow (f). In this case (f) is being represented as an effective friction slope. The basis of this derives from the du Buoys equation, which assumes one-dimensional, steady uniform flow, locally as follows:

$$\tau = \rho g R S_f, \qquad (16.3)$$

where R is the hydraulic radius of the flow (equal to A/P, where A is the cross-sectional flow area and P is the wetted perimeter). In specifying (f), equation (16.3) is being expressed using one of a range of resistance formulae, in this case using the Darcy–Weisbach uniform flow equation. This leads us to the essence of the problem of modelling natural river channels. We must assume that the flow is approximately hydrostatic and that the downstream component of mass flux is significantly greater than the vertical and cross-stream components. We also have introduced the term f in equation (16.2), and this has a very poor physical basis.

To apply these equations to a natural river channel, as opposed to a channel with a straight rectangular cross-section, we have to introduce the natural geometric variability of river channel morphology. If we consider equations (16.1) and (16.2), the only physically based parameter that we have for representing this variability is in the discretisation of the equations in the x-direction: as x tends towards very low values, so we will have a better representation of downstream variability in channel shape. There are two limitations to this. The first is practical and relates to the density with which cross-sections may be measured and hence specified in the model. The second is theoretical. Equations (16.1) and (16.2) assume negligible mass and momentum flux in both the vertical and lateral directions and they also introduce a roughness term that represents the friction losses at the channel perimeter. The most common way of dealing with geometric variability is through the roughness term. In addition to the skin friction associated with shear between the fluid and the channel bed, the roughness term is augmented, normally implicitly, to represent

(i) the effects of aspects of channel geometry that are not represented through channel discretisation,

(ii) the effects of vertical and lateral components of mass and momentum flux upon the downstream flux, and

(iii) the effects of turbulent velocity fluctuations upon the extraction of momentum from the mean flow, and its dissipation at smaller spatial scales.

Both (ii) and (iii) reflect the basic assumption that three-dimensional flow and turbulence can be treated as a net increase in the effective boundary roughness in a one-dimensional hydraulic model.

16.2.2 Roughness in one-dimensional models

There are two distinct approaches in which boundary roughness may be specified under these assumptions. The first involves the use of roughness as a calibration parameter. Approximate estimates of starting values, derived from methods considered below, are used to obtain a set of preliminary estimates. Roughness values are then perturbed in order to get a good fit between field observations and model predictions. In the second approach, roughness is specified using a range of empirical or semi-quantitative approaches. The most straightforward empirical approach involves the inversion of velocity formulae, such as that for Manning's n, or the Darcy–Weisbach friction factor f, given by (16.4b), assuming steady uniform flow (see, for example, Wolman, 1954), namely

$$n = \frac{R^{2/3} S_0^{1/2}}{V}, \tag{16.4a}$$

$$f = \frac{8gRS_0}{V^2}. \tag{16.4b}$$

In these cases, the roughness (n or f) can be estimated from properties of the channel (S_0 and R) and of the flow (i.e., V and R). In addition to questions over the validity of expressions (16.4a) and (16.4b), the main problem here is that both f and n vary as a complex and nonlinear function of the flow. However, for practical river management purposes, notably associated with floodplain mapping and flood forecasting, we need to be able to predict flow properties on the basis of a given channel geometry and bed roughness. Thus, n and f are strictly speaking measures of effective hydraulic roughness and not of bed roughness itself. Furthermore, both expressions (16.4a) and (16.4b) have a strong dependence upon R. Knight (2001) notes that this can result in an odd situation when water levels reach bank full, and floodplain flow begins. At this point, there is a sudden increase in the wetted perimeter which results in an effective reduction in R, and hence n or f, when floodplain flow results in a net increase in flow resistance due to

(i) lateral shear between the main channel and the floodplain flow, see, for example, Thornton *et al.* (2000), and

(ii) the higher relative roughness of floodplain flows.

The second approach to the specification of n and f is based upon the concept that roughness is additive. This follows from the observation above. Following Cowan (1956), it involves augmenting a roughness due to skin friction or individual grains (n_0) with that due to bed geometry (n_1), cross-section morphology (n_2), obstructions in the flow such as boulders or islands (n_3), and vegetation (n_4). In addition, these values may be scaled by m to represent the effects of channel curvature as follows:

$$n = (n_0 + n_1 + n_2 + n_3 + n_4)\, m. \qquad (16.5)$$

There is a clear logic to this expression in that this recognises the effective roughness of a river depends upon the scale over which it is measured. However, the question remains as to how n is estimated. The third approach addresses this in part by considering the relationship between channel bed material and the roughness parameter, see, for example, Strickler (1923). The final approach reflects all of these methods and is based upon photographs of river reaches of known roughness, often determined through expressions (16.4a) and (16.4b), see, for example, Barnes (1967) and Hicks and Mason (1991).

A recent study by the Environment Agency (R & D Technical Report *Scoping Study for Reducing Uncertainty in River Flood Conveyance*) consulted a range of twenty-six practitioners and academics as to how they specified roughness in 1D hydraulic models. This found that 88% used tables and photographs at some stage, 38% used the Cowan method at some stage and 88% also listed 'experience' as important in roughness specification. Only one consultee felt that they had a high confidence in how to estimate roughness.

Equation (16.5) introduces the idea that roughness values depend upon the scale over which they measured. Equation (16.4) implies that roughness depends upon the hydraulics of the flow, as well as the channel surface itself. Both of these factors have resulted in the suggestion that different parts of the river–floodplain complex should be treated separately within a model, with appropriate interfacing between them, i.e., mass and momentum transfer. Thus, cross-sections may be divided into sub-areas and conveyance of each sub-area is calculated and then summed to give the total conveyance, see, for example, Knight (2001). The most extreme illustration of the need for this is in the case of out-of-bank flow, when there is water in both the main channel and on the floodplain, and where the frictional retardation of the floodplain surface is significantly greater than that of the main channel due to the presence of vegetation on the flood plain and/or the greater relative roughness due to the shallower floodplain flows. Furthermore, empirical research has shown, see, for example, Thornton *et al.* (2000), that the apparent shear that results between the in-bank and out-of-bank portions of the flow can act as a significant resistance to flow, with associated implications for conveyance.

In theory, this appears a logical development. However, in practice, it emphasises the need for higher dimensionality analysis (2D and 3D) of river channel flows in order to help justify the way in which channel capacity is estimated in one-dimensional models.

16.2.3   Roughness in higher dimensionality models

What emerges from this review, and the above observations from the EA study, is the clear lack of knowledge that we have of how roughness should be treated in 1D numerical models. In recent years, there has been a progressive move in the research community to 2D, see, for example, Bates *et al.* (1992, 1995), Lane *et al.* (1994, 1995) and Lane and Richards (1998), and 3D numerical modelling, see, for example, Hodskinson and Ferguson (1998), Sinha *et al.* (1998), Gessler *et al.* (1999), Lane *et al.* (1999), Bradbrook *et al.* (2000a, 2000b), Booker *et al.* (2001) and Nicholas (2001), of natural river channel flows. It is with these developments that the limitations of traditional approaches to roughness terms become clear. The majority of 2D models continue to rely upon simple roughness parameters developed for 1D uniform flow in their application to the 2D case, see, for example, Bates *et al.* (1992) and Lane *et al.* (1995). Clearly, the validity of the associated assumptions in 2D must be questioned. Similarly, little is known about how to specify effective roughness parameters in the 2D case. In 2D, roughness varies spatially. Thus, optimisation of model predictions through roughness parameterisation becomes seriously complex as there is the possibility of adjusting both the absolute values of roughness and their spatial distribution.

Again, in the move to 3D little is known about how to deal with all of the spatial scales of river channel geometry. Following from the observation made with respect to 2D models that we have a range of spatial scales of topography that must be included in the numerical model, 3D models of natural river channels typically have a basic representation of river channel morphology, based upon the channel outline and either cross-sections or distributed data points sampled from within the channel. This involves a discrete sample of a continuous surface, and the roughness parameter is required to represent topographic scales that are not included in the topographic data that is used to define mesh geometry. This is illustrated conceptually in Figure 16.2.

The roughness parameter is commonly identified in a fundamentally different way to that which is commonly used in 1D and 2D models: it is based upon specification of a numerical roughness height that controls the vertical variation of velocity with elevation above the channel bed. In this case, roughness is strictly a property of the surface. Indeed,

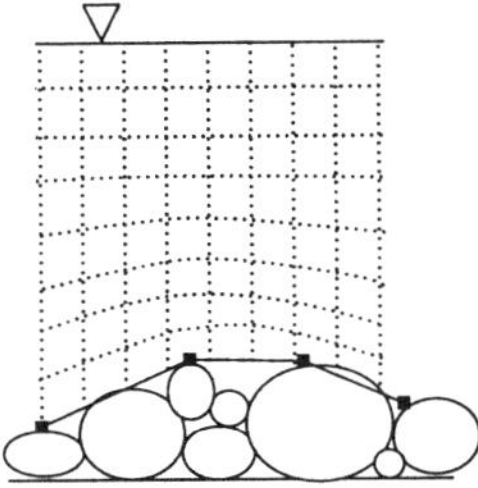

Figure 16.2   *Boundary fitted coordinates applied to sampled topographic data*

a better label for the roughness terms in Section 16.2.2 would be conveyance factors as they represent aggregating measures that essentially control the rate at which discharge moves through a river channel system. In 3D models, roughness terms needs to be spatially variable but temporally fixed, unless there is sediment entrainment and movement, and the shape of the channel changes, or the roughness of individual grains changes, e.g., due to periphyton growth, see Godillot *et al.* (2001). However, determination of this spatial variability is not straightforward.

There are a number of ways in which a roughness height may be specified. First, a skin friction associated with grain surfaces is identified. Second, this is multiplied upwards, see, for example, Clifford *et al.* (1992), to represent the effects of sub-grid-scale topography, e.g., sand dunes, grain surface morphology, grain interactions, as frictional retardation of the flow at the bed. The multiplication occurs up to the scale of topographic information that is contained within the numerical model. Third, the roughness height is applied to either an equilibrium or a non-equilibrium version of the law of the wall. Launder and Spalding (1974) recommend a non-equilibrium law-of-the-wall in which shear velocity (u_*) is replaced by the square root of the turbulent kinetic energy per unit mass, k, as the characteristic velocity scale, namely

$$\frac{u_y \sqrt{k}}{u_*^2} = \frac{1}{c_\mu'^{0.25} \kappa} \ln \left(c_\mu'^{0.25} \frac{\sqrt{k}}{u_*} \frac{y}{y_0} \right), \tag{16.6}$$

where u_y is the flow velocity at elevation y above the bed, y_0 is the roughness height, c_μ' is an experimentally determined empirical constant, and κ is von Karman's constant. In theory, provided the mesh is designed properly at the boundary, such that expression (16.6) is only applied to boundary cells, and the boundary is smooth, the main concern becomes the validity of the turbulence model. In practice, k may be determined from the transport equation in a two-equation turbulence model, with diffusion of energy to the wall assumed to be zero, production of what is expressed in terms of the shear velocity ($u_*^2 u_z / 2y$) and the boundary condition for ε set as

$$\varepsilon = \frac{c_\mu'^{3/4} k^{3/2}}{2\kappa y} \ln \left(c_\mu'^{0.25} \frac{\sqrt{k}}{u_*} \frac{y}{y_0} \right). \tag{16.7}$$

Whilst there are significant concerns over turbulence parameterisation in these near wall treatments, research has demonstrated that, in natural river channels with any sort of micro-scale topography, e.g., sand or gravel dunes, pebble clusters etc., much greater uncertainty is introduced into predictions of the three-dimensional velocity field due to poor knowledge and treatment of topographic variability than is introduced by uncertainty over turbulence treatments at the wall, see Lane *et al.* (1999). As with the 1D roughness parameters, much of the debate relates to how to determine y_0, and how to multiply it up to reflect larger scales of topography that are not represented in the mesh but which are clearly natural artefacts of the river channel, see, for example, Hey (1979), Bray (1982), Whiting and Dietrich (1990), Wiberg and Smith (1991) and Clifford *et al.* (1992).

Research has shown that it is possible to acquire high quality topographic data using close range remote sensing, see, for example, Butler *et al.* (2002). Data acquired at this scale has bee used to justify and to apply this multiplication, see, for example, Clifford *et al.* (1992). However, some studies have noted problems in terms of numerical stability and solution accuracy for flows characterized by high relative roughness, see, for example, Nicholas and Sambrook-Smith (1999), with these difficulties resulting from the existence of an upper limit of k_s for a given near-bed cell thickness, see Nicholas (2001). The implication of this is that the thickness of the near-bed cell limits the maximum shear velocity at the bed, so that near-bed velocities may be over predicted in field situations with high relative roughness, see Nicholas (2001). In addition, there is the basic problem of setting the reference height of the bed in a numerical mesh: it is normally implicitly assumed that the effective bed surface in mass conservation terms is the same as the bed surface sampled during field survey. Figure 16.2 shows how blocked grid cells are not effectively blocked in a boundary fitted coordinate treatment, even if the drag term can be effectively specified: there will be mass conservation errors arising from cells that are not blocked but which should be, and cells that are blocked that shouldn't be. It also demonstrates the real uncertainty in the value that the multiplier of roughness length should take. The grid cells in Figure 16.2 are smaller than the average topographic spacing. Thus the multiplier effect has to represent both sub-grid-scale topographic roughness and a roughness component that is lost because of the coarse spacing of topographic data collection.

In practice, the problems presented in Figure 16.2 arise from the coarse sampling of topographic data. The main alternative to using a multiplier of roughness length is to begin to include topographic data in the model. If actual data is available then roughness due to the difference in sampling density and grid density may be represented through the introduction of topographic variability into the boundary fitted coordinates. For instance, Nicholas (2001) attempted to include bedform roughness, e.g., particle clusters through using a random elevation model to introduce topographic variability into the CFD mesh. This reduces the roughness problem to sub-grid-scale topographic variability but assumes that the mesh distortion that arises from using the boundary fitted coordinate approach does not result in significant numerical diffusion or instability. The need for careful investigation of numerical diffusion associated with grid specification, the accuracy of discretisation, see, for example, Wallis and Manson (1997), and convergence problems associated with fine grids in finite volume discretisations, see, for example, Cornelius *et al.* (1999), are all well established in the 3D case.

This review implies that whilst rivers that are not vegetated may be conceived as solid boundary problems, dealing with those boundaries requires the development of new approaches. After a review of the vegetated case, we consider how some of these problems may be addressed through the treatment of the channel boundary as a porous media.

16.3 VEGETATION IN RIVERS AND ON FLOODPLAINS

Most natural river channels, except in arid regions, contain some form of vegetation. The presence of vegetation has two key effects. First, it reduces the volume of the channel that can be occupied by water, and hence affects the mass conservation equations. Second, it affects the momentum equations by acting as a source and a sink for turbulent velocity fluctuations. Both of these increase the effective drag upon the water flow, affect river channel conveyance, and hence have implications for water levels and flood routing. However, these same processes also result in the significant enhancement of river channel habitat, by creating a complex flow structure, environmental refugia, and the source of important components of the aquatic food chain. Whilst the traditional engineering approach to the management of river vegetation focused upon its removal, the growth of a more holistic approach to environmental management has questioned the extent to which this is sustainable. There remains considerable uncertainty about the effects of vegetation upon river channel flow processes and this makes understanding the role of vegetation increasingly more important.

16.3.1 Conveyance and multipliers of roughness parameters

Most research into the interactions between in-channel and/or floodplain vegetation and the flow of water has been concerned with the effects of vegetation upon flood conveyance. The basic assumption here is that the flood conveyance is reduced by the presence of vegetation and that this reduction can be represented through an increase in a roughness parameter. Indeed, it has been demonstrated that vegetation can have a far greater effect upon conveyance than other components of roughness, e.g., bed material, in vegetated channels, see, for example, Kutija and Hong (1996).

In practice, augmentation of n has varied in its sophistication. The simplest approach has been based upon addition or multiplication of roughness or roughness length to take into account the presence of vegetation, as per n_4 in expression (16.5). A more quantitative approach, and one that is still widely used, is based upon empirical relationships for channels with different vegetative characteristics, normally related back to basic hydraulic parameters such as the vR product, see, for example, Watson (1987) and Bakry (1992). Strictly speaking, see Chen (1976), this is an $n\text{--}Re$ relationship, where Re is the flow Reynolds number given by

$$Re = \frac{vR}{\nu} \tag{16.8}$$

and ν is the viscosity of water. For submerged grasses, the $n\text{--}vR$ relationship is normally negative, see Wu *et al.* (1999), and associated with plant bending as the vR product increases, resulting in the channel becoming hydraulically smoother.

There are two main problems with these empirical approaches. First, the derivation can appear to be circular. As n has only a very poor physical basis, it is commonly estimated from properties of the flow. Discharge, water surface slope and hydraulic radius measurements are used to determine n, which is then plotted against parameters like the hydraulic radius, or the velocity as derived from the discharge divided by the cross-sectional area of the flow. This problem is noted by Kouwen and Fathi-Maghadam (2000) who plotted a vegetative friction factor with a $1/V^2$ dependence on V, the section-averaged velocity. They observed the importance of making sure that the form of the relationship is dominated by factors other than that imposed by the use of V on both sides of the relationship. This is an issue that is commonly overlooked in n–VR studies.

The second problem is the dependence upon R. The hydraulic radius varies as a function of the ratio A/P, which is conventionally determined as $wh/(w + 2h)$. Thus, changes in R are dependent upon the nature of interaction between w (width) and h (flow depth), which depends upon the shape of the channel and the rate of change of h. This reduces the generality of curves derived experimentally for channels of particular shapes, and n has been shown to be strongly dependent upon h. Pre-submergence, n tends to increase with h, see Chow (1959), but this depends upon the plant physiology, i.e., its resistance to bending.

With this in mind, it may be preferable to estimate vegetative roughness using a more physically-based analysis. A general form for the augmentation of roughness can be derived by considering the force exerted on a plant of exposed area A_v, which may also be referred to as the momentum absorbing area, see, for example, Fathi-Maghadam and Kouwen (1997). The drag on a plant per unit area (F_v) will be defined by

$$F_v = 0.5\,\rho m C_{vd} v^2, \tag{16.9}$$

where C_{vd} is the drag coefficient associated with vegetation and m is a multiplier that reflects the exposed area of vegetation. A direct analogy here with flow around piers, after Hsieh (1962), leads to

$$F_v = 0.5\,yd C_{vd}\rho v^2, \tag{16.10}$$

where y is the water depth and d is the stem diameter.

If expression (16.10) is expressed as an augmentation to the du Buoys bed shear stress, and we substitute in expression (16.4a), then we get the increased n due to vegetation as

$$n = \sqrt{n_0^2 + \frac{mC_{vd}}{2gL}R^{4/3}}, \tag{16.11}$$

where L is the channel reach length and n_0 is the channel roughness in the absence of vegetation. There have been various suggestions for the value that m should take. For instance, Petryk and Bosmajian (1975) use the effective exposed plant area, namely

$$n = \sqrt{n_0^2 + \frac{C_{vd}A_v}{2gAL}R^{4/3}}. \tag{16.12}$$

Similarly, relationships can be developed for the Darcy–Weisbach friction factor (16.4b) that are similar in form to expression (16.12), see, for example, Pasche and Rouvé (1985) and Fathi-Maghadam and Kouwen (1997). The use of an effective exposed plant area term is useful as it can be specified independently to flow parameters or dependent upon flow parameters, according to the type of vegetation being studied. For instance, vegetation can bend when subject to shear, see Kouwen and Li (1980) and Darby (1999), and so it is important to make a distinction between flexible and non-flexible plants. Roughness with flexible plants will need a strong flow dependence. For instance, Fathi-Maghadam and Kouwen (1997) considered flexible unsubmerged vegetation. They showed that the frictional response is dependent upon changes in both flow velocity (which results in a streamlining effect) and the submerged momentum absorbing area (which depends upon flow depth). For subcritical, turbulent flow, they show that the vegetative drag coefficient can be expressed as

$$C_{vd}\left(\frac{A_v}{ah_n}\right)h_v = F\left(\frac{\rho v^2 h_n^4}{J}\right),\qquad(16.13)$$

where F is the friction factor or resistance parameter in non-submerged isolated plant flow, h_n is the normal flow depth, assumed to be equal to the hydraulic radius for wide shallow channels, a is the horizontal area of the bed covered by vegetation, h is the average vegetation height, and J is the flexural rigidity of the plants, equal to the modulus of elasticity times the cross-sectional moment of inertia of the tree (see below). The precise form of the relationship, i.e., F and J, depends upon the vegetation species and their age. Kouwen and Fathi-Maghadam (2000) generalised this to a relationship of the form

$$f = \alpha\left(\frac{v}{\sqrt{\xi E/\rho}}\right)^{\beta}\left(\frac{h_n}{h}\right),\qquad(16.14)$$

where α and β are species-dependent empirical constants, and ξE is a vegetation index that is unique for all specimens of a tree species and related to the resonant frequency, mass and length of vegetation. Experimental results suggested that β was negative, confirming the expected result that the friction factor is inversely proportional to velocity, i.e., a flexure and streamlining effect, and positively proportional to flow depth, i.e., an increase in momentum absorbing area effect, up to the point that vegetation becomes submerged.

16.3.2 Roughness height treatments

As considered in Section 16.2.3, it can be more useful to consider roughness as a roughness height. This is certainly required in 3D models, but has also been the basis of derivation of parameters like n and f for conveyance calculations. Darby (1999) summarises this for both flexible and non-flexible vegetation. In the case of f, it is common to use a relationship of the form, see, for example, Darby (1999),

$$\frac{1}{\sqrt{f}} = a + c\ln\left(\frac{R}{y_0}\right),\qquad(16.15)$$

where a and c are empirical constants. Kouwen and Li (1980) conducted experiments over plastic strips to determine an empirical expression for y_0 that might be suitable for flexible vegetation as follows:

$$y_0 = 0.14\,d\left[\left(\frac{MJ}{\tau}\right)^{0.25}\bigg/d\right]^{1.59}, \qquad (16.16)$$

where d is the local height of the strips, τ is the boundary shear stress, M is the vegetation stem density, and J is the product of E, the stem's modulus of elasticity, and I, the stem area's second moment of inertia. Measurement of the individual components of MJ is difficult, see Darby (1999), and thus Kouwen (1988) treated it as a combined parameter in which stem density, M, and stem stiffness, EI, are thought to have an equivalent effect. Experiments by Temple (1987) justified this as they showed a strong correlation in laboratory experiments between vegetation height and the MEI parameter.

In the case of non-flexible vegetation, see Darby (1999), an alternative approach is required. Non-flexible vegetation has a stronger effect upon the generation of turbulence. One solution here, see, for example, Li and Shen (1973) and Thompson and Roberson (1976), is to determine wake velocities and to use these to determine the friction factor. For instance, Darby (1999) used the Thompson and Roberson (1976) wake velocity equations to estimate the wake velocities from the ratio of vegetation spacing and vegetation diameter. The wake velocities were then substituted for the section-averaged velocities in expression (16.4b) to give a value of f. This emphasises an important point, namely, close attention needs to be given to the interaction between vegetation and turbulence, and especially the generation of turbulence, see, for example, Tsujimoto (1999). It is insufficient to introduce vegetation simply as a source of drag, and the turbulence treatment, especially in 2D and 3D models, also needs to be developed.

16.3.3 Application in numerical models

Both conveyance and roughness height treatments are currently being used in numerical models. Morin *et al.* (2000) used a conveyance-type approach in a two-dimensional depth-averaged model of flow in Lake Saint François, in the St Lawrence River, Canada. This used a similar relationship to (16.11), but with the actual value of n set as a function of the maximum possible value of n. The latter was defined for vegetation at its maximum growth stage, maximum height and maximum density. This maximum value was determined through distributed parameterisation of the 2D model to optimise model predictions in relation to distributed velocity observations. Morin *et al.*'s work introduces an additional problematic issue for roughness estimation in the presence of vegetation. Whereas roughness due to river channel perimeter characteristics remains relatively constant, that due to vegetation shows a sensitive dependence upon season.

The main problem with conveyance approaches in higher-dimensionality models is their dependence upon effective parameterisation and their failure to account for the interaction of vegetation-induced drag and vegetation-induced turbulence. The facts that 2D and 3D models are more sensitive to turbulence treatments, see Lane and Richards (1998), and that a nonlinear plant species dependence between drag and turbulence is to be expected, emphasise the need to consider treatments that explicitly incorporate both drag and turbulence treatments separately. Fischer-Antze *et al.* (2001) introduced a drag term based upon projected plant area and a drag coefficient as an additional momentum sink in the 3D momentum equations and this successfully reproduced velocity patterns in a compound section with the floodplain containing rigid vegetation. López and Garcia (2001) went one stage further and in addition to introducing a vegetation-related drag term, also modified the turbulent kinetic energy and dissipation rate equations in a two-equation turbulence model. They found that this resulted in a good agreement between predicted and measured turbulent flow properties.

Although recent years have seen considerable developments in modelling flow around vegetation in rivers, there is still a relative paucity of modelling attempts. The treatment of vegetation in numerical models is still largely empirical, see Sections 16.3.1 and 16.3.2, although there is some important research seeking to define conveyance parameters in terms of generic models, see, for example, Darby (1999) and Kouwen and Fathi-Maghadam (2000), defined by properties of the vegetation, rather than the extension of solid boundary roughness parameters to the vegetated case. Progress in understanding atmospheric flows has been much more extensive, especially in relation to numerical modelling. Thus, the next section of this paper considers research undertaken in atmospheric flows in an attempt to identify potentially fertile avenues for research in rivers.

16.4 ANALOGIES WITH ATMOSPHERIC FLOWS

The flow boundaries associated with atmospheric flows are directly analogous with river flows. There can be quite complex fixed boundaries, e.g., buildings, and dynamic boundaries, e.g., vegetation. As noted in Section 16.3, the problem of vegetation in river channels has conventionally been dealt with using the extension of roughness parameters designed to represent non-vegetated cases to vegetated situations using empirical evidence. Treatment of atmospheric flows has progressed in a much more rigorous manner, based largely upon roughness height treatments and less so on conveyance-based parameters.

Traditionally, the main motive for studying turbulent flow in the plant environment has been to understand the processes governing momentum, heat and mass exchange between the atmosphere and the biologically active canopy, see Raupach and Thom (1981). As in fluvial studies, the means of representing vegetation has been conditioned by the spatial scale of application. Also as in fluvial studies, a single parameter is unlikely to be able to represent the heterogeneous and complex boundary structures in the atmospheric case. This has been confirmed further when the processes operating with a plant canopy are considered. Raupach and Thom (1981) highlight the processes that require representation when this surface interacts with the air flow and within it. The interactions include

(i) momentum absorption from the flow by both form and skin friction acting on individual drag elements,

(ii) the transport of momentum and scalar properties by turbulent diffusion,

(iii) the generation of turbulent wakes, which convert the mean kinetic energy of the flow into turbulent kinetic energy at length scales characteristic of the elements, and

(iv) oscillatory plant movement in relation to the flow passing around and through vegetation, with the plants storing mean kinetic energy as strain potential energy to release it as turbulent kinetic energy half a waving cycle later.

The aerodynamic drag on the foliage is consequently the cause of two characteristics of flow through canopies, namely

(i) an unstable inflected velocity profile, and

(ii) a spectral short cut mechanism that removes energy from large eddies and diverts it to finer scales, where it is rapidly dissipated, bypassing the inertial eddy cascade, see Finnigan (2000).

This forms a complex dynamic environment where the total dissipation rate is often very large as a result of the fine-scale shear layers that develop around the foliage. The velocity moments scale with single length and time scales throughout the layer, rather than being dependent on height, see Finnigan (2000).

Despite these complexities, the treatment of atmospheric flows over vegetation is surprisingly similar to roughness length treatments of flows in rivers. A standard approach is the parameterisation of a drag coefficient, which is representative of the total shear stress on the rough surface or the downward flux density of streamwise momentum to the surface, see Raupach (1992). This is normally computed by the Monin–Obukhov similarity theory and specified as a constant surface roughness length, see Mahrt *et al.* (2001). In the simplest case, where there is no vegetation, steady flow over bare soil can be described by the application of the logarithmic law, see Monin and Yaglom (1971), as follows:

$$u_y = \frac{u_{*g}}{k} \ln\left(\frac{y}{y_{0g}}\right), \tag{16.17}$$

where u_y is the horizontal velocity at height z, u_{*g} is the friction velocity for a bare soil and y_{0g} is the roughness length of bare soil. Expression (16.17) is similar in structure to expression (16.6), as discussed earlier in relation to representation of roughness in a fluvial environment. However, as the vegetative surface is more complex, the equation is modified to

$$u_z = \frac{u_*}{k} \ln\left(\frac{y - d_s}{y_{0g}}\right), \tag{16.18}$$

where d_s is the displacement height defined as the mean height in the vegetation on which the bulk aerodynamic drag acts, see Thom (1971). This is a different sort of treatment of the boundary to that used in fluvial flows. In expression (16.18), the vegetated canopy is effectively treated as a displacement (d) of the reference height at which the velocity becomes zero. As with the fluvial case discussed in Section 16.2, expression (16.18) has a relatively poor physical meaning, and under represents the processes operating in the environment. For example, a plant canopy consists of numerous elements such as leaves, stems and branches, aggregated into one complex structure, see Raupach and Thom (1981), and there can be significant flow within the structure. Further, the roughness length for canopies has been shown to vary as a function of both wind direction and wind speed. The drag coefficient for individual leaves is sometimes formulated to be inversely proportional to wind speed, v^{-n}, where $n = 0$ for high Reynolds number flow (no dependence on wind speed) and $n = 0.5$ for low Reynolds number flow, see Raupach and Thom (1981). The large drag coefficients observed at weak winds were due to the importance of viscous boundary layers at low Reynolds numbers, see Deacon (1957). Similarly, Brunet *et al.* (1994) suggested that for weak winds, viscous stresses dominate for a greater portion of the canopy. As with river flows, the wind speed also determines the architecture of the canopy, where the canopy can become more streamlined with increasing wind speed. For a weak wind, the canopy drag coefficient is at its maximum as the leaf surface area perpendicular to the wind velocity. However, the leaf orientation will alter due to wind stresses. The canopy drag coefficient has been observed to decrease by a factor of two with increasing wind speed for conifers, see Mayhead (1973) and Johnsen *et al.* (1982). The response of the canopy to turbulence depends on several factors including the density of shrubs, see Dudley *et al.* (1998) and Grant and Nickling (1998), the architecture of the canopy and the stiffness and sway damping coefficients, see Kerzenmacher and Gardiner (1998). Due to these complex effects, researchers have found that individual plant species require individual drag coefficients, see, for example, Gillies *et al.* (2001), mirroring the observations of Kouwen and Fathi-Maghadam (2000) for flows in rivers.

What emerges as more important is that for the roughness sub-layer above a vegetative surface, there is significant departure of the wind profile from that predicted by the logarithmic relationship, see Shaw and Pereira (1982), Wilson *et al.* (1982) and Sellars *et al.* (1986). Given these observations, numerical modelling has developed away from the standard roughness height approach. Standard free-air Reynolds equations have been adapted for use in canopies by the *ad hoc* addition of a source drag term, which was regarded as a smooth function of space, see Finnigan (2000). However, when attempts to write second-order turbulence closure models of canopy flow were made, the limitation of this approach quickly became apparent, see Finnigan (2000). At second-order, the drag term appeared as a strong non-Newtonian viscous damping of the turbulence, augmenting regular viscous dissipation and some orders of magnitude larger. However, in reality, the interaction of the wind field with the foliage ought to produce large amounts of fine-scale turbulence in the wake of canopy elements. Wilson and Shaw (1977) demonstrated that a rigorous spatial averaging procedure was required which produced equations for the area averaged wind field that contained the required source and drag terms as well as terms corresponding to the production of fine-scale wake turbulence. The spatial averages used by Wilson and Shaw (1977) and further developed by Raupach and Shaw (1982) were averaged over a horizontal plane while more general volume averages were subsequently introduced by Finnigan (1985).

One of the interesting aspects of both fluvial and atmospheric flows is that most of the emphasis in numerical modelling has been upon treatments of the momentum equations. One of the most exciting developments in this area involves work on '*honami*', the wavy motion of a flexible canopy, see Ikeda *et al.* (2001), which was first studied by Inoue

(1963) over the top of rice plants. Ikeda *et al.* (2001) present work studying flow through a *honami* employing a two-dimensional large eddy simulation (LES). The modelling system is set up so that there are two different grids. Primarily there is a spatial discretisation which the LES simulation is calculated upon. In addition to this, there is a secondary grid, defined as the plant grid. In this grid movement of the plant is simulated by a flexible cantilever, in which the average displacement of plants is calculated. The mass of the plants is automatically conserved in the plant grid system while the drag due to the plant and its movement is formulated in proportion to the square of the relative velocity between flow and motion of plant in momentum equations for a turbulent flow field. This represents a combined treatment of both mass and momentum. A mass treatment may be crucial at high densities of vegetation, something that is illustrated implicitly by Godillot *et al.* (2001) for flow over periphyton in a river. They found that including the thickness of the periphyton matrix was crucial for effective prediction of the vertical variation of velocity. This is where we begin to see the potential for treating rivers as porous flows, and this is explored in Section 16.5.

In summary, the representation of vegetation in atmospheric flows has seen less emphasis upon conveyance parameters and more emphasis upon the interaction between the canopy and both the mean and turbulent properties of the flow field. In some senses, this is not surprising: flows in rivers are clearly depth-limited. However, changes in the position of the water surface is of major concern for river management as it controls flood inundation. Given the strong flow dependence of these conveyance parameters, and the fact that models must be used in a predictive sense, the interaction between vegetation and water remains an area that needs considerable additional research.

16.5   THE MATHEMATICAL BASIS OF POROSITY IN RIVERS

In this section, we argue that a much more effective treatment of river channels, in terms of both complex topography and vegetation, is through an explicit treatment of the effects of blockages upon mass conservation using a numerical porosity.

Figure 16.3 revisits the problem considered in Section 16.2, in which roughness heights are required to represent the effects of complex topographic variability, see Figure 16.3(a). As noted above, serious levels of grid deformation may be required if boundary fitted coordinates are used over these surfaces.

One alternative to this approach is to develop the method identified by Olsen and Stokseth (1995). This involves the use of regular structured grids, in which all control volumes are orthogonal, in both computational and Cartesian space, with the bed topography specified using cell porosities (P_0): $P_0 = 1$ for cells that are all water, $P_0 = 0$ for cells that are all bed, and $1 < P_0 < 0$ for partly blocked cells. Hardy *et al.* (2002) have developed the Olsen and Stokseth method to address issues that were not resolved in the original Olsen and Stokseth study, notably through the inclusion of relevant drag terms, see, for example, Wen *et al.* (1998), in the momentum equations in combination with the porosity

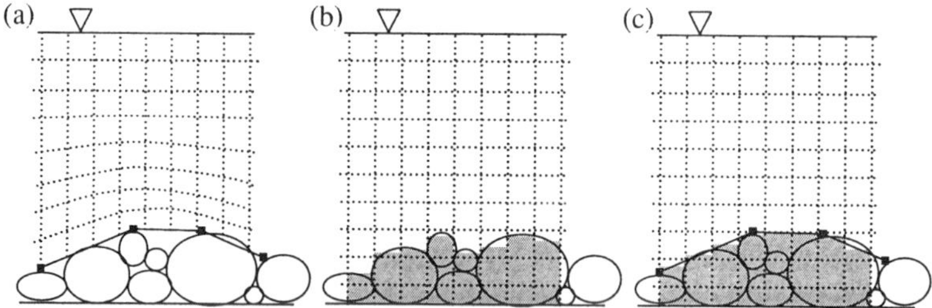

Figure 16.3 *(a) Boundary fitted coordinates applied to sampled topographic data. (b) Topographic representation using a porosity treatment with high resolution topographic data. (c) Topographic representation using a porosity treatment with sampled topographic data. Figures (a) to (c) represent an idealized, rough, gravel-bed surface. Black squares represent field sampled data points*

treatment. Hardy *et al.* also validate this approach using a dataset based upon particle image velocimetry as applied to a regular cuboidal surface.

Figure 16.3(b) shows application of this method to the case where high resolution topographic data, for example Figure 16.4, is available. The development of digital photogrammetric methods for the precise and high resolution measurement of river bed topography, see Butler *et al.* (1998, 2002), Lane (2000, 2001) and Lane *et al.* (2001), means that such data are becoming readily available. This is normally in the form of a digital elevation model (DEM) which comprises a regular grid of elevation data. This leads to the representation of continuous elevation values as a series of stepped elevation values, which is acceptable if the grid spacing is small as compared to local topographic variability.

Figure 16.4 *(a) Photograph and (b)* 3 mm *resolution digital elevation model of a gravel-bed river surface*

Thus, under Figure 16.3(b), the numerical mesh is defined as a regular grid, with planform dimensions identical to those used in the DEM. In this case, porosity can be mapped directly into the numerical grid. The effective drag terms are specified as vectors in the 3D momentum equations according to low-flow direction, i.e., an initial approximation of near boundary flow direction is obtained and used to define the local directions in which drag must be augmented by the effective cell area. As an example of the potential of this approach, Figure 16.5 shows a long profile of predictions of flow using a full 3D solution of the Navier–Stokes equations. This uses a $0.002\,\text{m}^3$ grid, inlet boundary conditions specified from detailed measurement of inlet velocities, and a horizontal rigid lid across which normal velocity resolutes are set to zero. This gives us a flow field resolution that has never been seen before. Lane *et al.* (2002) report on this method in detail, including validation.

Figure 16.3(c) shows a scenario more common in practice, where high resolution topographic data is not available. Even in this situation, use of a porosity treatment will result in a numerical solution that is more stable and subject to less numerical diffusion as a result of the use of orthogonal grid cells. The porosity treatment is especially amenable to roughness parameterisation: with basic knowledge of the scaling properties of a rough surface, see, for example, Butler *et al.* (2002), it is possible to simulate a surface with the same roughness characteristics at all spatial scales. This is currently being used to explore the effects of rough topography upon conveyance parameters avoiding the sort of circularity that can emerge in traditional methods of conveyance determination, see

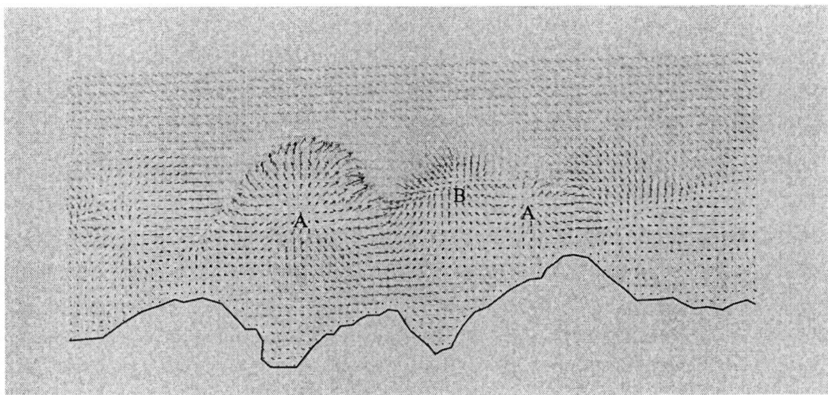

Figure 16.5 *Flow vectors modelled over a cross-section gravelly material. Fluid flow is into picture. The flow is depth limited (water depth 0.24 m). The model is finite volume, and grid cells are cuboidal (0.002 m length). The figure shows the bottom 0.080 m of the model predictions. The solid line is the bed surface. The image shows rapid divergence (labelled A) around gravel grains further downstream from this cross-section and convergence (labelled B) into areas of low pressure downstream of grains upstream of this cross-section*

Section 16.2.2. It also avoids the need to arbitrarily scale drag coefficients in situations where there is complex bed microtopography. Similar treatments may be suitable for dealing with vegetation, and the next stage of this research will extend this approach to the vegetated case. This will need careful consideration of drag terms, and certainly either modification of traditional turbulence models (as per López and Garcia, 2001) or a large eddy simulation treatment as per Ikeda *et al.* (2001).

16.6   CONCLUSIONS

The aim of this chapter has been to highlight the basic limitations of traditional treatments of complex bed geometries and vegetation in traditional models of river channels. Many of the methods used to deal with bed topography and vegetation suffer from one or more of the following problems:

(i) a strong dependence upon parameters that have a poor physical basis and which are only readily determined using empirical means;

(ii) a poor conceptual basis, in terms of the way they represent the effects of topography and vegetation upon the flow; and

(iii) a tendency to introduce problems of numerical diffusion and numerical stability, especially in higher dimensionality numerical models.

The use of numerical porosity may lead to a much improved representation of a range of open channel flow processes. Preliminary experiments over complex river gravels are producing very encouraging results, and may provide the means for an improved representation of vegetation in higher dimensionality numerical models. Problems of determining the geometry of the bed and the morphology of plants will inevitably mean that application of these methods in many practical situations may prove to be unfeasible. However, experimental investigation of river channel processes using a numerical porosity approach may result in a better justification and more reliable identification of the conveyance parameters needed for flood identification and habitat characterisation.

Acknowledgements

This research was supported by NERC Grant GR9/5059 awarded to SNL, Professor Derek B. Ingham and Dr Lionel Elliott. Dr Kate Bradbrook (JBA Consulting) provided useful comments on an earlier version of the text.

REFERENCES

Bakry, M. F. (1992). Effect of submerged weeds on the design procedure of earthen Egyptian canals. *Irrig. Drainage Syst.* **6**, 179–188.

Barnes, H. H. (1967). Roughness characteristics of natural channels. USGS Water Supply Paper 1849.

Bates, P. D., Anderson, M. G., Baird, L., Walling, D. E., and Simm, D. (1992). Modelling floodplain flows using a two-dimensional finite element model. *Earth Surface Proc. Landforms* **17**, 575–588.

Bates, P. D., Anderson, M. G., and Hervouet, J. M. (1995). Initial comparison of two 2-dimensional finite element codes for river flood simulation. *Proc. Inst. Civ. Eng. – Water, Maritime, Energy* **112**, 238–248.

Booker, D. J., Sear, D. A., and Payne, A. J. (2001). Modelling three-dimensional flow structures and patterns of boundary shear stress in a natural pool–riffle sequence. *Earth Surface Proc. Landforms* **26**, 553–576.

Bradbrook, K. F., Lane, S. N., and Richards, K. S. (2000a). Numerical simulation of time-averaged flow structure at river channel confluences. *Water Resources Res.* **36**, 2731–2746.

Bradbrook, K. F., Lane, S. N., Richards, K. S., Biron, P. M., and Roy, A. G. (2000b). Large eddy simulation of periodic flow characteristics at river channel confluences. *J. Hydraul. Res.* **38**, 207–216.

Bray, D. I. (1982). Flow resistance in gravel-bed rivers. In *Gravel-Bed Rivers* (eds R. D. Hey, J. C. Bathurst, and C. R. Thorne), pp. 109–133. Wiley, Chichester.

Brunet, Y., Finnigan, J., and Raupach, M. R. (1994). A wind tunnel study of air flowing in waving wheat: single point velocity statistics. *Boundary-Layer Meteorol.* **70**, 95–132.

Butler, J. B., Lane, S. N., and Chandler, J. H. (1998). Assessment of DEM quality characterising surface roughness using close-range digital photogrammetry. *Photogramm. Record* **16**, 271–291.

Butler, J. B., Lane, S. N., Chandler, J. H., and Porfiri, K. (2002). Through-water close-range digital photogrammetry in flume and field environments. *Photogramm. Record.* In press.

Chen, C. L. (1976). Flow resistance in broad, shallow, grassed channels. *J. Hydraul. Div.* **102**, 307–322.

Chow, V. T. (1959). *Open Channel Hydraulics.* McGraw–Hill, New York.

Clifford, N. J., Robert, A., and Richards, K. S. (1992). Estimation of flow resistance in gravel-bedded rivers: a physical explanation of the multiplier of roughness length. *Earth Surface Proc. Landforms* **17**, 111–126.

Cornelius, C., Volgmann, W., and Stoff, H. (1999). Calculation of three-dimensional turbulent flow with a finite volume multigrid method. *Int. J. Numer. Meth. Fluids* **31**, 703–720.

Cowan, W. L. (1956). Estimating hydraulic roughness coefficients. *Agricult. Eng.* **7**, 473–475.

Darby, S. E. (1999). Effect of riparian vegetation on flow resistance and flood potential. *J. Hydraul. Eng.* **125**, 443–454.

Deacon, E. (1957). Wind profiles and the shearing stress: an anomaly resolved. *Quart. J. Roy. Meteorol. Soc.* **78**, 537–540.

Dudley, S. J., Fischenich, J. C., and Abt, S. R. (1998). Effect of woody debris entrapment on flow resistance. *J. Amer. Water Resources Assoc.* **34**, 1189–1197.

Fathi-Maghadam, M. and Kouwen, N. (1997). Non-rigid, nonsubmerged, vegetative roughness on floodplains. *J. Hydraul. Eng.* **123**, 51–57.

Finnigan, J. (1985). Turbulent transport in flexible plant canopies. In *The Forest Atmosphere Interaction* (eds B. A. Hutchinson and B. B. Hicks), pp. 443–480. Reidel, Dordrecht.

Finnigan, J. (2000). Turbulence in plant canopies. *Ann. Rev. Fluid Mech.* **32**, 519–571.

Fischer-Antze, T., Stoesser, T., Bates, P., and Olsen, N. R. B. (2001). 3D numerical modelling of open-channel flow with submerged vegetation. *J. Hydraul. Res.* **39**, 303–310.

Gessler, D., Hall, B., Spasojevic, M., Holly, F., Pourtaheri, H., and Raphelt, N. (1999). Application of 3D mobile bed, hydrodynamic model. *ASCE J. Hydraul. Eng.* **125**, 737–749.

Gillies, J. A., Lancaster, N., Nickling, W. G., and Crawley, D. M. (2001). Field determination of drag forces and shear stress partitioning effects for a desert shrub (Sarcobatus vermiculatus, greasewood). *J. Geophys. Res. – Atmos.* **105**, 24871–24880.

Godillot, R., Caussade, B., Ameziane, T., and Capblanco, J. (2001). Interplay between turbulence and periphyton in rough open channel flow. *J. Hydraul. Res.* **39**, 227–240.

Grant, P. F. and Nickling, W. G. (1998). Direct field measurement of wind drag on vegetation for application to windbreak design and modelling. *Land Degrad. Devel.* **9**, 57–66.

Hardy, R. J., Lane, S. N., Elliott, L., Ingham, D. B., Best, J. L., and Lawless, M. (2002). Development of a method for numerical modelling of flows over complex surfaces using structured grids. Under review.

Hey, R. D. (1979). Flow resistance in gravel-bed rivers. *ASCE J. Hydraul. Div.* **105**, 365–379.

Hicks, D. M. and Mason, P. D. (1991). *Roughness Characteristics of New Zealand Rivers.* Water Resources Survey, DSIR Marine and Freshwater, Christchurch, New Zealand.

Hodskinson, A. and Ferguson, R. I. (1998). Numerical modelling of separated flows in river bends: Model testing and experimental investigation of geometrical controls on the extent of flow separation at the concave bank. *Hydrolog. Proc.* **11**, 1323–1338.

Hsieh, T. (1962). *The resistance of piers on high velocity flow.* Masters thesis. State University of Iowa.

Ikeda, S., Yamada, T., and Toda, Y. (2001). Numerical study on turbulent flow and honami in and above flexible plant canopy. *Int. J. Heat Fluid Flow* **22**, 252–258.

Inoue, E. (1963). On the turbulent structure of air flow with crop canopy. *J. Meteorol. Soc. Japan* **1**, 317–346.

Johnsen, R. C., Ramey, G., and O'Hagan, D. (1982). Wind induced forces on trees. *J. Fluids Eng.* **104**, 25–30.

Kerzenmacher, T. and Gardiner, B. (1998). A mathematical model to describe the dynamic response of spruce to the wind. *Trees* **12**, 385–394.

Knight, D. W. (2001). *Conveyance in 1D River Models.* Report to HR Wallingford and the Environment Agency.

Kouwen, N. (1988). Field estimation of the biomechanical properties of grass. *J. Hydraul. Res.* **26**, 559–568.

Kouwen, N. and Fathi-Maghadam, M. (2000). Friction factors for coniferous trees along rivers. *ASCE J. Hydraul. Eng.* **126**, 732–740.

Kouwen, N. and Li, R.-M. (1980). Biomechanics of vegetative channel linings. *J. Hydraul. Div.* **106**, 1085–1103.

Kutija, V. and Hong, H. (1996). A numerical model for assessing the additional resistance to flow introduced by flexible vegetation. *J. Hydraul. Res.* **34**, 99–114.

Lane, S. N. (2000). The measurement of river channel morphology. *Photogramm. Record* **16**, 937–961.

Lane, S. N. (2001). The measurement of gravel-bed river morphology. In *Gravel-Bed Rivers V* (ed. M. P. Mosley), pp. 291–320. New Zealand Hydrological Society, Wellington.

Lane, S. N. and Richards, K. S. (1998). Two-dimensional modelling of flow processes in a multi-thread channel. *Hydrolog. Proc.* **12**, 1279–1298.

Lane, S. N., Bradbrook, K. F., Richards, K. S., Biron, P. M., and Roy, A. G. (1999). The application of computational fluid dynamics to natural river channels: Three-dimensional versus two-dimensional approaches. *Geomorph.* **29**, 1–20.

Lane, S. N., Hardy, R. J., Ingham, D. B., and Elliott, L. (2002). Numerical modelling of flow of rough, gravel surfaces using structured grids. Under review.

Lane, S. N., Porfiri, K., and Chandler, J. H. (2001). Monitoring river channel and flume surfaces with digital photogrammetry. *ASCE J. Hydraul. Eng.* **127**, 871–877.

Lane, S. N., Richards, K. S., and Chandler, J. H. (1994). Distributed sensitivity analysis in modelling environmental systems. *Proc. Roy. Soc., Series A* **447**, 49–63.

Lane, S. N., Richards, K. S., and Chandler, J. H. (1995). Within reach spatial patterns of process and channel adjustment. In *Rivers* (ed. E. J. Hickin), Chapter 6, pp. 105–130. Wiley, Chichester.

Launder, B. E. and Spalding, D. (1974). The numerical computation of turbulent flows. *Comp. Meth. Appl. Mech. Eng.* **3**, 269–289.

Li, R.-M. and Shen, H. W. (1973). Effect of tall vegetation on flow and sediment. *ASCE J. Hydraul. Div.* **99**, 793–814.

López, F. and Garcia, M. H. (2001). Mean flow and turbulence structure of open channel flow through non-emergent vegetation. *J. Hydraul. Eng.* **127**, 392–402.

Mahrt, L., Vickers, D., and Sun, J. (2001). Determination of the surface drag coefficient. *Boundary-Layer Meteorol.* **99**, 249–276.

Mayhead, G. (1973). Some drag coefficients for British forest trees derived from wind studies. *Agricult. Meteorol.* **12**, 123–130.

Monin, A. S. and Yaglom, A. M. (1971). *Statistical Fluid Mechanics: Mechanics of Turbulence*, Vol. 1. MIT Press, Cambridge, Massachusetts.

Morin, J., Leclerc, M., Secretan, Y., and Boudreau, P. (2000). Integrated two-dimensional macrophytes-hydrodynamic modelling. *J. Hydraul. Res.* **38**, 163–172.

Nicholas, A. P. (2001). Computational fluid dynamics modelling of boundary roughness in gravel-bed rivers: An investigation of the effects of random variability in bed elevation. *Earth Surface Proc. Landforms* **26**, 345–362.

Nicholas, A. P. and Sambrook-Smith, G. H. (1999). Numerical simulation of three-dimensional flow hydraulics in a braided channel. *Hydrolog. Proc.* **13**, 913–929.

Olsen, N. R. B. and Stokseth, S. (1995). Three-dimensional numerical modelling of water flow in a river with large bed roughness. *J. Hydraul. Res.* **33**, 571–581.

Pasche, E. and Rouvé, G. (1985). Overbank flow with vegetatively roughened floodplains. *J. Hydraul. Eng.* **111**, 1262–1278.

Petryk, S. and Bosmajian, G. (1975). Analysis of flow through vegetation. *J. Hydraul. Div.* **101**, 871–884.

Raupach, M. R. (1992). Drag and drag partition on rough surfaces. *Boundary-Layer Meteorol.* **60**, 375–395.

Raupach, M. R. and Shaw, R. H. (1982). Averaging procedures for flow within vegetation canopies. *Boundary-Layer Meteorol.* **22**, 79–90.

Raupach, M. R. and Thom, A. S. (1981). Turbulence in and above plant canopies. *Ann. Rev. Fluid Mech.* **13**, 97–129.

Reiser, D. W. (1998). Sediment in gravel bed rivers: Ecological and biological considerations. In *Gravel-Bed Rivers in the Environment* (eds P. C. Klingemann, R. W. Beschta, P. D. Komar, and J. B. Bradley), Chapter 10, pp. 199–225. Water Resources Publications, Colorado.

Sellars, P. J., Mintz, Y., Sud, Y. C., and Dachler, A. (1986). A simple biosphere model (SiB) for use with general circulation models. *J. Atmos. Sci.* **43**, 505–531.

Shaw, R. H. and Pereira, A. R. (1982). Aerodynamic roughness of plant canopy: A numerical experiment. *Agricult. Meteorol.* **26**, 51–65.

Sinha, S. K., Sotiropoulos, F., and Odgaard, A. J. (1998). Three-dimensional numerical model for flow through natural rivers. *ASCE J. Hydraul. Eng.* **124**, 13–24.

Strickler, A. (1923). Beitrage zur Frage der Geschwindigheitsformel und der Rauhigkeitszahlen fur Strome, Kanale und Geschlossene Leitungen. *Mitteilungen des Eidgenossischer Amtes fur Wasserwirtschaft.* Bern, Switzerland.

Temple, D. M. (1987). Closure of 'Velocity distribution coefficients for grass-lined channels'. *J. Hydraul. Eng.* **113**, 1224–1226.

Thom, A. S. (1971). Momentum absorption by vegetation. *Quart. J. Roy. Meteorol. Soc.* **97**, 414–428.

Thompson, G. T. and Roberson, J. A. (1976). A theory of flow resistance for vegetated channels. *Trans. Amer. Soc. Aeronaut. Eng.* **19**, 288–293.

Thornton, C. I., Abt, S. R., Morris, C. E., and Fischenich, J. C. (2000). Calculating shear stress at channel-overbank interfaces in straight channels with vegetated floodplains. *J. Hydraul. Eng.* **126**, 929–936.

Tsujimoto, T. (1999). Fluvial processes in streams with vegetation. *J. Hydraul. Res.* **37**, 789–803.

Wallis, S. G. and Manson, J. R. (1997). Accurate numerical simulation of advection using large time steps. *Int. J. Numer. Meth. Fluids* **24**, 127–139.

Watson, C. C. (1987). Hydraulic effects of aquatic weeds in UK rivers. *Regulated Rivers: Res. Manag.* **1**, 211–227.

Wen, X., Ingham, D. B., Dombrowski, N., and Foumeny, E. A. (1998). The generation of an orthogonal grid by the use of a boundary integral technique. *Eng. Anal. Boundary Elements* **21**, 197–205.

Whiting, P. J. and Dietrich, W. E. (1990). Boundary shear stress and roughness of mobile alluvial beds. *ASCE J. Hydraul. Eng.* **116**, 1495–1511.

Wiberg, P. L. and Smith, J. D. (1991). Velocity distribution and bed roughness in high gradient streams. *Water Resources Res.* **27**, 825–838.

Wilson, J. D. and Shaw, R. H. (1977). A higher-order closure model for canopy flow. *J. Appl. Meteorol.* **16**, 1198–1205.

Wilson, J. D., Ward, D. P., Thurtell, G. W., and Kidd, G. E. (1982). Statistics of atmospheric turbulence within and above a corn canopy. *Boundary-Layer Meteorol.* **24**, 495–519.

Wolman, M. G. (1954). A method of sampling coarse river bed material. *Trans. Amer. Geophys. Union* **35**, 951–956.

Wu, F., Shen, H., and Chou, Y. (1999). Variation of roughness coefficients for unsubmerged and submerged vegetation. *J. Hydraul. Eng.* **125**, 934–942.